GW00762478

ADVANCED MACHINING:

The Handbook of Cutting Technology

IFS

ADVANCED MACHINING:

The Handbook of Cutting Technology

Graham T. Smith

IFS Publications, UK
Springer-Verlag
Berlin · Heidelberg · New York
London · Paris · Tokyo

Graham T. Smith
Dept of Engineering and Naval Architecture
Southampton Institute of Higher Education
East Park Terrace
Southampton SO9 4WW

British Library Cataloguing in Publication Data

Smith, Graham T., *1947* –
Advanced machining – The handbook of cutting technology
1. Materials. Machining
I. Title
670.42'3

ISBN 1-85423-022-0 IFS Publications Ltd
ISBN 3-540-50650-0 Springer-Verlag Berlin
ISBN 0-387-50650-0 Springer-Verlag New York

Phototypeset by Area Graphics, Letchworth
Printed and bound by Short Run Press, Exeter, U.K.

To my wife
BRENDA
who has been of great support to me
throughout my industrial and academic years

Contents

Preface

There are numerous reasons why a book on cutting tool technology for turning and machining centre applications is necessary. Anyone with experience of using such equipment knows that these machine tools dominate the high production metal cutting industry and, therefore, I believe a review of the present state of these technologies is overdue. Most engineers have been involved in some form of metal removal techniques most of their working lives; if this is so, then why write about a subject in which a company already has experience through years of familiarisation? The answer to this question lies at the heart of the problem and can be neatly summed-up by the well-known saying, 'familiarity breeds contempt'.

When a company has built up considerable experience in cutting tool applications over many years, it is often loath to change a 'good thing'. The company wrongly surmises that if new tooling concepts are introduced it will only produce, at best, marginal productivity gains, at the penalty of high financial outlay. This form of tooling apathy was 'brought home' to me recently when visiting a company involved in the manufacture of wrought components — produced from the solid, for use in the aerospace industry. The company in question had recently acquired a machining centre which was tooled-up with cutters of dubious vintage by today's standards. However, even more remarkable than this were the feeds and speeds used for stock removal. The cutting data would have been familiar to any engineer reading a tooling manual 40 years ago! It seemed such a waste of production potential to shackle a new, highly efficient machine tool in such a manner with an old-fashioned cutting tool approach. In fact, it would have been obvious to anyone with a working knowledge of the new tooling concepts that even though the company had sent its own employees on the machine tool training course, it had ignored the tooling lectures and reverted to type by continuing with its previous manufacturing methods. Unfortunately, this anecdote would be quite humorous but for the fact that – by degrees – it is relevant to quite a few companies. One wonders how long this out-dated approach to cutting tool technology will survive in the highly competitive field of manufacture practised by most companies today. Even more remarkable than this is the question: 'How did they last this long?'. The anecdote was only mentioned because it infortunately summarises the worst aspects of the 'ostrich' approach to tooling philosophy and, even now, some companies still regard tooling applications as a 'black art'.

Returning to the book, it was written with the help and assistance of a wide range of cutting tool and associated companies, in order to highlight the latest cutting tool technology trends, hopefully with an informal treatment. The aim was not to develop a highly mathematical treatment with abstract cutting tool equations to reinforce the text; this would only have produced a rather daunting and ponderous text for the reader. Diagrams and photographs have been used to emphasise points in the text whenever possible and, to a certain degree, to liven a possibly 'dry' subject as it occurs. It would have been easy to spend much longer explaining certain tooling aspects – such as chip-breaker theory, for example – but comments were confined to a concise and generalised treatment throughout.

After the introductory remarks on some of the cutting tools available, their material development and geometries, a chapter devoted exclusively to modular quick-change tooling is included. This fast method of manual/automatic tool changing is becoming a popular concept. Initially developed for use on turning centres, where it has rapidly expanded in popularity, it can now be applied to machining centre cutting applications, or it may be used on conventional machine tools.

A chapter is also included on tool monitoring, the need for which was highlighted once a form of un-manned manufacturing operations had been developed. Although certain aspects of tool monitoring have been around for some time, in the early days it was principally the concern of adaptive control systems. Latterly, new systems have been incorporated using tool collision and wear protection and so on, all of which are covered in the chapter.

With companies developing larger and larger tooling libraries in recent years, a system of tool management has become of critical importance. The chapter which addresses this topic discusses the means of rationalising, controlling and distributing tooling throughout a company. Systems range from highly sophisticated software-based methods, through to a simple, but effective tool control system. The method of application depends on the company's product and range of machine tool plant.

Although not, strictly speaking, a cutting tool subject, the application of cutting fluids to metal cutting operations is relevant and, therefore, needs adequate coverage. It is essential that the correct usage of these cutting fluids is achieved in order to combat the problems of tool wear and improve the surface condition of the workpiece.

The last section of the book discusses the importance of assessing the machinability rating of specific materials, by a brief review of the subject. There are other inter-related subjects which may be coupled to machinability asssessment such as methods used in determining machined surface characterisation and using surface integrity techniques. The chapter discusses the damage caused to a machined surface and the likely effects it could have on the in-service life, when coupled to a poor surface topography.

Also covered are methods of characterising a surface.

The final comments are reserved expressly for you, the reader. In my mind's eye, I have formed a vision of a harrassed, production engineer, supervisor, planning/application engineer, as well as a designer, who needs to know concisely – or indeed to confirm his own judgement – the latest methods and developments in the application of cutting tool technology. This book is aimed principally at such a person in industry. But, it is also hoped that academics and students alike will gain some benefit from this book. The book has been written with the serious intention of stimulating discussion through thought-provoking assumptions and, to a certain extent, contentious statements. If just some of these ideas are implemented by a company, then the effort required in writing the book will have been worthwhile and, I hope, prove to give productivity and economic savings!

Graham T. Smith
Broughton
Hampshire
August 1988

Acknowledgements

Any book with the accent on the latest applications of cutting-tooling and related subjects must, of necessity, feature the most recent products in the industrial marketplace. Therefore I am indebted to the many companies who have given of their time and provided literature in support of the book.

First and foremost, I must extend my sincere thanks to Steve Chapman of Rocol Ltd, Leeds, for his great contribution to Chapter 5, on cutting fluids. Without his help, this chapter would have been much poorer and less succinctly written.

I should also like to thank the other major contributors to large sections of the book: Paul Brohan of Sandvik (UK) Ltd., for the time and effort he spent in discussing the company's philosophy of the latest tooling developments and for supplying vast amounts of technical and support information, assisted by photographic support from M. Stokes; Bill Kennedy of Kennametal Inc. (USA), for his speedy collation of information and comprehensive technical support on cutting-tool technology, and B. Williams of Kennametal Erickson (UK); A. John Hale of Seco Tools (UK) Ltd., for kind permission to use extracts from their training manuals and general literature on tooling; Alan Street of Karl Hertel Ltd., for information and photographs; Terry Reynolds of Krupp Widia (UK) Ltd., for information on tool-condition monitoring and related subjects; and Lyndon Mortimer of Stellram Ltd., for tooling information.

A whole host of other companies, too numerous to mention here, also willingly supplied technical information, and their help is gratefully acknowledged; a complete list of them is to be found in the company addresses at the end of the book.

Lastly, I should like to acknowledge the help and assistance of Anna Kochan of IFS Publications, who has provided enthusiasm and support throughout the book's compilation.

Graham T. Smith
August 1988

Chapter 1
Cutting-tool technology

1.1 An introduction to cutting-tool technology

At a seminar a few years ago, a cutting-tool specialist told an anecdote that summarised well the way that cutting-tool technology is regarded by manufacturing industry: 'Imagine it is the day of a Formula One Grand Prix. On the front row of the grid sits the world champion in a car costing approximately £200 000. The team manager decides that in order to cut costs he will put a set of retread tyres onto the car . . . and still try to win the race!' The probable outcome of this poor judgement is obvious. Not that any team manager would do this in reality, of course – far from it.

So, why does cutting-tool technology suffer from this 'Cinderella syndrome'? It is not unusual for a company investing in a machining centre, say, to spend more than £300 000 after spending months on a feasibility study, and only then to think about tooling it up. The same thing occurs at the other end of the scale when a small subcontract shop buys a computer-numerically controlled (CNC) lathe for £30 000 and does not budget for the tooling. The tooling costs for machining and turning centres can easily approach a tenth of the cost of the original machine, so why is the same care not taken over their purchase? This is a particularly valid point, since it is only when a machine is cutting metal that it makes any money! To press the point further, a small CNC milling machine correctly tooled-up can be more efficient in terms of productivity than a machining centre with inadequate tooling.

So why has this penny-pinching attitude been sustained? There are several factors. Firstly, there is the engineers' ignorance of the latest tooling advances, in conjunction with the argument 'We have always used these cutters and they perform satisfactorily, so why change?' Secondly, the senior management may not release funding for tooling, on the basis that 'The stores are full of tools' and other similar excuses. These attitudes must change if a company is to compete on an equal footing with its competitors.

A recent US survey by a leading cutting-tool manufacturer showed that increases in productivity of around 30% were possible in the average plant of the companies studied without increasing capital investment. If this increase in productivity is possible, how can it be achieved with the minimum of disruption to current production? First the company must analyse its manufacturing environment. We shall consider some of the problems and solutions that might arise from such a survey.

The first and most obvious thought might be to speed up every operation, but this does not lead to increased productivity, as bottlenecks will still occur. For some operations the inclination may be to maximise the number of pieces per hour, whereas for others it might be to reduce costs. Of prime importance however, is the overall increase in productivity and efficiency, not the perfection of any one operation. From these considerations, it can be seen that the production flow through the shop will highlight any bottlenecks, where components pile up at a machine for a certain operation. Every such bottleneck is an opportunity to improve productivity, either at the machine or over the whole line. Where bottlenecks do not occur, then it is often possible to reduce the cost per part, or influence costing by other means. The ideas that 'no machine is an island' and that the shop should be thought of as one big harmonious machine and not a lot of independent problems, will lead the way to productivity increases.

One of the most obvious ways of reducing costs is to rationalise the company's tooling inventories. This fulfils two objectives, firstly of decreasing unnecessary tooling, and secondly of saving money. In the US survey, it was found that 45%–55% of the tooling listed on the companies' inventory was obsolete.

Any tooling that remains after rationalisation should be consolidated. Reducing the number of insert grades by half, for example, often proves to have little effect on capability. Grouping the inserts by size, shape and nose radius will eliminate many of the less-used items and make bulk purchasing possible, thus reducing costs further. By such consolidation of tooling, it may be possible to purchase high-performance grades that meet wider ranges of applications. These will also have longer tool life and can be used at higher speeds which almost eliminates their extra cost. If fewer grades are stocked, the tooling engineers will know them more thoroughly, which will result in more effective usage.

A boost to productivity can also occur if the depth of cut can be maximised and, as a result, the number of passes minimised; however, this has little effect for components with little stock to be removed. It can be argued that increasing the depth of cut leads to a reduction in tool life (in terms of minutes of cutting per edge). However, there are fewer cuts per part, so each part requires fewer minutes' cutting overall, and many more parts per edge are produced. More important than this is the fact that cycle times for roughing operations are reduced: a reduction in the number of roughing passes from two to one, for example, results in a 50% reduction in cycle time. This increase in productivity may justify the decreased tool life on the basis that it removes a bottleneck. To extract the maximum productivity from today's high-performance grades they must be worked hard and be pushed to their fullest capabilities.

If tool life is reduced by increasing the depth of cut there are several ways in which this loss can be minimised. For example, it is known that the size of the nose radius of an insert has a pronounced effect on tool life; thus a doubling of the depth of cut can largely be compensated for by using a larger nose radius – assuming that part geometry allows this. When an increase in nose radius cannot be used, then increasing the insert's size helps to compensate for higher wear-rates by better dissipation of heat.

The accepted practice when roughing out is that no more than half the cutting edge length of an insert should be used; as the depth of cut approaches this value, a larger insert size must be considered. If large depths of cut are used in combination with high feedrates, a roughing-geometry insert produces longer tool life than a general-purpose insert. Often, using a single-sided insert rather than a double-sided one for roughing cuts brings the twin benefits of increased productivity and longer tool life (in terms of pieces per edge). Normally, single-sided inserts are recommended whenever the depth of cut and feedrate are so high that surface speed must be reduced below the grade's normal range in order to maintain an adequate tool life. They should also be considered if erratic breakage occurs.

The fact that some materials produce curled chips when being cut has shown that the highest-temperature region on the insert is not at the cutting edge, but some distance back where the crater forms. The position of this region can vary depending upon the feedrate: if the feedrate increases, the high-temperature region moves away from the cutting edge, and if the feedrate is reduced it moves towards the edge. This means that if the feedrate is too low for the geometry and edge preparation of the insert, heat will be concentrated too near the cutting edge and abrasive wear will be accelerated. Increasing the feedrate moves the maximum heat zone away from the cutting edge and increases tool life in terms of minutes per edge as a result; each part will be produced in less time at the higher feedrate, and the tool life in terms of parts per edge will increase.

Incorrect feedrates produce a number of symptoms: when they are too low, chip-control problems occur, and when they are too high, tool life is extremely short and edge chipping and insert breakage are likely to occur. Once the insert grades have been consolidated and their geometries chosen, it is relatively easy to determine the feedrates for the workpiece materials. Suppliers of the inserts can help by recommending an inititial selection of grade, speeds and feedrates. Juggling the grades and geometries marginally around these specified values may allow the feedrates to be increased and provide a significant pay-off in terms of productivity with little, or no, capital investment.

If the speed is increased instead of the feed, a point is reached where increased surface speed results in decreased productivity. In other words, cutting too fast will mean spending more time changing tools than making parts! By cutting too slowly, on the other hand, tools will last considerably longer, but this will be at the expense of the number of parts produced. So, what is the 'right' surface speed?

Returning to the 'no machine is an island' theory and treating the shop as one big machine, it is clear that every shop should determine its own particular objectives when considering cutting speeds and tool life. Typical objectives for tool life might be completing a certain number of parts before indexing the insert, or indexing after one shift, or after completing part of a shift. If very expensive parts are being machined, the main goal is to avoid catastrophic failure of the insert, which would necessitate scrapping the part. When very large parts are being machined, the objective may simply be to complete one part per insert, or just one pass of a part. When small parts are being made, then the tool life may be controlled in order to minimise size variations over time and to reduce the need for offset compensations. One idea is shared by all these approaches, namely that the surface speed must be adjusted to meet the overall production objectives. As a consequence of this, there is no 'correct' surface speed for any specific combination of material and insert grade.

Long production runs are ideal because they allow experimentation to find the optimum speed for the particular objective required. However, sometimes it is not possible to find a speed to meet the production requirements, and a change of grades to one of the higher-technology materials may be in order. Short production runs often rule out any experimentation with insert grades, and consultation with the client (who may have previous experience of problems) or the use of reference works may be of assistance.

Care must be taken when using published recommendations; generally they are only guidelines, to help get the job into production. Compared with a starting point within the recommended range for specific conditions, interruptions in the cut, large depth of cut, high feedrates, very long continuous cuts, surface scale, the absence of coolant or a less than rigid set-up of the workpiece would all suggest reductions in surface speed. Conversely, conditions such as smooth uninterrupted cuts, short lengths of cut, light depths of cut, low feedrates, clean preturned materials, flood coolant or a very rigid set-up would suggest that the recommended ranges for the insert could be exceeded whilst still maintaining an acceptable tool life.

As well as using today's much improved cutting-tool materials to their full potential, the full capabilities of machines must also be employed. In the US survey referred to earlier, it was found that one sub-contract shop used nearly 300 special individually-ground triangular inserts for grooving applications. The company switched to using grooving inserts, which were sufficiently rigid for light profiling cuts on turning centres; just as importantly, they exploited the ability of their machines to generate complex shapes. These changes reduced their threading and grooving inventory to 50 inserts and saved time in regrinding 'specials'.

Finally, it must be remembered that the requirement is for an overall increase in production, not perfection. After the analysis, when the tooling inventory has been consolidated, there will be fewer and more versatile grades and geometries of inserts. Then a survey of speeds and feedrates in conjunction with variations in depth of cut will determine how specific production objectives may be attained. With the optimum machining parameters now defined, and with full utilisation of capital equipment, large increases in overall production efficiency will result.

From these discussions it can be seen that the parameters of tool life, cutting speeds and feedrates form a complex relationship, which is shown in Fig. 1.1a. If you change one parameter, it will affect the others, so a compromise has to be reached to obtain the optimum performance from a cutting tool. Ideally a cutting tool should have superior performance in five specific areas: hot-hardness, resistance to thermal shock, lack of affinity between tool and work, resistance to oxidation, and toughness (Fig. 1.1b). We shall consider these factors in turn.

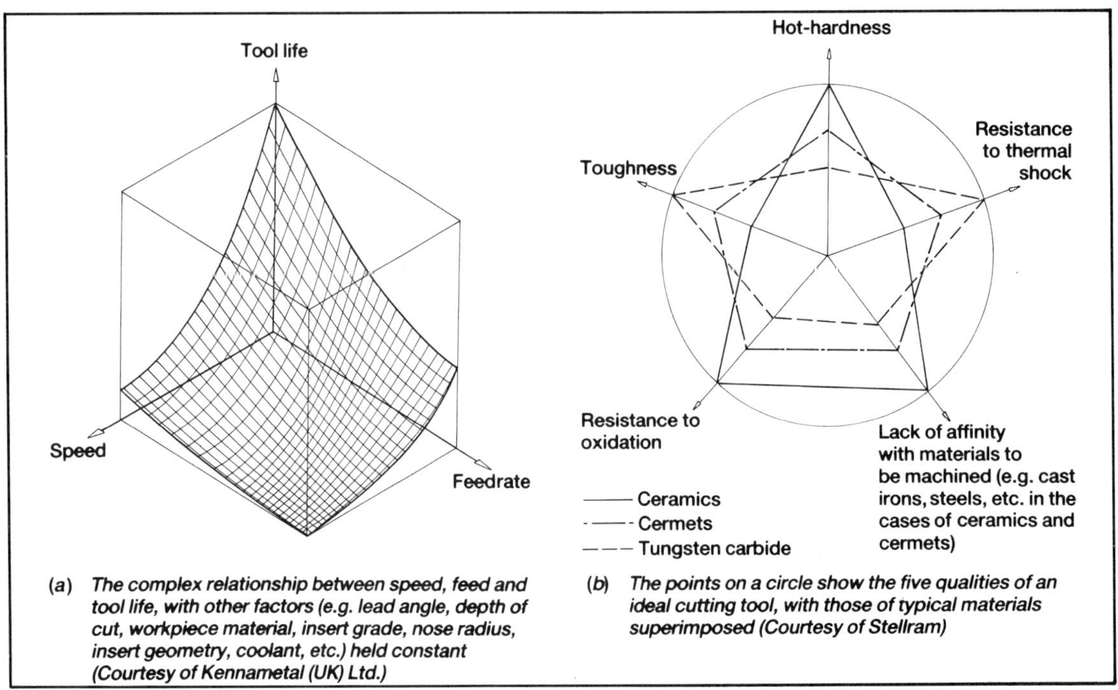

(*a*) *The complex relationship between speed, feed and tool life, with other factors (e.g. lead angle, depth of cut, workpiece material, insert grade, nose radius, insert geometry, coolant, etc.) held constant (Courtesy of Kennametal (UK) Ltd.)*

(*b*) *The points on a circle show the five qualities of an ideal cutting tool, with those of typical materials superimposed (Courtesy of Stellram)*

Fig. 1.1 Factors affecting the life of a cutting tool

A cutting tool requires hot-hardness in order to maintain a sharp and consistent cutting edge at the elevated temperatures that occur whilst machining. If the hot-hardness is not sufficient for the temperature generated at the tip, then it will degrade quickly and be useless.

Resistance to thermal shock is necessary in order to overcome the effects of the continuous cycle of heating and cooling that is typical of a milling operation, or of intermittent cutting on a lathe (such as in eccentric turning, etc.). If this shock resistance is too low, then rapid wear-rates of the insert occur. Inserts with low shock resistance may only be used in continuous cutting operations (and, even here, with some reservations).

Ideally, a lack of affinity between the tool and workpiece is to be encouraged, since any degree of affinity will lead quickly to the formation of a built-up edge. The result of this is that the tool geometry is modified, leading to poorer chip-breaking ability, higher forces being generated and a poorer workpiece surface finish, as well as several other undesirable effects. Thus an insert should ideally be inert to any reaction with the workpiece.

Another desirable feature is a high resistance to oxidation, in order to reduce the debilitating wear that oxidation produces.

Lastly, there is the insert toughness to be considered. Toughness allows the insert to absorb the forces and shock loads produced whilst cutting; it is particularly relevant in intermittent cutting operations. If an insert is not sufficiently tough, then induced vibrations alone (in, say, continuous cutting) will eventually cause the edge to shatter.

By careful balancing of these five factors, the tool manufacturers can produce grades of inserts which vary to the last degree, allowing a wide range of materials to be machined through choice of the correct insert grade for a particular material type. In fact, recent progress by the tool manufacturers has resulted in wider ranges of workpiece-cutting ability from fewer types of inserts. The manufacturers are confident of reducing the number of inserts still further, so that in the future a few inserts will be sufficient for successful machining of a wide range of materials with differing speeds and feeds, which will allow a company to reduce tooling inventories further.

This introductory section has indicated how correct tooling can lead to increased production rates, but when and how did the present advanced state of tooling develop? The advances that contributed to it will be discussed next.

1.2 The evolution of cutting-tool materials

Fig 1.2 illustrates the evolution in terms of the time taken to machine a standard component. Before 1870 all lathe tools were made from plain carbon steel, with a typical composition of 1% carbon and 0.2% manganese (the remainder being iron). This material had low hot-hardness (withstanding temperatures only up to 250°C) and the tools could only be used at speeds of approximately 5m/min. Quench cracks were frequently present as materials were hardened by water quenching. By 1870 Mushet (working in England) had introduced a steel containing 2% carbon, 1.6% manganese, 5.5% tungsten and 0.4% chromium. This was an air-hardened steel which retained its hardness at higher temperatures than previously, so that it could be used at speeds of up to 8m/min. This modified material was used for turning tools until the turn of the century, with chromium gradually replacing manganese.

In 1901 F. W. Taylor and M. White greatly improved the stability of these tools, allowing them to reach cutting speeds of 19m/min. The material they used became known as

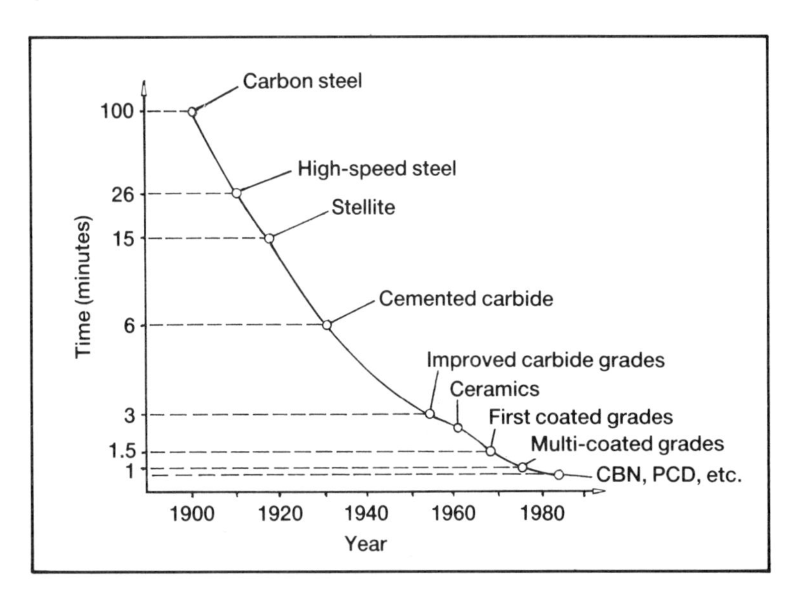

Fig. 1.2 The evolution of cutting-tool materials (Courtesy of Sandvik (UK) Ltd.)

high-speed steel (HSS); this was basically a new heat-treatment procedure, and not a new material as is often thought. A typical composition was 1.9% carbon, 0.3% manganese, 8% tungsten and 3.8% chromium. It differed from Mushet's steel by an increased amount of tungsten and further replacement of mangenese by chromium. During the next ten years a rapid development of HSS took place, and by 1904 the carbon content had been reduced, making the steel more easy to forge. Increases in the tungsten content improved the hot-hardness and Dr J. A. Matthews found that additions of vanadium improved the material's resistance to abrasion. By 1910 the content of tungsten had increased to 18%, with 4% chromium and 1% vanadium. The well-known 18 : 4 : 1 HSS had arrived, and it continued with only marginal modifications over the next 40 years. The most notable modifications to HSS occurred in 1923 when the so-called 'super' HSSs were developed, although they did not become commercially available until 1939, when Gill reduced the tungsten content to allow the material to be hot-worked. In the present M2 HSS, introduced in the USA in 1950, some of the tungsten content has been replaced by molybdenum. This HSS has a composition of approximately 0.8% C, 4% Cr, 2% V, 6% W and 5% Mo. It will withstand temperatures approaching 650°C and still maintain a cutting edge. This is the most recent advance in the development of HSS.

In 1915 an alloy called stellite was produced, which had high contents of cobalt, chromium and tungsten, and which could withstand temperatures up to 800°C. Another major cutting-tool material, cemented tungsten carbide, was developed in Germany in 1928 using a powder-metallurgy processing route. These early 'carbides' suffered from a chipping tendency (brittleness), there were difficulties in brazing and grinding them, and most lathes were not sufficiently powerful or rigid to use them adequately. There was also a tendency to cratering, especially when machining steels. Over the next 25 years improved grades of carbides were produced with greater shock resistance; however, their use was limited to machining cast irons and non-ferrous materials. During this period, diamond grinding wheels were also developed, which allowed tool preparation to be done more easily. The problem of cratering when machining steels was alleviated by additions of titanium and tantalum carbides, and this also improved other mechanical properties (Fig. 1.3).

In the early to mid-1950s, ceramic cutting tools (Fig. 1.4) were developed in England, Germany and Russia. These were made from finely sieved α -alumina and processed by sintering, without the use of binders or additives. The tools were harder and more

Fig. 1.3 A range of carbide inserts, showing their diverse shapes and geometries (Courtesy of Sandvik (UK) Ltd.)

Fig. 1.4 A typical range of ceramic inserts, illustrating the various shapes available (Courtesy of Sandvik (UK) Ltd.)

refractory than those made from carbides, but they still suffered from being brittle. To impart a degree of toughness to ceramics, a small grain size of alumina is used, and they are sintered at maximum density. The cost of the alumina is low, but the manufacturing processing costs tend to be high, mainly because of the cost of first slicing large blocks of alumina into small pieces with diamond saws.

The relative brittleness of ceramics demands very stable machines and clamping, which has led to a slow acceptance of them by industry. It is normal to use these materials at speeds up to 900m/min when turning steels. Typical composite ceramics e.g. those of alumina and titanium carbide (cermets), are less prone to brittleness, with transverse rupture strengths greater than 690MPa.

Alumina inserts can be either cold-pressed (white) or hot-pressed (black). The cold-pressed types are not so strong as the hot-pressed ones, but are cheaper. A typical particle diameter of the hot-pressed types is less than 2μm; when it is possible to use smaller particles and higher densities, superior properties result.

Ceramics consist of inorganic non-metallic elements in combination with metallic elements and have good atomic bonding properties. They have high melting points, hardnesses and chemical stabilities (i.e. low reactivity), coupled with low thermal expansion and conductivity, which allows them to be used where a continuous cutting action occurs. Intermittent cutting conditions promote impact stresses and lead to cutting-edge chipping, or thermal cracking.

New composites are continually being developed; typical of these are Kyon (Fig. 1.5) and Sialon (Fig. 1.6), which have the superior impact resistance necessary for intermittent cutting. Kyon is an alumina-based ceramic that is reinforced with silicon carbide 'whiskers', which are elongated single crystals of silicon carbide. They act to disperse mechanical shock, whilst increasing fracture toughness and improving thermal conductivity. Thus these inserts are ideal for machining applications on nickle-based alloys, and they can machine Inconel 718 productively. They have a typical hardness of 45HRC and can be used at speeds up to ten times those for tungsten carbide.

Fig. 1.5 Kyon, an alumina-based ceramic with whiskers of silicon carbide as reinforcement, can turn nickel-based alloys with speeds of up to 600m/min and feeds of 0.4mm/rev (Courtesy of Kennametal (UK) Ltd.)

Fig. 1.6 A silicon nitride (Sialon) insert turning an Inconel nickel-based alloy (Courtesy of Kennametal (UK) Ltd.)

Coated tools (Fig. 1.7) were developed in the late 1960s. The first ones were coated with titanium carbide (TiC) to a depth of 0.005mm by vapour-phase deposition. These coatings reduced the tendency of cemented carbide tools to cratering and enabled machining to be conducted at higher speeds and feeds. At present the most widely-used coatings are TiC, titanium nitride (TiN) and alumina (Al_2O_3). The inert nature of TiC and Al_2O_3) coatings reduces cratering, whereas TiN offers a lower tool friction. The nature of the coating process makes these inserts more susceptible to the effects of interrupted cutting, and they are usually reserved for the machining of materials that are prone to cratering. Multi-coated grades appeared during the 1970s; these offer a combination of the properties of both types of coating in conjunction with various substrates. A typical example of this hybrid range can be seen in Fig. 1.8.

During the period 1975 to 1984, the materials polycrystalline diamond (PCD) and cubic boron nitride (CBN) were introduced by the major tool companies. These basically consisted of a thin coating of PCD or CBN on a substrate of sintered tungsten carbide, with a cobalt binder. The layer of PCD or CBN has a small grain size; it is subjected to extremely high pressures and temperatures, which allows a small amount of cobalt to diffuse to the surface and act as a binder phase, producing a very hard coating. In fact, the PCD synthesised coatings are an intergrown mass of randomly-oriented diamond crystals, produced by high-temperature sintering. In most cases, the finer the diamond

Fig. 1.7 A range of TiC coated-carbide milling inserts for steels (Courtesy of Kennametal (UK) Ltd.)

Fig. 1.8 A TiC/TiN cermet grade insert turning stainless steel (Courtesy of Kennametal (UK) Ltd.)

size, the better the cutting edge. PCD inserts can turn and mill very abrasive non-metallic and non-ferrous materials. Owing to the affinity between carbon and iron, however, the PCD insert have an inherent weakness on any ferrous metals and the CBN inserts are required. These are made from cubic boron nitride particles that are carefully selected, then sintered. Using CBN inserts it is now possible to turn and mill components that used to be ground in the hardened state. This has obvious productivity benefits, assuming that any surface-finish requirements are satisfied. These exotic insert mixes allow for much longer tool life on 'difficult' materials, but they are rather expensive.

Various methods are used to deposit coatings onto cutting tools; typical ones are shown in Fig. 1.9. The coatings for most carbide turning tools are applied by the chemical vapour deposition (CVD) process. This process imposes limitations when milling, particularly at low to moderate chip loads. The CVD process is typically effected at a temperature of around 1000°C. This temperature causes degradation of the substrate and an eta-phase is formed at the interface between the coating and substrate. This brittle phase decreases the cutting tool's toughness and causes detrimental effects at sharp corners; as a result, most CVD-coated cutter inserts and tools for milling are now honed or chamfered (or both) prior to the coating process. The brittle eta-phase can be minimised in milling cutters by careful choice of the process parameters of CVD and of the coating and substrate compositions, and by applying a thinner coating than for an equivalent turning tool.

However, another process is available for milling-cutter coatings; this is known as the physical vapour deposition (PVD) process. This process is carried out at much lower temperatures than CVD, typically 500°C. These lower temperatures circumvent the problems posed by eta-phase formation, and give good edge sharpness. The PVD process is only now gaining acceptance. Owing to the belief that PVD coatings do not

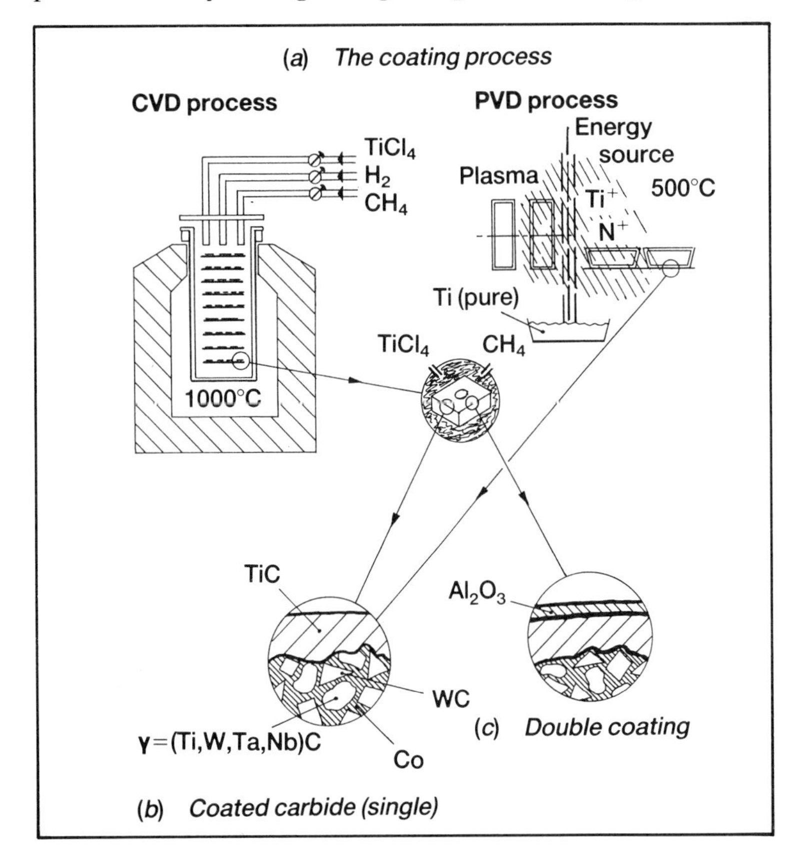

Fig 1.9 Insert-coating methods and metallurgy: thicknesses of typical coatings are 5μm for TiC and 1μm for Al_2O_3

adhere well to the carbide substrates and that CVD ones outperform them in use. This belief is false; indeed, a large supplier of coated tools has demonstrated that there are positive advantages in using the PVD process. The coated surface of a PVD insert is much smoother, appearing shallow-dimpled and with a finer grain size than the crystalline, blocky-grained appearance of inserts coated by the CVD process. This smoother surface reduces problems to do with surface morphology, such as tendencies for thermal cracking, which lead to edge chipping and premature failure. It also improves resistance to repeated mechanical and thermal stresses, thus controlling edge chipping and promoting slower flank wear whilst minimising interface friction.

The CVD process for coating the inserts is illustrated in Fig. 1.9a *left*. Oxygen is removed from the vacuum-type crucible by flushing it with argon. Then the container is heated to 1000°C and flushed with pure dry hydrogen to remove the oxides from the substrate surfaces. Titanium tetrachloride ($TiCl_4$) and methane (CH_4) are introduced through separate lines, with hydrogen as the carrier gas. As a result of this procedure, the coating grows very slowly. The treatment continues for several hours, until the necessary thickness of coating is achieved. The alternative PVD process is shown in Fig. 1.9a *right*. Here, a furnace ionises a pure titanium source to form a plasma stream which combines with nitrogen and is deposited onto the inserts. Both of the processes are quite time-consuming, and the inserts are generally costly to produce at present; however, they have many advantages over the uncoated carbide grade inserts.

So far, we have discussed the historical development of cutting tools up to the present. Now we shall consider the manner in which the most popular types of inserts are classified.

1.3 The classification of carbide inserts

Considering the number of permutations of metallurgical content, coatings, geometrical shapes, etc, the choice of cemented carbide inserts available from manufacturers seems limitless! In classifying this variety, the International Standards Organisation (ISO) faced the decision of whether to adopt a standard based on the properties of the carbide grades (such as hardness, transverse rupture strength, density, etc) or one which classified them according to their application or use. After debating these issues, the ISO compiled Recommendation No. 513, *Classification of Carbides According to Use*, which has now been adopted by most leading industrial nations.

The classification groups are designated P, M and K (Fig. 1.10):

- P refers to use with long-chipping materials (e.g. steel, etc.).
- K refers to short-chipping materials (e.g. cast iron, etc.).
- M represents an intermediate application, between P and K, and includes steel castings, malleable cast irons, and similar materials.

Within each classification group there are several sub-groups, designated on a numerical scale from 01 to 50. The higher the sub-group number, the higher will be the toughness: however, it should be said that this toughness is normally at the expense of hardness (i.e. wear resistance). In the advanced coated carbide grades, though, where a tough core exists, surface hardness can be combined with shock resistance, giving a broad range of machining applications.

The K-group has a high percentage of tungsten carbide, which imparts good resistance to wear, in particular to flank wear when machining cast iron. If crater wear occurs whilst

Distinguishing colours	Designation	Material to be machined	Application	Direction of increase in characteristic: Of cut	Direction of increase in characteristic: Of carbide
BLUE	P01	Steel, steel castings	Finish turning and boring; high cutting speeds, small chip section, accuracy of dimensions and fine finish, vibration-free operation	↑ Increasing speed ↓ Increasing feed	↑ Wear resistance ↓ Toughness
	P10	Steel, steel castings	Turning, copying, threading and milling, high cutting speeds, small or medium chip sections		
	P20	Steel, steel castings Malleable cast iron with long chips	Turning, copying, milling, medium cutting speeds and chip sections		
	P30	Steel, steel castings Malleable cast iron with long chips	Turning, milling, planing, medium or low cutting speeds, medium or large chip sections, and machining in unfavourable conditions		
	P40	Steel Steel castings with sand inclusion and cavities	Turning, planing, slotting, low cutting speeds, large chip sections, with the possibility of large cutting angles for machining in unfavourable conditions		
	P50	Steel Steel castings of medium or low tensile strength, with sand inclusion and cavities	For operations demanding very tough carbide; turning, planing, slotting, low cutting speeds, large chip sections, with the possibility of large cutting angles for machining in unfavourable conditions		
YELLOW	M10	Steel, steel castings, manganese steel Grey cast iron, alloy cast iron	Turning, medium or high cutting speeds; small or medium chip sections	↑ Increasing speed ↓ Increasing feed	↑ Wear resistance ↓ Toughness
	M20	Steel, steel castings, austenitic or manganese steel, grey cast iron	Turning, milling; medium cutting speeds and chip sections		
	M30	Steel, steel castings, austenitic steel, grey cast iron, high-temperature-resistant alloys	Turning, milling, planing; medium cutting speeds, medium or large chip sections		
	M40	Mild free-cutting steel, low tensile steel Non-ferrous metals and light alloys	Turning, parting off		
RED	K01	Very hard grey cast iron, chilled castings of over 85 Shore, high silicon–aluminium alloys, hardened steel, highly abrasive plastics, hard cardboard, ceramics	Turning, finish-turning, boring, milling	↑ Increasing speed ↓ Increasing feed	↑ Wear resistance ↓ Toughness
	K10	Grey cast iron over 220 Brinell, malleable cast iron with short chips, hardened steel, silicon–aluminium alloys, copper alloys, plastics, glass, hard rubber, hard cardboard	Turning, milling, drilling, boring, broaching		
	K20	Grey cast iron up to 220 Brinell, non-ferrous metals (copper, brass, aluminium)	Turning, milling, planing, boring, broaching, demanding very tough carbide		
	K30	Low-hardness grey cast iron, low tensile steel, compressed wood	Turning, milling, planing, slotting, for machining in unfavourable conditions and with the possibility of large cutting angles		
	K40	Soft wood or hard wood Non-ferrous metals	Turning, milling, planing, slotting, for machining in unfavourable conditions and with the possibility of large cutting angles		

When applying coated inserts it must be borne in mind that these are basically intended for turning operations, and will in many cases be multi-purpose covering a number of groups and sub-groups

Fig. 1.10 Classification of carbides according to use (Courtesy of Seco)

machining at high cutting speeds, then other grades must be adopted. Inserts from the P-group may be applicable, as they contain titanium, tantalum and niobium carbide. Combined with tungsten carbide, these provide high hardness at elevated temperatures in conjunction with good resistance to crater wear.

The American C-system starts its grouping with C1 to C4 for cast-iron type grades (corresponding to the K-classification) and continues with C5 to C8 for steel grades (corresponding to P). The M-range is not covered.

Achieving the optimum machining results is often a compromise, but the likelihood of achieving this aim is enhanced if the correct carbide insert grade is selected. Having identified the correct insert, what other factors must be considered before machining commences? These factors will now be discussed.

1.4 Milling techniques, insert geometry and forces

The direction in which the cutter approaches the work is a basic factor which has important ramifications in terms of the forces exerted, the chip morphology, and residual stresses induced into the machined surface.

Fig. 1.11a illustrates the conventional, or up-cut, method of milling, where the milling cutter rotates against the feed direction. In up-cut milling, a 'cold' cutting edge enters the work and a chip builds up during the cut. Before the insert starts to cut, burnishing takes place as the insert glides over the surface; this tends to separate the cutter from the workpiece as a result of induced cutting forces. The burnishing action produces cold plastic deformation at the machined surface, resulting in residual compressive stresses. During further contact, the heat in these chip surface layers increases, and they are removed by the next insert.

When the milling cutter rotates with the feed direction, it is known as either climb or down-cut milling (Fig. 1.11b). With down-cut milling, the cutting process starts at maximum chip thickness, which causes an impact stress. During the contact, the temperature will rise and this heat will be transmitted into the machined surface, producing tensile stresses.

In up-cut milling, the direction of the cutting force is away from the work, whereas in down-cut milling it is towards the workpiece. Generally, down-cut milling should be encouraged in any CNC application, as a more efficient cutting action occurs, less spindle power is required, and lower forces are generated. The resultant force generated by up or down-cut milling can cause work-holding problems or cutter push-off, and extreme care must be taken to avoid these.

In face milling (Fig. 1.11c), where the cutter is symmetrically positioned over the workpiece, the two conditions balance one another.

With regard to cutter geometries, there are more cutting angles to consider in milling than in turning. The simplest geometry occurs in peripheral milling, where the rake angle affects the power requirements and chip formation: the larger the rake angle, the lower are the cutting forces and power requirements. If tough 'sticky' materials such as light alloys are to be machined, then a large rake angle is required: conversely, for hard brittle materials the angle is smaller. Clearance angles are designed to provide clearance between the insert and workpiece; they may sometimes be designated as primary or secondary. Ductile materials, for example copper, require larger clearance angles than harder materials. Also, flank wear develops more slowly as the clearance angle is

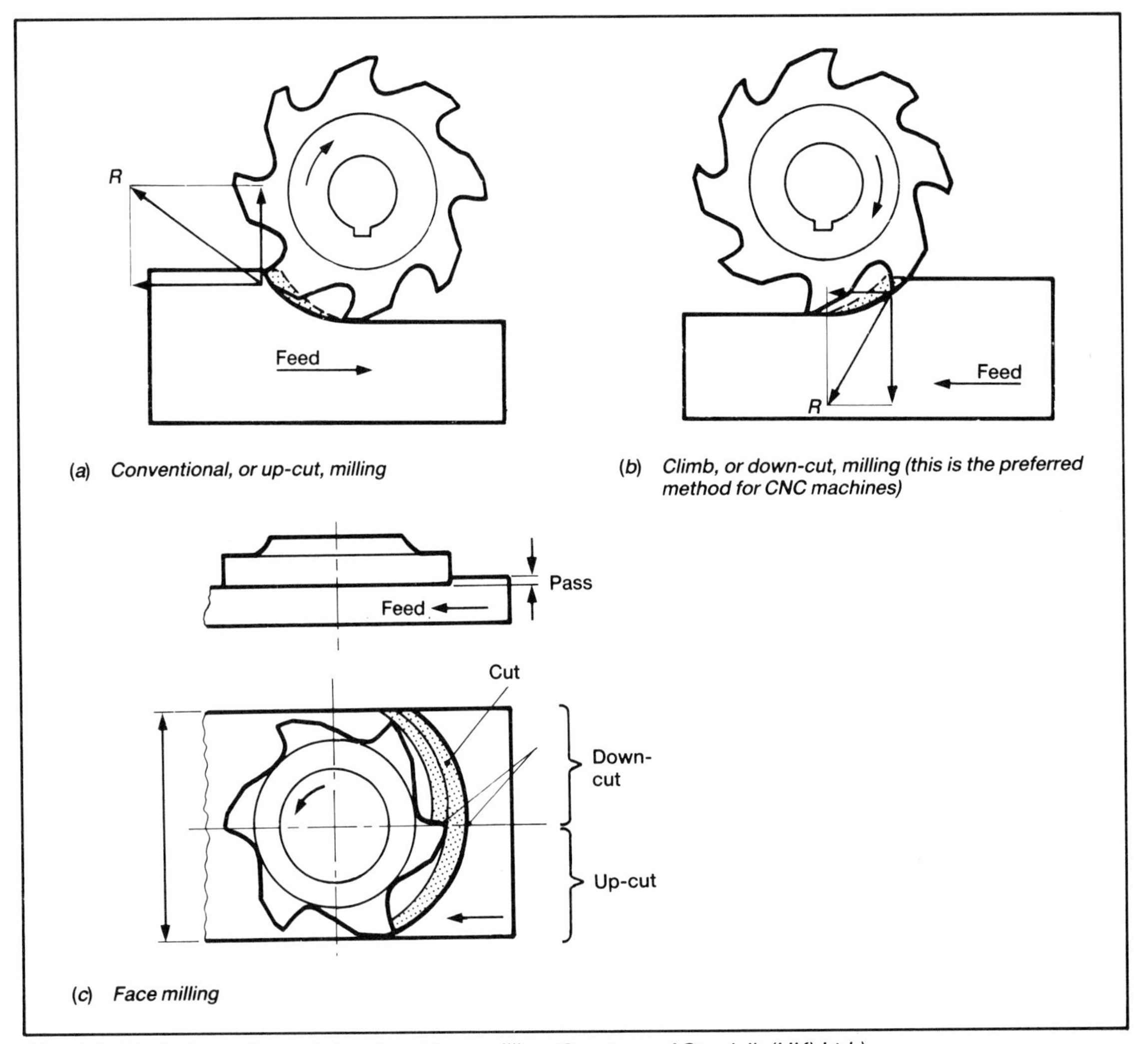

Fig. 1.11 Techniques for peripheral and face milling (Courtesy of Sandvik (UK) Ltd.)

increased. The combination of the rake and clearance angles determines the wedge angle, which greatly influences the insert's strength. Carbide inserts are usually reinforced by rounding the cutting edge, or by providing a negative primary land to avoid fracture. The helix angle brings the cutting edge progressively into cut, resulting in a quieter run; it also gives rise to an axial component of the cutting force. This force has a tendency to either pull the cutter out of the spindle or push it towards it, depending on whether it is right or lefthand.

There are basically three types of cutters so far as the geometry of the axial and radial rake angles is concerned: double-positive, double-negative and positive/negative. Examples of these are illustrated in Fig. 1.12. We shall consider the effects of these insert geometries in turn.

The double-positive insert geometry (Fig. 1.12a) has a single-sided positive insert. This geometry allows for a freer cutting action than with double-negative cutters. Lower cutting forces are produced, owing to the reduced chip thickness and the shorter length of contact. Therefore less spindle power and lower insert strength are required, compared with those for a typical double-negative cutter. The way that chip formation occurs is beneficial, in that spiral chips are formed which can easily be removed from chip pockets. When machining materials such as aluminium, ductile steels, and some stainless and heat-resistant steels, where there is a tendency to form built-up edge, the

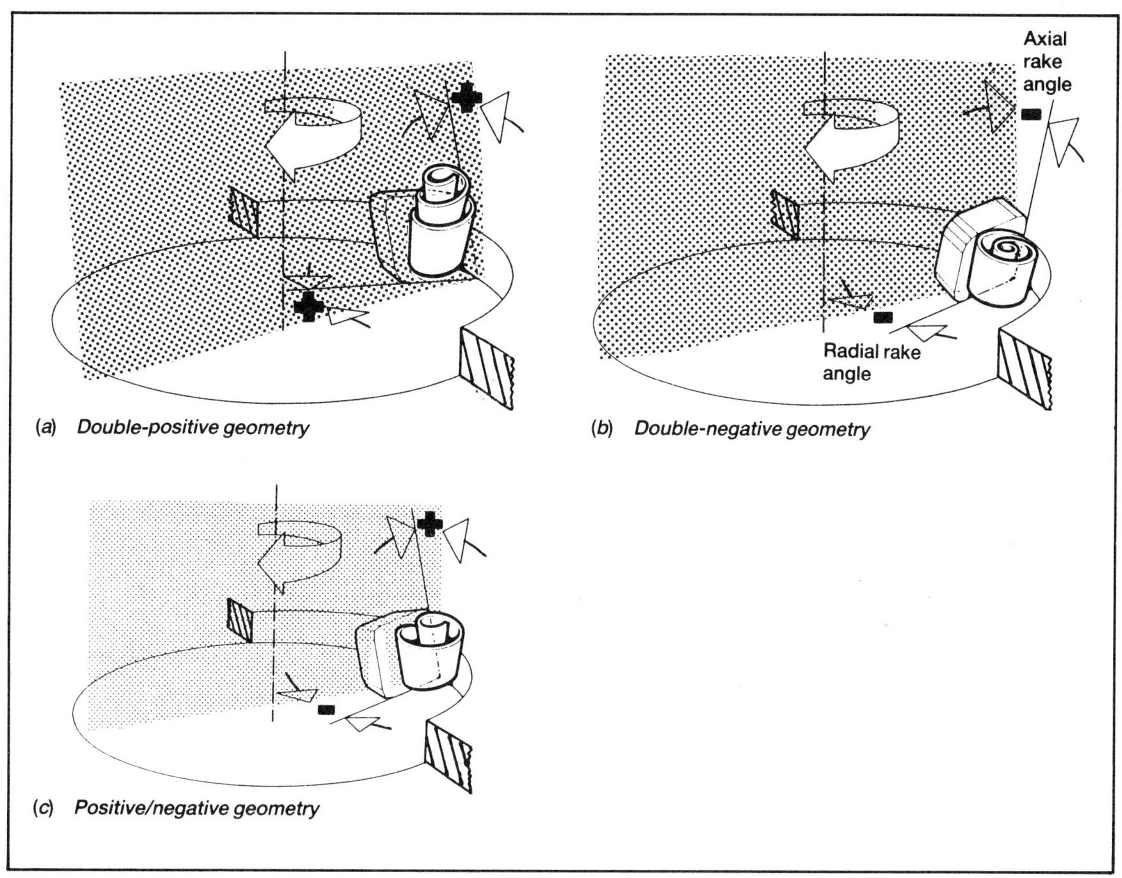

Fig. 1.12 Insert geometry for milling applications (Courtesy of Sandvik (UK) Ltd.)

double-positive geometry is often the only solution. Alternatively, if the workpiece has an unstable tendency or the spindle power is low, then the double-positive insert geometry is again the most suitable.

The double-negative insert geometry (Fig. 1.12b) occurs when the radial and axial angles are both negative. A negative insert is used, and clearance is obtained by tilting the insert. This has an added benefit economically, since both sides of the insert can be used, allowing more cutting and resulting in stronger edges. This geometry is suitable for materials or machining conditions involving heavy impact stresses associated with hard steels and cast iron, or if the machine has poor axial bearings. The demands on power and stability are considerable, owing to the large cutting forces associated with this geometry and the large chip thicknesses.

A combination of positive axial angle and negative radial angle occurs in the positive/negative insert geometry (Fig. 1.12c). Although the basic form of the insert may be negative, the edge on its end face must be positive in order to give a positive axial rake. The power requirements of this geometry are a compromise between the lower demands of the double-positive geometry and the higher ones of the double-negative. High feeds per tooth combined with large depths of cut can be achieved with positive/negative geometry, as the negative radial rake provides high insert strength whilst the positive axial rake offers good chip formation and directs chips away from the cutter. This can be considered to be an excellent general-purpose cutter geometry.

Of equal importance to the insert geometries mentioned above are the approach, or entering, angles of the inserts. Fig. 1.13 shows typical approach angles to the workpiece, and the effects of these entering angles will now be considered.

Inserts with entering angles of 90° are known as 'square-shoulder' face mills, as they are usually used for cutting up to a shoulder. These cutters have some limitations:

- The chip thickness will be at a maximum for a given feed and, as a result, insert loads are high.
- The chip flow may be hampered.
- The high radial force (in relation to the axial force) produces unfavourable loads on the spindle and creates vibration tendencies.
- Positive triangular inserts are usually used, which weakens insert corners. For more difficult operations, the use of rhomboid-shaped inserts can offer much stronger cutting edges.

(*a*) *Problems associated with milling-cutter approach angles*

Effects of cutter approach angle on swarf thickness

Milling cutters of 90° approach angle have an effective tooth load equal to the selected feed per tooth. Cutters with less than 90° approach have tooth loads of the selected feed per tooth multiplied by the sine of the approach angle.

90° approach cutters produce minimum downward pressure and give the best results on thin-section components.

The action of 90° cutting, produces a 'knocking' effect, so cutters of less than 90° approach are smoother in performance. Because the effective tooth load is less, these cutters produce a degree of downwards or tangential force.

For general milling applications, the use of milling cutters with approach angle of less than 90° is recommended, providing the specific application allows. The following table gives effective tooth load applicable to the approach angle.

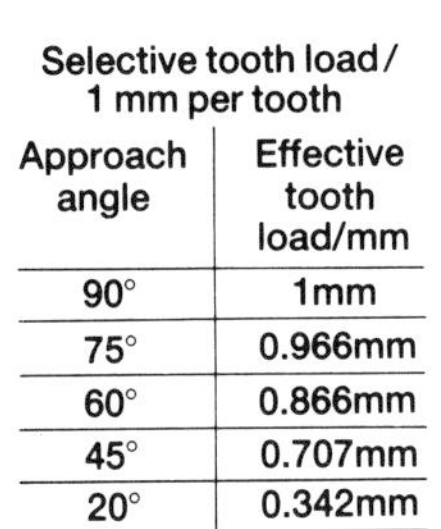

Selective tooth load / 1 mm per tooth

Approach angle	Effective tooth load/mm
90°	1mm
75°	0.966mm
60°	0.866mm
45°	0.707mm
20°	0.342mm

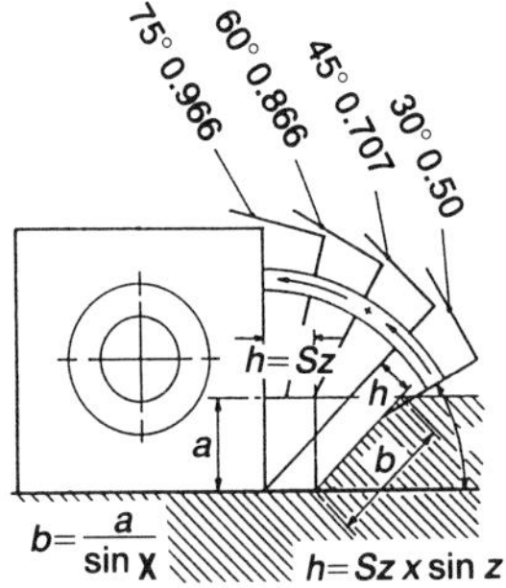

Minimising edge-breakout condition

Edge breakout is of prime importance when machining such materials as cast iron. If a 90° approach angle is used, edge frittering can occur (see below).

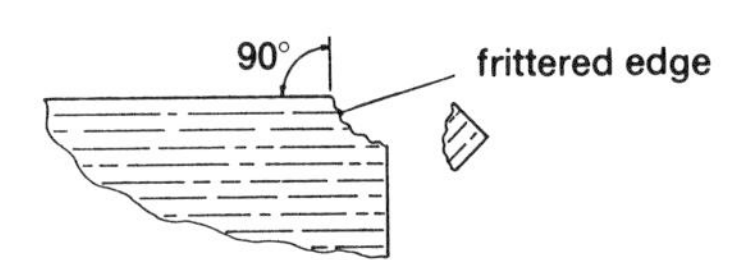

Edge breakout can be eliminated by reducing the approach angle below 90°

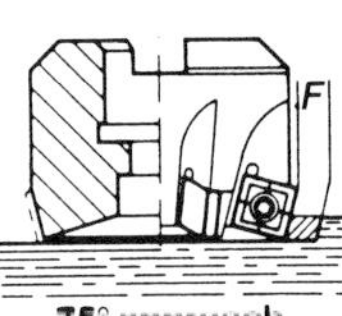

F=feed per tooth

Due to reduced approach angle, chip thickness is reduced thus leaving the base of the metal adjacent to component edge stronger. In order to minimise edge breakout, the following factors will have the greatest influence:

1 Low approach angle
2 Low feed per tooth
3 Low cutting force

(*b*) *Typical problems and remedies*

Problems \ Remedy		Use harder carbide	Use tougher carbide	Increase cutting speed v	Use lower cutting speed v	Increase feed/per tooth s_z	Reduce feed per tooth s_z	Change cutter (geometry)	Change cutter (coarse pitch)	Parallel land size	Reduce approach angle	Increase stability	Check cutting mounting	Edge condition
Insert fracture			X		X	X	X			X				
Cutting-edge spalling		X	X	X			X	X				X		X
Rapid flank wear		X			X	X								X
Rapid cratering					X		X							
Edge build-up				X		X								
Chip clogging								X						
Vibration								X	X					
Poor surface finish							X	X		X		X	X	
Workpiece frittering				X			X	X		X	X	X	X	

Useful calculations

$$\text{RPM} = \frac{\text{Cutting speed (m/min)} \times 1000}{\pi \times d \text{ (mm)}}$$

$$\text{Cutting speed (m/min)} = \frac{\pi \times d \text{ mm}}{1000} \times \text{RPM}$$

Table feed = RPM × Feed per rev (m/min) *or* RPM × No. of teeth × Feed per tooth

$$\text{Power} = \frac{\text{Width of cut} \times \text{Depth of cut} \times \text{Feed per min}}{F \times 1000}$$

where *F*=

EN1 – 3	= 14.7
EN4,5,6,8 & 42	= 13.1
EN18,21	= 11.5
EN24	= 8.2
Aust. St. Stl	= 6.5
Stain Steel	= 13.1
Cast iron	= 32.8
Sg iron	= 24.6
Alum	= 49.1

Fig. 1.13 Milling-cutter approach angles (Courtesy of Stellram)

Intermediate approach angles, such as 75°, are common for general-purpose milling operations; they offer good edge strength combined with a favourable relationship between insert size and cutting depth. An entering angle of 45°, on the other hand, will spread the cut and its load over a longer cutting edge. This 45° geometry gives good chip flow for long-chipping materials, radial forces are low in comparison with the axial force, there is a strong insert edge, and higher feedrates may be utilised. The machining of cast iron is improved with this entering angle geometry as there is a lower radial force, which reduces the frittering that normally occurs at the end of the cut (i.e. breakout at the corners).

If rough cuts are required, or if difficult materials requiring strong insert edges are machined, then a round insert might be the answer. In general, round inserts have a positive geometry; they offer more cutting edges than other inserts. This type of cutter usually exerts high axial forces and may cause a rough finish. A variation of the round insert geometry occurs in an American cutter, in which the insert's geometry is such that it is, in a sense, free to rotate whilst cutting. This rotation increases the cutting edge enormously and allows high feeds and speeds to be selected with good surface finishes.

Some typical face-milling cutters are shown in Fig. 1.14, which illustrates some of the range of sizes and insert geometries that are available. In fact, inserts can be fitted to almost every type of milling cutter; some of the different types are shown in Fig. 1.15.

Fig. 1.14 Inserted-tooth face-milling cutters for machining centres, etc., illustrating variations in size and in the number of inserts present (Courtesy of Sandvik (UK) Ltd.)

Fig. 1.15 Milling cutters using inserts: a staggered-tooth side and face cutter, siot drills and a porcupine cutter (Courtesy of Sandvik (UK) Ltd.)

Fig. 1.16 A typical face-milling operation: using an inserted-tooth TiC-grade milling cutter for milling steels (Courtesy of Kennametal (UK) Ltd.)

Inserted-tooth milling cutters should allow for easy evacuation of swarf, as any blockages can affect the cutting action or the application of a coolant. An efficient design is shown in Fig. 1.16, in which the swarf can be seen flowing freely away from the machined surface and the cutter.

In this section we have covered some of the major requirements of inserted-tooth milling, for which the choice of insert geometry and grade for the so-called optimum machining conditions depends upon many factors and will at best be a compromise. Now we shall consider the insert requirements of another major metal-removal process – turning.

1.5 Turning geometries, forces and insert selection

In turning, the insert geometry can be extremely varied, with a wide range of shapes, nose radii, plan approach angles, angular inclinations, etc.; two typical insert geometries are show in Fig 1.17a. Inserts can have their top and bottom faces the same size, in which case they are known as 'negative'; such inserts can sometimes be turned over to offer extra cutting edges, or they may be single-sided. If the top face is larger than the bottom, the insert is classified as 'positive'; positive inserts usually have cutting edges on one face only.

A large wedge angle occurs using negative-geometry inserts (Fig. 1.17b), they must be inclined in order to give a clearance angle and avoid rubbing on the workpiece. This inclination produces a strong blunt edge and causes the cutting forces to be increased accordingly, so a rigid machine-tool structure and higher power are required than if a positive-geometry insert is used. Positive inserts allow a smaller wedge angle to be utilised, resulting in lower cutting forces and less tendency for vibration because the shear plane is reduced. The main drawback with positive inserts is that, owing to the smaller wedge angle, tool life is reduced if heavy roughing cuts are taken.

Depending on the plan approach angle (Fig. 1.18), the cutting action of an insert will be orthogonal or oblique. So, for a given depth of cut, the chip zone may be extended or reduced depending upon whether the insert has a large or small lead angle. An orthogonal cutting action occurs when the insert is 'square' to the job (i.e. the entering angle is 90°, so the lead angle is 0°); this means that the load on the tool is immediately at a maximum as soon as the insert starts to cut. Orthogonal tools are ideal for turning long shafts of small diameter (i.e. sliding) as only one force is produced (if we ignore the effects of the tangential force and the small radial force generated by the tool's nose

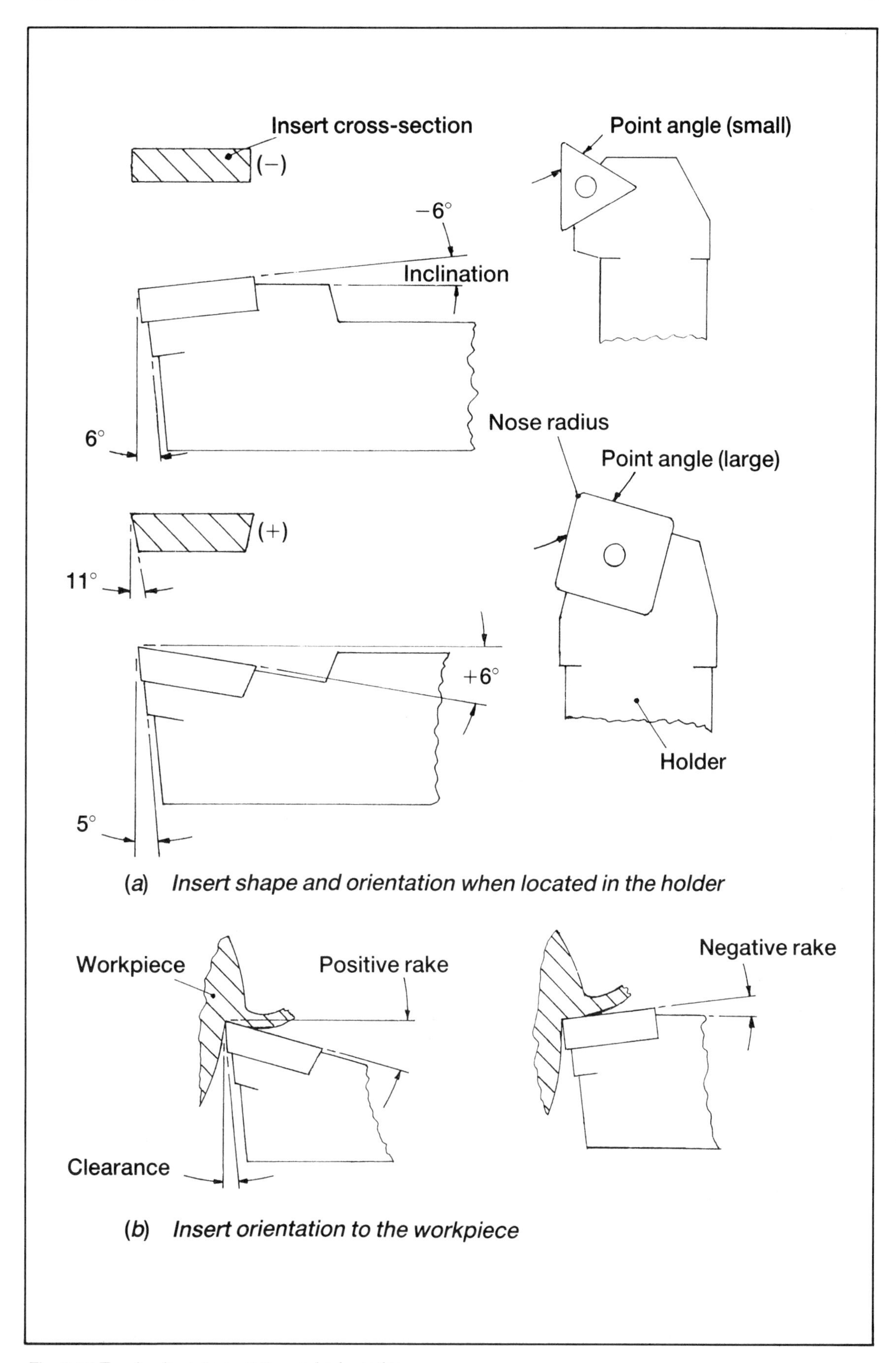

Fig. 1.17 Turning insert geometry and orientation

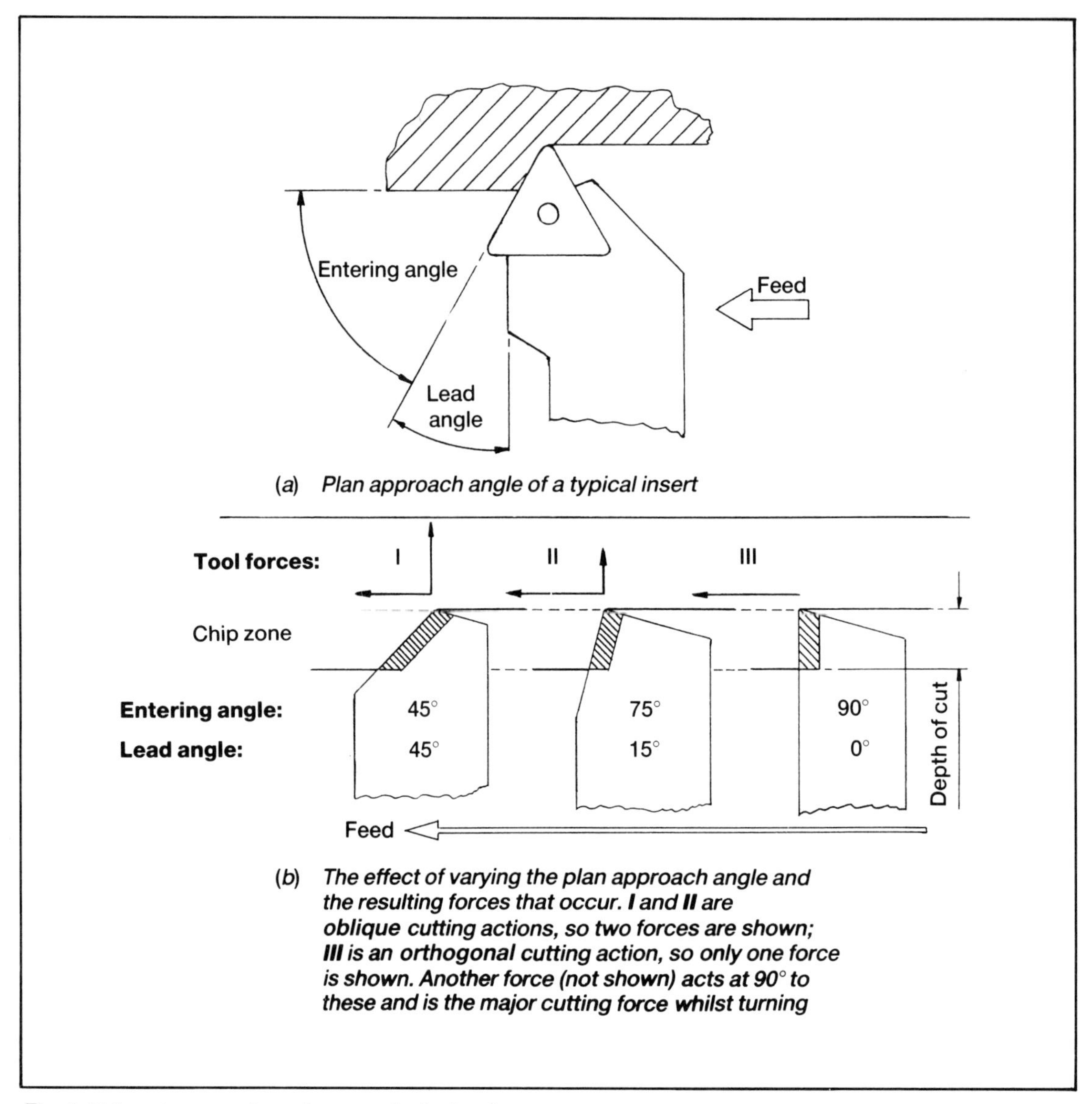

Fig. 1.18 Insert approach angle geometry for turning

radius). Thus there is no 'push-off' half-way along the bar which would cause a barrel-shaped effect (i.e. a larger diameter at the bar's centre and tapering at either end). Oblique inserts can have either positive or negative lead angles. Some positive lead-angled inserts are shown in Fig. 1.18b, which illustrates that the chip zone is extended along a longer cutting edge. This means that, for a given depth of cut, the load per unit area is reduced using oblique tools, compared with an orthogonal cutting condition; also the maximum cut depth is achieved more gradually. Therefore, a large plan approach angle increases the tool's life. However, it also produces a large radial force, and the greater the lead angle, the larger this becomes. As a result, oblique inserts should only be used on large-diameter work, or on relatively short workpieces (if unsupported by the tailstock), in order to minimise the influence of the radial force component.

Tools with positive plan approach angles are ideal for sliding operations (i.e. turning diameters), but are not so effective for facing operations (i.e. facing to length). If a combined turning and facing tool is required, then a negatively-inclined plan approach angle is used; this gives a clearance to the insert whilst facing-off and whilst turning diameters.

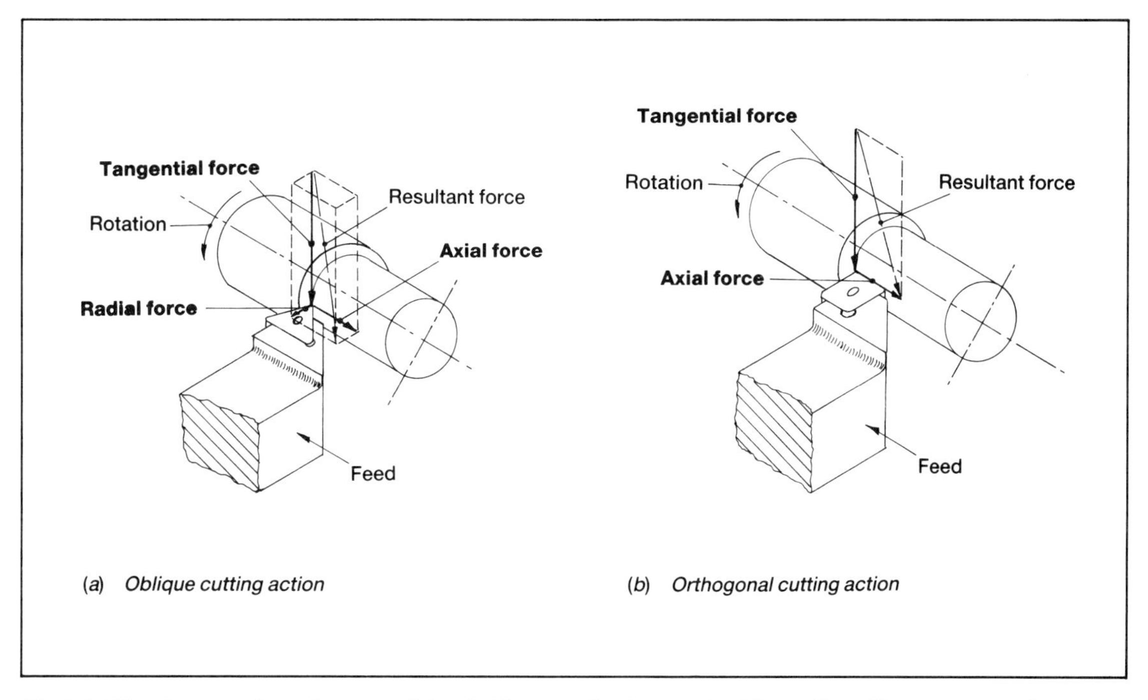

Fig 1.19 The three- and two-force models of oblique and orthogonal cutting actions (the component forces are scaled to give an indication of their magnitudes)

The forces produced by oblique and orthogonal insert geometries can be seen in Fig. 1.19a and b respectively. In both oblique and orthogonal cutting conditions, the tangential force generated by the bar's rotation is by far the largest force; it will vary in magnitude depending upon whether a positively or negatively-inclined insert is used. If a positive insert is in use, then a lower force will result, and vice-versa. The axial force is much smaller than the tangential component, and is determined by the feedrate: if the feedrate is increased, there will be a higher axial force. Thus the feedrate will influence the axial force – but not in isolation, since it will also lead to increases in the tangential force. The radial force is usually assigned to the plan approach angle: as this increases, so will the radial force, as illustrated in Fig 1.18b. The radial force will occur with both positive and negative plan approach angles. The nose radius of the insert will also produce a small radial force. 'Balanced turning' using twin turrets on slender long components equalises the radial forces of the two tools, so the radial force can be ignored with this configuration.

All of the factors mentioned above must be considered if the correct selection of an insert for a given application is to be made. However, there are many other variables that must be considered before the final insert may be chosen; these are shown in Fig. 1.20. In general, the fixed conditions cannot be modified, but by juggling with the variable ones it is possible to optimise the cutting conditions to obtain the best compromise insert for the job.

Fig. 1.21 gives a practical example of how changing only one variable – the geometry – can influence an insert's application in turning. The shape of the insert will be either strong or weak, and this must be considered depending on whether a roughing or a finishing operation is undertaken. The shape will either be prone to vibration, or else its life will be reduced, depending upon which of the extreme shapes it approaches. Variable conditions such as the indexable insert's geometry affects other parameters, and the same is true for most insert factors, so a compromise will always occur in any machinability testing.

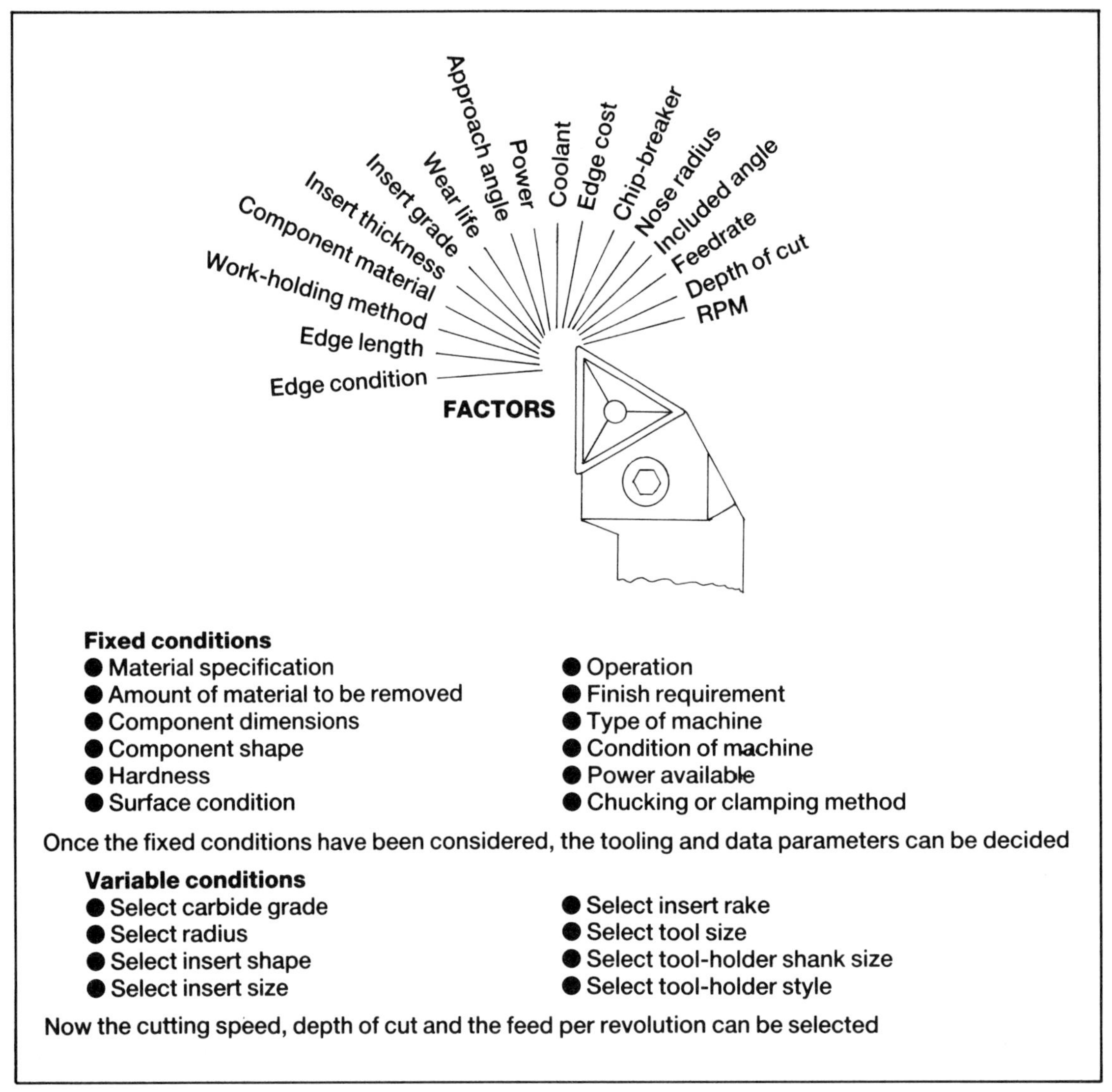

Fig. 1.20 The factors that must be considered before commencing turning when using indexable inserts (Courtesy of Seco)

Cutting tools may be classified into three basic types: unqualified, semi-qualified or qualified. Unqualified tools are basically 'one-offs' made specifically for a certain operation, for example form tools, where all the dimensions are variable. Semi-qualified tools usually have one or more standard dimensional characteristics with the others variable, whereas qualified tooling has standard values for all the dimensional data. Two examples of qualified tooling are shown in Figs. 1.22 and 1.23, which illustrate the ISO classification of indexable inserts and tool-holders.

In Fig. 1.22, the indexable inserts are classified (ISO 1832, or BS 4193 Part 1) according to an alphanumeric code. The first seven symbols are compulsory, the next two optional, and the remainder are used for the tool company's own internal classification. A typical example of an insert's designations is shown in Fig. 1.22; this is how an appropriate insert is ordered from a cutting-tool supplier.

Tool-holders and cartridges can also be ordered using standard coding, as shown in Fig. 1.23; these codes too are based upon an ISO standard (ISO 5608, or BS 4193 Part 6). Once again, an alphanumeric code is used to determine the appropriate holder or cartridge for the application under consideration. (The main difference between a tool-holder and a

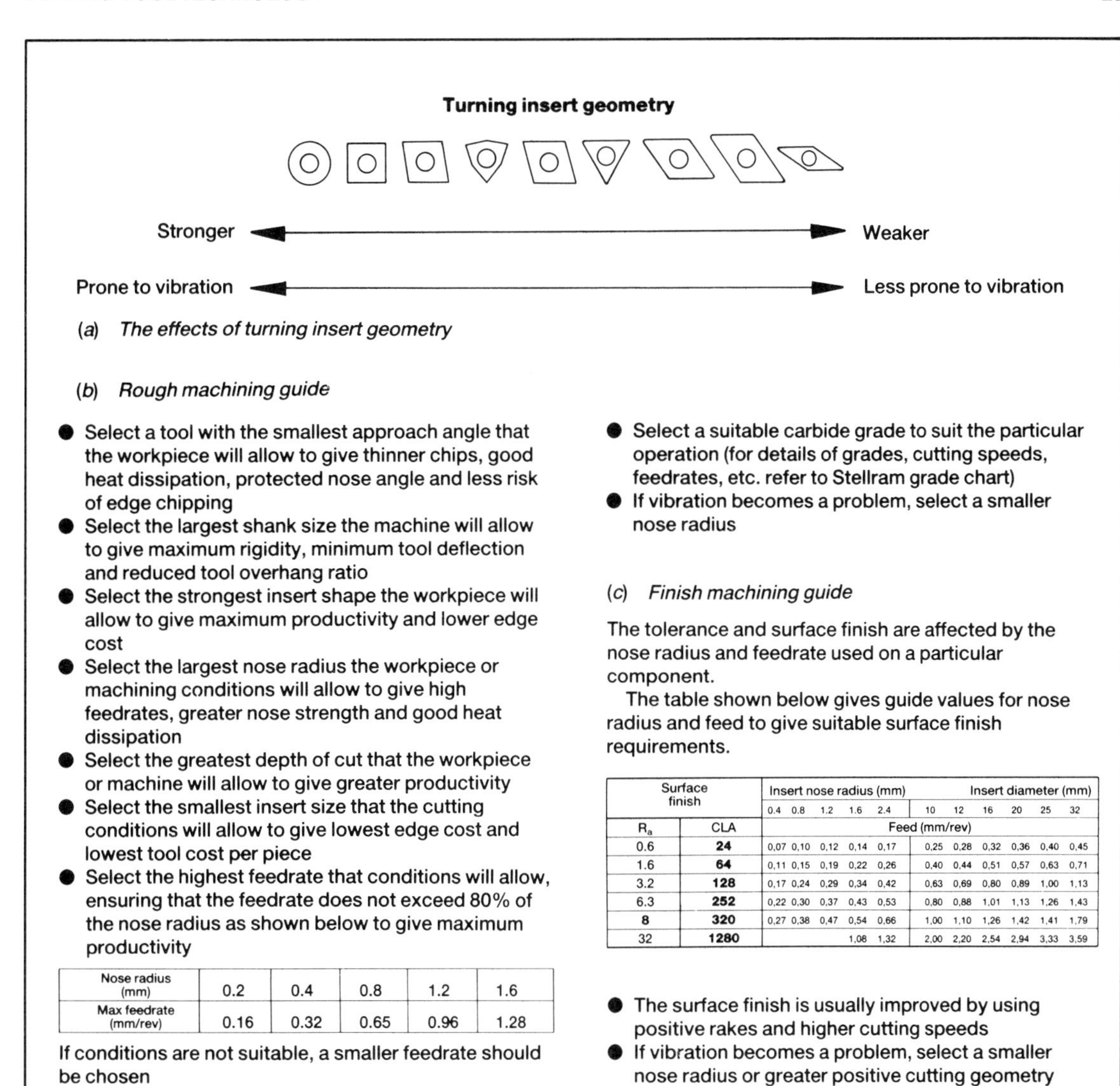

Turning insert geometry

(*a*) *The effects of turning insert geometry*

(*b*) *Rough machining guide*

- Select a tool with the smallest approach angle that the workpiece will allow to give thinner chips, good heat dissipation, protected nose angle and less risk of edge chipping
- Select the largest shank size the machine will allow to give maximum rigidity, minimum tool deflection and reduced tool overhang ratio
- Select the strongest insert shape the workpiece will allow to give maximum productivity and lower edge cost
- Select the largest nose radius the workpiece or machining conditions will allow to give high feedrates, greater nose strength and good heat dissipation
- Select the greatest depth of cut that the workpiece or machine will allow to give greater productivity
- Select the smallest insert size that the cutting conditions will allow to give lowest edge cost and lowest tool cost per piece
- Select the highest feedrate that conditions will allow, ensuring that the feedrate does not exceed 80% of the nose radius as shown below to give maximum productivity

Nose radius (mm)	0.2	0.4	0.8	1.2	1.6
Max feedrate (mm/rev)	0.16	0.32	0.65	0.96	1.28

If conditions are not suitable, a smaller feedrate should be chosen

- Select a suitable carbide grade to suit the particular operation (for details of grades, cutting speeds, feedrates, etc. refer to Stellram grade chart)
- If vibration becomes a problem, select a smaller nose radius

(*c*) *Finish machining guide*

The tolerance and surface finish are affected by the nose radius and feedrate used on a particular component.

The table shown below gives guide values for nose radius and feed to give suitable surface finish requirements.

Surface finish		Insert nose radius (mm)					Insert diameter (mm)					
		0.4	0.8	1.2	1.6	2.4	10	12	16	20	25	32
R_a	CLA	Feed (mm/rev)										
0.6	**24**	0,07	0,10	0,12	0,14	0,17	0,25	0,28	0,32	0,36	0,40	0,45
1.6	**64**	0,11	0,15	0,19	0,22	0,26	0,40	0,44	0,51	0,57	0,63	0,71
3.2	**128**	0,17	0,24	0,29	0,34	0,42	0,63	0,69	0,80	0,89	1,00	1,13
6.3	**252**	0,22	0,30	0,37	0,43	0,53	0,80	0,88	1,01	1,13	1,26	1,43
8	**320**	0,27	0,38	0,47	0,54	0,66	1,00	1,10	1,26	1,42	1,41	1,79
32	**1280**				1,08	1,32	2,00	2,20	2,54	2,94	3,33	3,59

- The surface finish is usually improved by using positive rakes and higher cutting speeds
- If vibration becomes a problem, select a smaller nose radius or greater positive cutting geometry

Fig. 1.21 Selecting inserts for turning (Courtesy of Stellram)

cartridge is that a cartridge is usually incorporated into a special-purpose body for performing multiple machining operations simultaneously, as illustrated later in this chapter, with radial and axial adjustments possible, whereas a tool-holder is used for conventional turning operations.) An example of the code for a typical tool-holder is also given in Fig. 1.23.

Before inserts were classified in this way, nearly every cutting-tool company had its own designations for tools and holders, which caused considerable confusion and unnecessary expense for the user.

Once a tool inventory is established, tools should be selected to cover the widest range of machining conditions possible, in order to ensure maximum utilisation of cutters and simultaneously reduce in-process idle-time. As a result of this careful selection, a certain amount of tool rationalisation will occur, enabling a few tools to machine a variety of turned features. An example of typical tooling is shown in Fig. 1.24, in which various insert geometries and tool-holder shapes are shown cutting features applicable to their shape. As may be appreciated from the examples in this figure, many of the tools can be used for more than one feature; this allows inventories to be further reduced and consolidated, with the financial savings this generates.

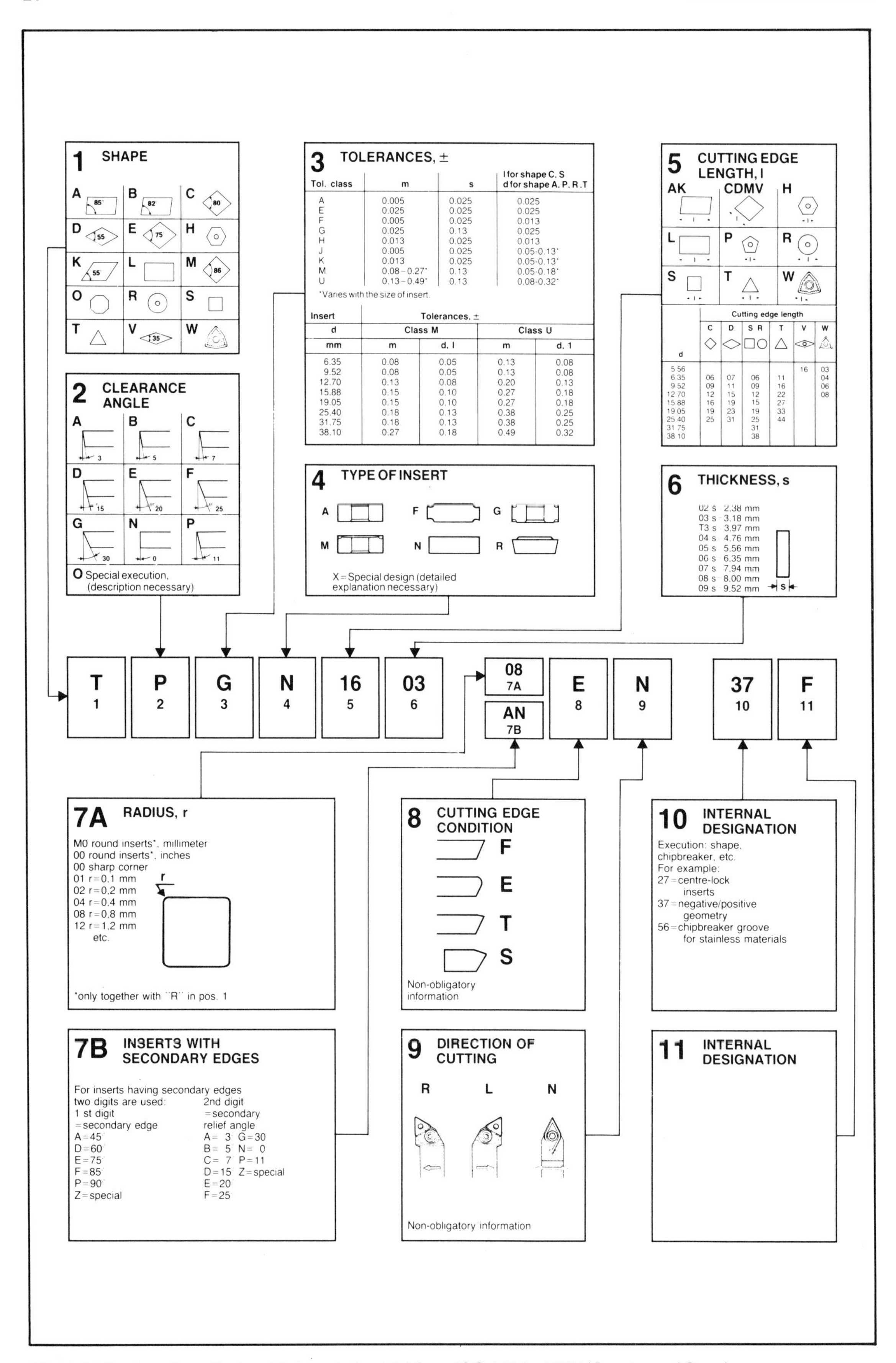

Fig. 1.22 Designation of indexable inserts (metric) from ISO 1832 : 1977 (Courtesy of Seco)

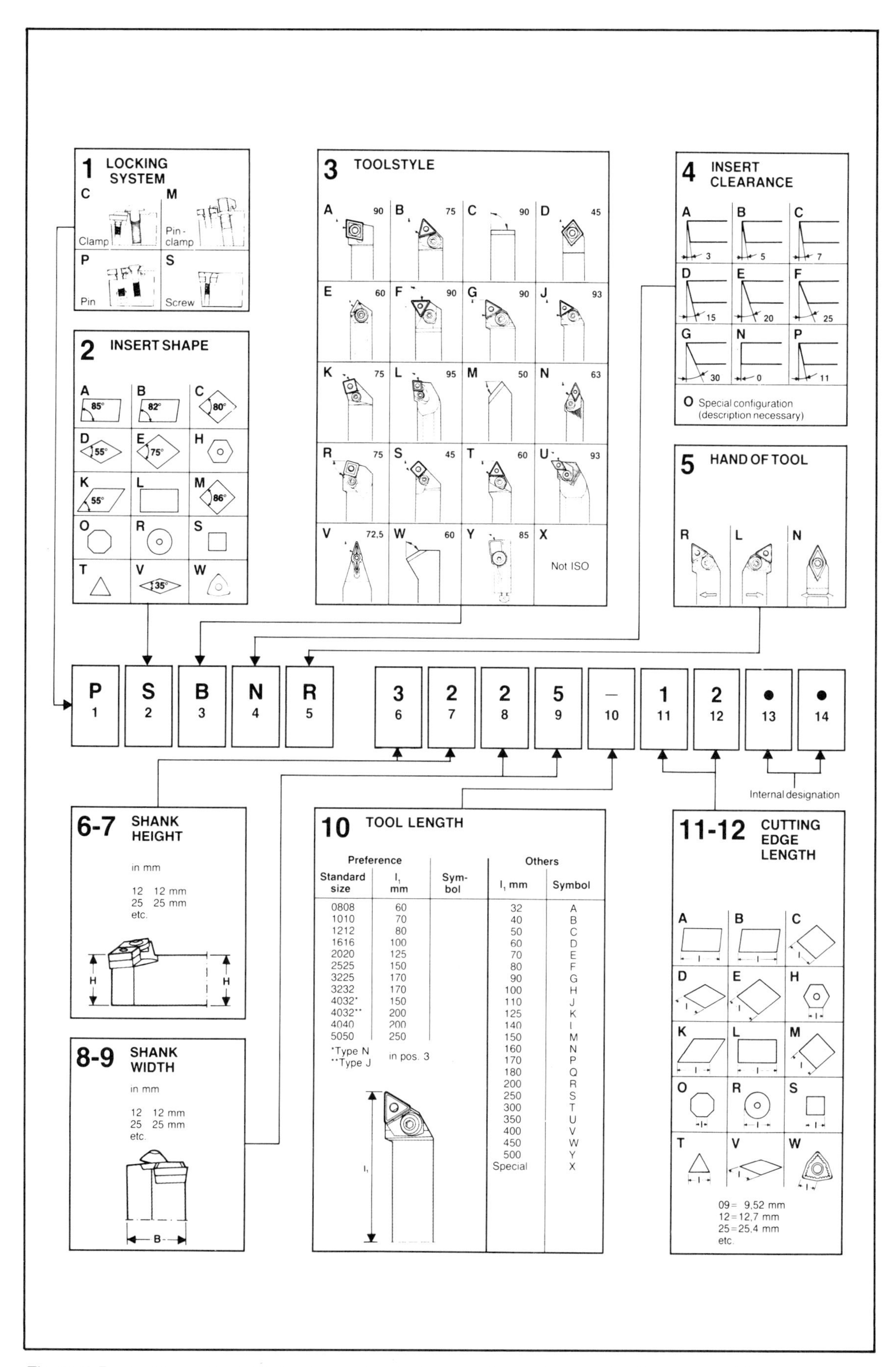

Fig. 1.23 Designation of tool-holders for turning tools according to ISO standard (Courtesy of Seco)

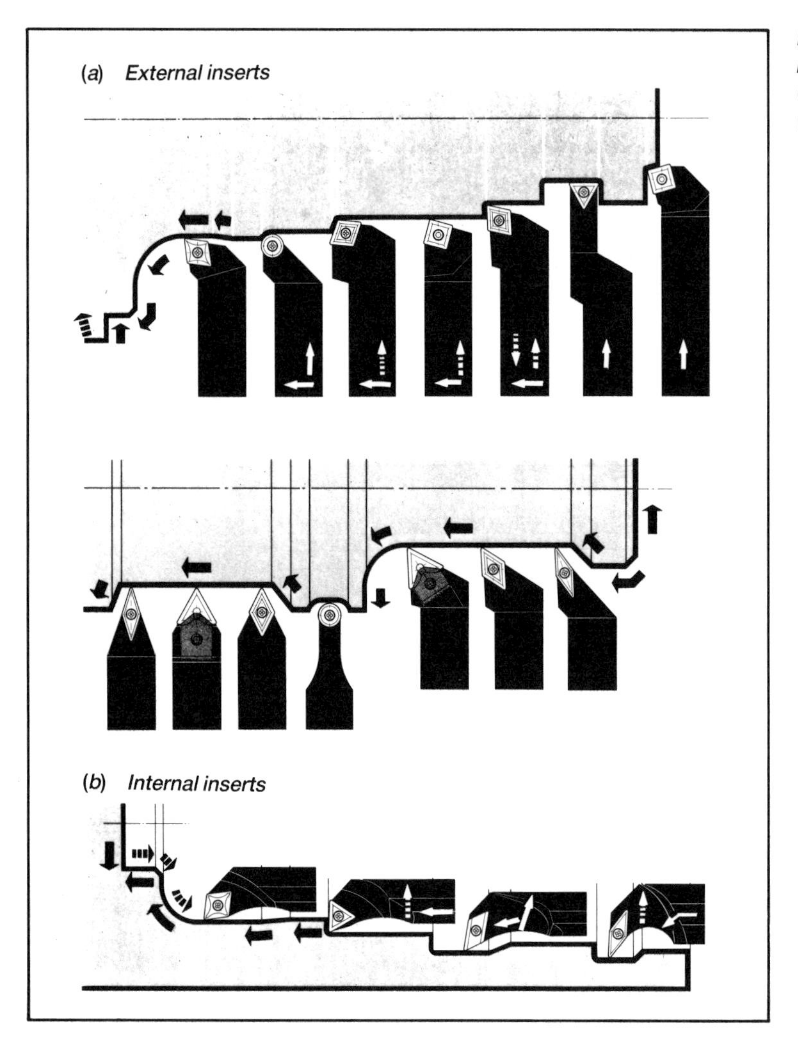

Fig. 1.24 Using typical turning insert geometries to machine different features on a workpiece (Courtesy of Stellram)

1.6 The principles of chip breaking

1.6.1 Swarf problems and remedies

A major problem of CNC turning is the large volume of swarf produced. The chip shape and its behaviour can cause a variety of problems. The degree to which swarf affects the turning operation will depend on operating conditions, but basically there are three requirements:

- The swarf must flow smoothly away from the cutting area without impairing the cutting action's efficiency.
- It must be of convenient shape and size to facilitate handling, storage, transportation and disposal.
- It must drop away to the swarf tray without snarling round the workpiece or tool, or interfering with other functions, e.g. automatic tool changing, component loading, etc.

In terms of priority, a smooth flow of swarf away from the cutting area is the most important factor in the efficient handling of swarf, although shape, size, handling, storage and lack of snarling have become more important with the advent of complex turning centres that use tool and part-delivery systems.

The cutting edge's primary function is to remove stock from the workpiece. Whether this is achieved by forming a continuous chip or by a flow of broken elemental chips will depend upon the properties of the workpiece material. The terms 'long-chipping' and 'short-chipping' are used when considering materials to be machined. Short-chipping materials, such as brass, cast iron, etc., do not present a chip-breaking problem for swarf disposal, so this section will concentrate on the long-chipping materials, with particular references to the steel 'family'.

Steels can be found with a wide variety of specifications, and their properties differ considerably. In addition, their method of processing (e.g. forging, casting, rolling, forming or sintering), together with the method of heat treatment, create further metallurgical variations that have greater influence on chip-breaking ability than the mere composition of the material. The strength and hardness values describe the material's character to some extent. But it should be borne in mind that it is the chip's mechanical strength that determines whether it can be broken easily. No absolute correlation exists between a steel component's strength and the mechanical strength of the chip produced.

A conventionally-turned chip is a rather frail product of serrated appearance (Fig. 1.25a). In order to promote good chip-breaking tendencies, so that short elements are formed, it is necessary to encourage this basic character by causing the serrations to be as deep as possible and the chip sections in between to be rigid, so that the chip is not flexible, and to steer the chip towards an obstacle that will provide a breaking force (Fig. 1.25b).

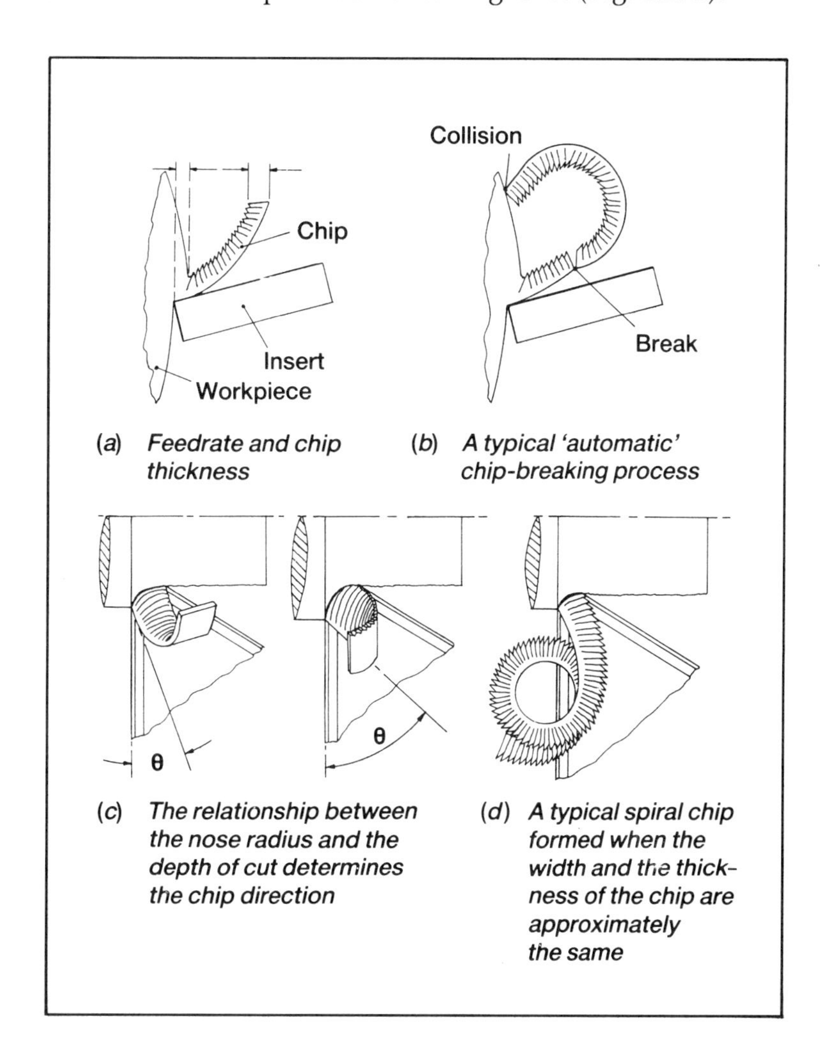

Fig. 1.25 The principles of chip breaking (Courtesy of Seco)

1.6.2 The chip thickness

It is a common misconception that feedrate and chip thickness are the same (Fig. 1.25a). At the instant when the chip detaches itself, the material is deformed, simultaneously shearing into a segmental form. For long-chipping materials (i.e. those of high ductility), a zone adjacent to the cutting-edge face will yield, intermittently producing segments that are held together by homogeneous materials. When conditions such as high cutting speeds and large rake angles are used, either separately or jointly, they cause the segmental structure to become less and less distinct. The yield zone usually occurs across the whole cross-section of the chip, which in this pure form may be known as a 'flow-chip'. The harder the chip compression, the less the segments will be connected and the thicker will be the chip, giving a lower chip strength. The degree of chip compression is increased by using a less positive, or more negative, rake angle, making it easier to obtain 'automatic' chip breaking (i.e. a self-breaking tendency). The term 'automatic' means that it can be achieved without the aid of a chip-breaker, using a flat insert to promote the formation of nicely broken chips at certain (high) feedrates.

1.6.3 Chip curvature

One of the contributory factors in obtaining a satisfactory chip-breaking tendency is the radius of curvature that the chip assumes at its initial formation. It is known that this radius decreases with more negative rakes and as the feedrate is increased. This results in a self-breaking tendency since the chip deflects upwards, hitting the workpiece above the point of cutting (Fig. 1.25b). This collision creates a bending force that breaks the chip.

1.6.4 The chip-breaker

It would be ideal if all the problems of chip-breaking could be resolved by choosing a more negatively-raked insert and high feedrates, but this is not the case. With certain types of materials this would lead to chip clogging, causing premature failure of the insert. Furthermore, the high cutting forces generated by some materials might make diametrical tolerances impossible to maintain. Other problems, such as high vibration tendencies, or requirements of surface finish, might also rule out the use of the 'automatic' chip-breaking method. An alternative method of obtaining chip breaking 'automatically' is to employ a chip-breaker geometry. This forces the chip to deflect at a narrower angle, causing it to break off, either immediately or just after the free end has hit the workpiece or tool flank before the first coil is formed. If such a collision does not occur, the result is a smaller diameter spiral chip, and it is very likely that the chip will not develop to a too-inconvenient length before it breaks. (This is because of the increasing chip mass and the effect of gravity on it, with or without any collisions.)

The direction of chip flow depends upon many factors, including the chip-breaker profile, back rake, setting angle, nose radius, depth of cut and feedrate. It is worth emphasising the last three factors further, since the nose radius, depth of cut and feedrate, and the relationship between them, often change during the course of the machining operation. Although the nose radius of the insert is fixed, its influence on the chip direction will be different for different depths of cut, depending on how much corner rounding is represented by the total edge length engaged (Fig. 1.25c). The feedrate also influences chip thickness; at different depths of cut and a constant feedrate, the form of the cross-section, i.e. the ratio of width to thickness of the chip, will change and this also affects the chip-breaking ability.

1.6.5 The helical formation of chips

In a recessing or parting-off operation, the chip typically forms a 'watch-spring' spiral; this only occurs when the cutting edge is parallel to the axis of rotation. Under all other turning operations, such as 'sliding', the chip is rolled up into a helix, simply because the edges are formed with different rotation radii (Fig. 1.25d). The two edges of the chip consume different quantities of material, so they must be of different lengths; coupled to the fact that there is a variation in cutting speed, this results in a helical chip formation. The chip helix's appearance will depend on the particular combination of cutting angle and other conditions used, and on the material specification and workpiece diameter; this means that it is extremely difficult to quantify.

The diameter of the most common type of helical chip is determined either directly by the initial curvature from its origin, or by additional bending produced by the chip-breaker. This helical chip type (Fig. 1.25c *left)* has its segments turned inwards; it is a desirable chip form when not fully developed (i.e. before the first coil has been completed). Whether or not the chip is of this form will already be determined even before it meets any chip-breaker; it depends upon its cross-section and the natural tendency to bend according to the 'line of least resistance'. If the chip's width is no larger than its thickness, for example, the resistance to bending in the segment-stiffened thickness direction is larger than in the width direction. So, unless this kind of chip is broken early, by colliding with either the workpiece or some part of the tool while it is still short and stiff (i.e. self-breaking), a helical chip will be formed. In this case, the barbed edge is turned outwards (see Fig. 1.25c *right*) and the chip is awkward to handle. This outward-curving helical chip also has weak sections in the serrations between the chip segments, but loads on it are readily absorbed by the spring action of the chip. Thus, no chip-breaker geometry can prevent such outward-curling helical chips, so combinations of feed and depth of cut that would result in the width being too small in relation to the thickness must be avoided.

A coolant may be used to enable close-tolerance machining to be undertaken; and will rapidly affect the temperature of the chips and can sometimes favourably influence chip-breaking, particularly when large cross-section chips are formed. In such chips, the heat content is larger and the cooling slower than in thin ones. Since the heat is retained longer, the chip's toughness increases, which is detrimental to its disposition to break. This is why use of a coolant may assist in chip breaking in such situations. However, experience has indicated that use of a coolant has the opposite effect in some cases, and adversely influences chip control. The coolant may lower the temperature at the cutting point to such an extent that a built-up edge begins to form. This welded-on material may alter the chip-breaker's profile temporarily, or it may have a more lasting effect. When carbide inserts are used, liquid cooling may involve the risk of edge failure due to thermal shocks; this explains why coolants are never recommended unreservedly by cutting-tool manufacturers.

1.6.6 Chip-breaker designs

There are three methods of achieving a satisfactory chip-breaking geometry on carbide tooling:

- A chip-breaker is ground directly into the tip; this is usually done with brazed tool types (Fig. 1.26a).
- A separate chip-breaker is positioned on top of the insert; this is usually done with the flat indexable varieties (Fig. 1.26b).

- A chip-breaking profile is pressed into the insert prior to sintering; these are referred to as a chip-breaker inserts (Fig. 1.26c).

This order reflects the historical development of inserts since the Second World War. The inserts were a result of the post-war boom in manufacturing which necessitated an increase in cutting speeds and meant that carbide tools were introduced rapidly to industry. The rising labour costs in the 1950s made companies more cost-conscious, the previous regrinding of tools was seen to be a waste of time, and the 'throwaway' principle was introduced along with the indexable insert philosophy. The first chip-breaking inserts were introduced in the early 1960s; the designs of these have become very complex and efficient recently.

We shall now look at these chip-breaking designs in more detail, begining with the ground-in types (Fig. 1.26a), which are still in use. By adapting the dimensions b and h as well as the angle θ (known as the chip-breaker edge angle) to the cutting data and of the work material, there is a likelihood that chip-breaking formation will be successful, even under difficult conditions. This technique is usually reserved for brazed tools, special-purpose indexable inserts and HSS tools (i.e. tool-bits).

The mechanical chip-breaker (Fig. 1.26b) is usually used on indexable inserts with a flat geometry, negative or positive. The design has been standardised with the chip-breaker edge parallel to the cutting edge (i.e. $\theta = 0°$). The b dimension can often be modified to cover a larger feed range. In practice, negative insert geometry produces poor overall economics, owing to problems with swarf, vibration or holding tolerances on small-diameter work. Swarf problems can also occur with some 'sticky' and tough materials. In

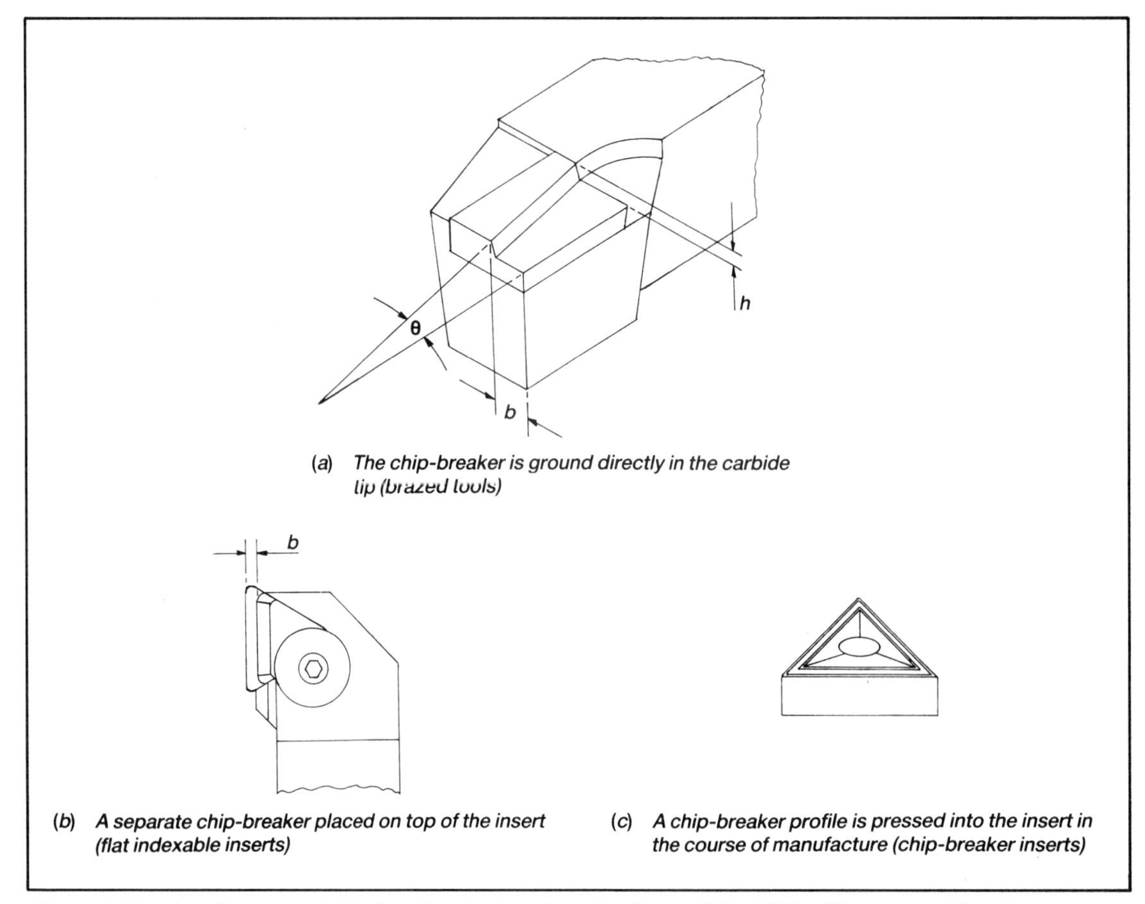

(a) The chip-breaker is ground directly in the carbide tip (brazed tools)

(b) A separate chip-breaker placed on top of the insert (flat indexable inserts)

(c) A chip-breaker profile is pressed into the insert in the course of manufacture (chip-breaker inserts)

Fig. 1.26 The development of chip-breaker design since the Second World War (Courtesy of Seco)

these conditions, an acute cutting edge (i.e. a large positive rake angle) is required in order to allow chips to flow away without clogging and blocking the cutting area. Chip clogging and hard chip compression often lead to insert failure, owing to the heavy pressure promoted by using negative-geometry inserts. So, although it seems paradoxical, the weaker positive-geometry inserts perform more satisfactorily in certain instances. For each degree that the cutting rake is altered in the positive direction, the cutting force will decrease by between 1% and 1.5%. Typical standard tools may have cutting rakes of -6° and +5° respectively, so the difference between them is 11°. This variation in the cutting rakes of standard tools can represent a net reduction or increase in the power requirements of between 10% and 15%. This margin of power variation might be especially important when the turning centre's power is limited.

Chip-breaker inserts (Fig. 1.26c) can have cutting rakes of up to +18° pressed into them, which is considerably more than the +5° provided by most flat positive inserts; they may be of double or single-sided design. Obviously the double-sided types of chip-breaker cannot withstand the higher feed forces and depths of cut that the single-sided varieties can, and they are intended for moderate use only. The single-sided types are recommended for normal use, as the reduced machining time produced by the higher feeds and depths of cut more than compensates for the extra cutting edges that would be obtained when using double-sided inserts. These single-sided inserts have completely flat bearing surfaces and can perfectly support the cutting edge, giving excellent chip-breaking qualities and allowing for a broader range of applications.

A problem associated with the chip-breaker inserts is the secondary wear of the insert face, caused by the chip helix breaking against it. This occurs when the helix attains such a diameter and pitch that its free end continually hits the non-cutting part of the insert's edge; it is known as 'chip hammering' and causes the edge to crumble. This can be alleviated by bearing in mind that a very slight increase of the helix diameter will cause the chip to break against the tool flank (below the edge), which is one of the favoured chip-breaking mechanisms.

If a flat insert is subjected to increasing crater wear, it progressively obtains a more positive cutting rake and becomes more free-cutting. Also, in its early development the back edge of the crater provides a chip-breaker action and aids chip flow. When the crater becomes deeper, this chip-breaking ability is lessened and chips become smaller and smaller, until eventually the swarf showers out onto the tray of the machine in an uncontrollable cascade, and at the same time the edge becomes weaker. This causes the crater to break through at the front edge, and the increased force will accelerate insert failure. Therefore, the chip-breaker's quality is diminished by the increasing crater wear.

In Fig. 1.27, the feedrate and depth of cut are represented along the two axes of a graph and the chip-breaking ability can be readily seen in terms of the areas occupied by the swarf generated. Where large depths of cut and small feedrates occur, it is apparent that a continuous chip form is present. If the cutting depth is maintained and the feedrate is increased, the chip-compression factor increases, leading to 'automatic' chip-breaking. If the cutting depth is now decreased whilst maintaining the feedrate, the chip-breaking is enhanced still further, with the added bonus that forces on the tool are lowered considerably; but this is at the expense of lower volumes of stock removed. As usual, a compromise is made, and an optimum chip-breaking effect is selected within the production constraints demanded by the product.

A good 'automatic' chip-breaking action can be seen in Fig. 1.28 where the chip-breaker (in combination with the feedrate and depth of cut) is causing the swarf to collide with

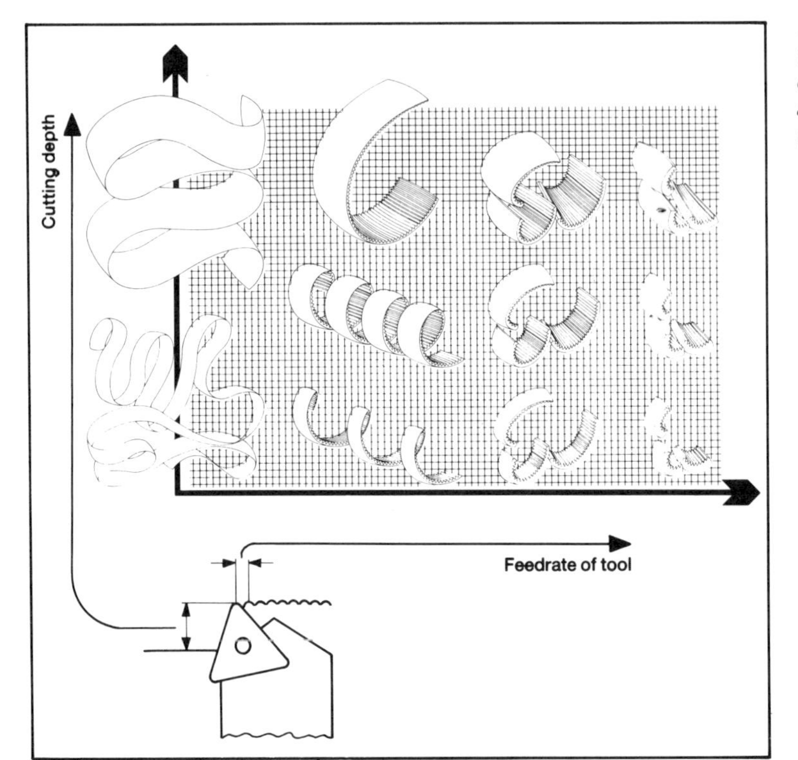

Fig. 1.27 The effect of increasing the feedrate and the depth of cut on an insert's chip-breaking ability (Courtesy of Sandvik (UK) Ltd.)

Fig. 1.28 A chip-breaking insert in a steel-turning operation, using TiC-coated grade (Courtesy of Kennametal (UK) Ltd.)

the insert. Another approach to good chip-breaking practice is illustrated in Fig. 1.29 where embossed dimples on the rake face lift and partially rupture the chip's underside. This in turn increases the curling of the chip, causing the swarf to be continually broken into convenient lengths. With this type of chip-breaker, lower feedrates and higher depths of cut can be maintained whilst still producing a chip-breaking action, whereas the conventional pressed-in chip-breaker would be producing a long unmanageable continuous swarf at this stage. Another added bonus of these latest chip-breaker designs is that, as the embossed dimples lift the chip away from the chip/tool interface much more quickly than the conventional pressed-in inserts and the area of contact is lessened, frictional effects are reduced and, as a result, the cutting forces are decreased considerably.

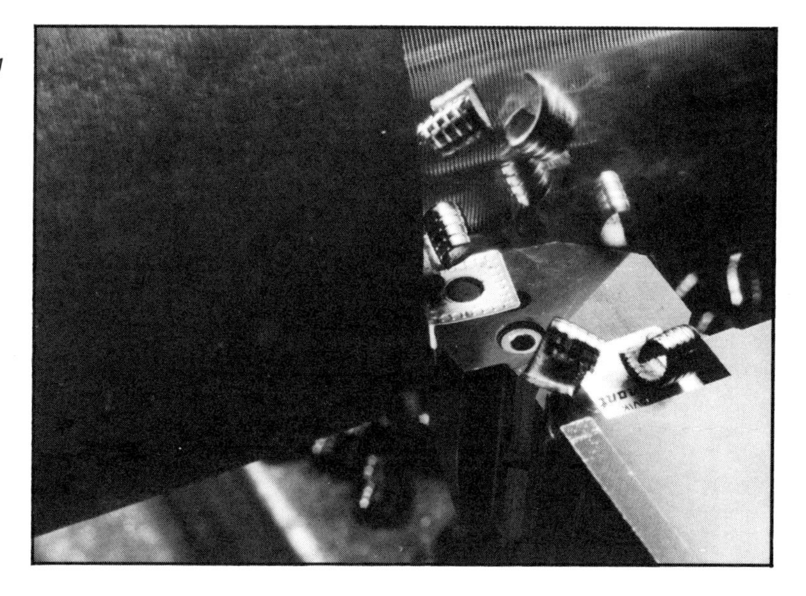

Fig. 1.29 A chip-breaking insert with embossed dimples, showing their influence on the swarf produced (Courtesy of Sandvik (UK) Ltd.)

This review of chip-breaking is by no means an exhaustive account; it is only meant as an explanation of ways of manipulating the variable factors in order to obtain the cutting conditions necessary to avoid the unacceptable entanglements with continuous ribbon-like swarf which would otherwise occur.

1.7 Drilling technology – an introduction

Hole making is by far the most prominent machining process, and it requires specialised techniques to achieve optimum cutting conditions. The term 'drilling' covers those methods used for producing cylindrical holes in the workpiece, including the machining of holes by counterboring. Two motions occur in drilling: rotation and linear feed, with either the tool being fed towards the workpiece or vice versa. Until the introduction of inserts in recent years, drilling technology had seen few advances since its original development; the traditional twist-drill design had hardly changed. In fact variations in twist-drill design have occurred latterly, including modifications to the cross-sectional area to increase strength whilst improving evacuation of chips. There have also been a variety of modifications to the design of drill points, including four-facet points, centre points and split points, whose function was to increase cutting efficiency as well as the drill's self-centring ability. The reasons why so much attention has been paid to the shape (cross-sectional area) and point angles of twist drills recently, is that for small holes (below 8mm in diameter) the indexable inserted type of drill cannot be utilised effectively.

1.8 Drill insert design and types of drill

If we disregard twist drills, drilling techniques with inserts can be classified in three ways (Fig. 1.30): solid drilling, trepanning and counterboring.

Solid drilling can be defined as the machining of a hole to a certain diameter in one operation, starting with a solid material. This is the most popular method of drilling.

Trepanning is also carried out in one operation. However, instead of cutting the complete hole, in trepanning only part of the hole is cut, leaving a core. This method is

(a) Solid drilling

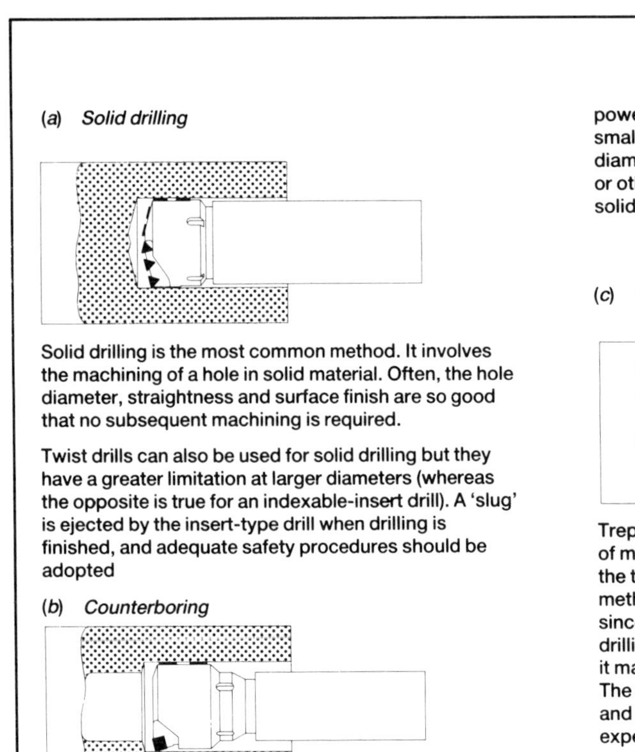

Solid drilling is the most common method. It involves the machining of a hole in solid material. Often, the hole diameter, straightness and surface finish are so good that no subsequent machining is required.

Twist drills can also be used for solid drilling but they have a greater limitation at larger diameters (whereas the opposite is true for an indexable-insert drill). A 'slug' is ejected by the insert-type drill when drilling is finished, and adequate safety procedures should be adopted

(b) Counterboring

Counterboring is a method for enlarging a hole, normally for the purposes of achieving a more accurate size, a better surface finish and/or a straighter hole. The method is used, for example, for counterboring forged, cast, pressed and extruded holes. Where machine power is insufficient, the hole can be predrilled with a smaller solid drill and then counterbored to the desired diameter. Hardening, tempering, stress relief annealing or other operations are sometimes performed between solid drilling and counterboring

(c) Trepanning

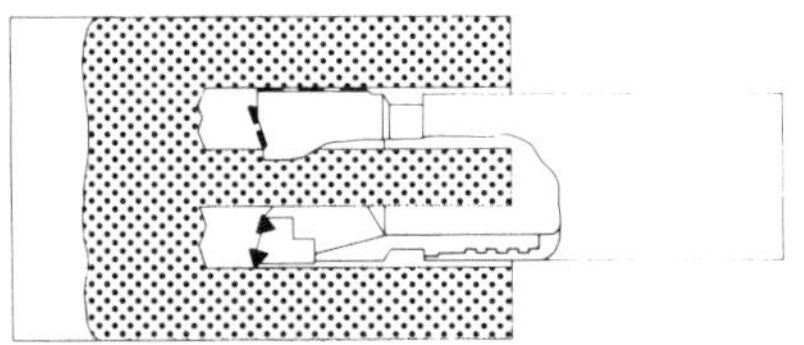

Trepanning is performed without predrilling, but instead of machining away all the material in the form of chips, the tool leaves a solid core in the middle of the hole. The method is chiefly used when machine power is limited, since the power requirement is lower than in solid drilling. In the case of large and expensive workpieces, it may be difficult to obtain suitable sample material. The core can then be used for tensile test specimens and material analyses. Particularly in the case of expensive materials, the core should be recovered and used for other purposes.

In the drilling of blind holes, a problem arises with cropping of the core. In deep holes, the core droops, which causes problems if insert-breakage occurs. The drill head must then be taken out of workpiece for insert replacement, and it is then difficult to guide the drill back into the drilled hole again

Fig. 1.30 The three different methods of drilling (Courtesy of Sandvik (UK) Ltd.)

particularly useful for large-diameter holes, as it requires less power and axial pressure than solid drilling. Restrictions occur with deep holes, in that they can cause problems in dealing with the core produced.

Counterboring is used to enlarge holes, either to give them an accurate dimension or for an improved surface finish; to a certain extent, axial inaccuracies can also be corrected. This technique can be performed using either counterboring heads or boring-bars.

These drilling operations can be performed on a variety of machine tools, such as machining and turning centres, transfer machines, or any type of machine capable of providing a relative rotation of the workpiece and the cutting tool.

Drilling is often further sub-divided into short and deep-hole drilling operations, with the design of the drill and positioning of the inserts being adjusted accordingly.

Short-hole drilling can be readily performed on both machining and turning centres and fulfils most hole-making requirements, whereas deep-hole drilling is usually confined to turning centres, horizontal machining centres or specialised machines. On turning centres, in which the drill's length is usually accommodated in the turret, care must be taken when indexing the turret so that the drill does not foul the machine. Similarly, on horizontal machining centres there must be enough clearance between the spindle nose and the workpiece when the automatic tool-changer is indexed, in order to avoid crashes.

The geometry of the insert will vary according to whether one is performing short or deep-hole drilling, and so will the positioning of the insert along the drill point and periphery. The arrangement shown in Fig. 1.31a is an example of complex insert geometry and positioning. Such multi-tipped drill heads are used, with coolant delivered through the body of the drill, for drilling larger diameters. As they rotate they generate a hole (Fig. 1.31b), so evacuation of chips from the cutting area is important. In them, the point has been displaced off-centre to relieve the high pressures that would otherwise occur on it as it rotates at extremely low cutting speeds compared with those at the periphery. The cutting fluid, which is fed under pressure, must fulfil a variety of functions: it must act as a coolant as well as a lubricant, and aid chip removal.

Chip-breaking ability is of prime importance with insert drills in order to avoid the 'bird's nest' effect around the tool. It is important to prevent the formation of excessively long and large chips, as they can become stuck in the chip ducts. But chips should not be broken harder than necessary, as chip-breaking is power-consuming and the heat generated increases wear-rates. The chip-breaking ability can be aided by the use of a 'peck-drilling' cycle. Normally, chips that are three to four times longer than they are wide are acceptable, provided that they can be evacuated effectively. Chip formation is affected by the work material, chip-breaker geometry, cutting speed and choice of cutting fluid. The chip-breaking ability of the insert varies with the cutting speed, which is particularly relevant since the cutting speed decreases from the periphery towards the centre of the drill. In a multi-tipped drill, for example, the central insert produces an over-hard breaking of the chip, whereas the peripheral insert gives satisfactory chip-breaking. In order to correct the hard breaking of the central insert, the feedrate should be reduced. Then the peripheral insert may produce excessively long chips, which can be remedied by simultaneously reducing the cutting speed.

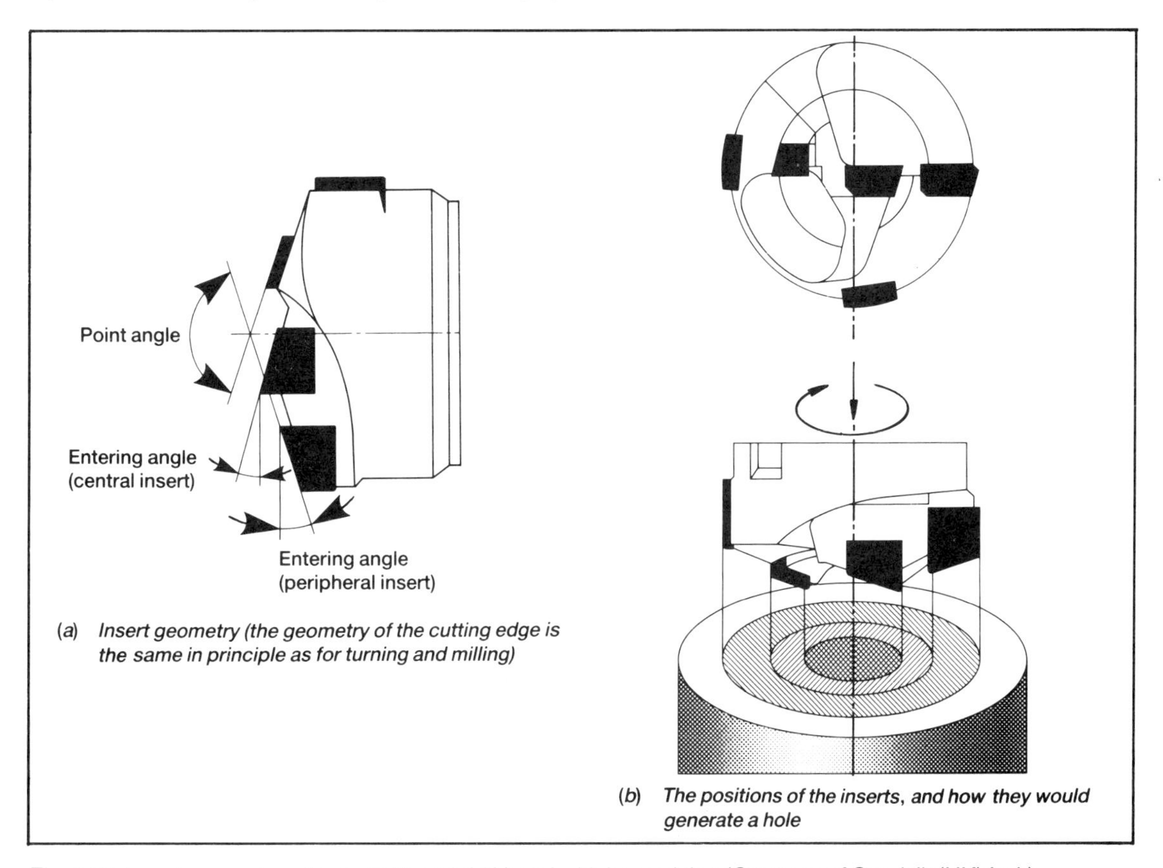

(*a*) *Insert geometry (the geometry of the cutting edge is the same in principle as for turning and milling)*

(*b*) *The positions of the inserts, and how they would generate a hole*

Fig. 1.31 Insert geometry of a multi-tipped drill head with brazed tips (Courtesy of Sandvik (UK) Ltd.)

It must be expected and accepted that a built-up edge will occur on the central insert with all drill types, and this may necessarily also be true for the intermediately-positioned inserts as well. The built-up edge formation becomes less of a problem with the coated grades of inserts. The effects of cutting fluids will be discussed in Chapter 5, and the influence of coolants on the tool work and swarf will be mentioned in Chapter 6.

Returning to insert geometry, it is worth stating that edge geometry in drilling is similar to those in milling and turning, with the angles on the inserts varying according to their position along the point and diameter. Normally, the angles at the drill point are different from those at the periphery.

Drills are also classified as either balanced or unbalanced, referring to the disposition of the cutting edges with respect to the axis. Typical balanced drills are twist drills, spade drills and some indexable drills, where the torque forces generated by the cutting edges cancel each other out. The unbalanced drills include gun drills and multi-tipped drills. To overcome a wandering tendency (due to imbalance of the torques on the inserts) guide pads are used on the longer unbalanced drills.

With some drills, particularly the non-rotating ones used on turning centres, it is possible to adjust the insert's position by off-setting it in order to obtain a better hole finish or size. The offset should be made with the cutting edges of the inserts parallel to the offset direction; this ensures that the small core always present at the centre of the tool does not become too large. If this core becomes larger than 0.5mm in diameter the drill's balance will become disturbed and the core may even damage the insert and the drill body. In order to achieve the required hole size when using a new insert it may be necessary to drill a trial hole first, adjusting its position afterwards to achieve the exact dimensions required around the nominal dimension.

With the insert-type drills, care must be taken when entering and emerging from the workpiece. If the face to be machined is inclined at an angle of not more than 5° to the plane of rotation, the U-type drills (Fig. 1.32) can normally be entered at full speed; if the angle is greater than 5°, the feedrate should be reduced until the drill has fully entered the workpiece. When the surface is convex (i.e. bulging outwards), the drill can be fed into the surface at full speed, as the centre insert is the first to make contact with the workpiece, causing minimal torque and vibration, and the hole size is not too much affected. However, if the surface is concave, the peripheral insert comes into contact with the surface first and it is safer to reduce the feedrate until full depth is reached. Some insert drill designs require a preliminary hole, or at least the start of one, to be made

Fig. 1.32 Insert geometry and location in U-drills (Courtesy of Sandvik (UK) Ltd.)

before the drill itself can be used; this is particularly the case with some of the unbalanced designs, as the self-centring tendencies of these are minimal.

Some of the most common types of drills for CNC machines are illustrated in Fig. 1.33, which shows the solid delta and U-drills, as well as the 'hollow' drills previously described. The wide range of inserts, drill diameters and lengths available caters for a wide variety of cutting conditions. The drill flutes can either be spiral, as shown in Fig. 1.34, or straight, as illustrated in Fig. 1.32. The spiral type, used in conjunction with a positive rake insert geometry, is ideal for use in the machining of aluminium or of materials that have a tendency to work-harden during cutting, such as stainless steels and high-temperature alloys. A feature of these drills is the replaceable pockets on the larger drills which extend the drill body's life. In the drills in both Fig. 1.32 and Fig. 1.34, coolant delivery to the insert is through holes in the body, which also help evacuation of chips. Such coolant channels can be clearly seen in Fig. 1.35a, which illustrates a different approach to insert design and shows how it is possible to overcome the problem of intermittent delivery of coolant to the cutting insert. This insert is of one-piece construction and non-indexable, but it is replaceable. Its design provides body clearance for the drill. Its application and limitations are also shown in Fig. 1.35b.

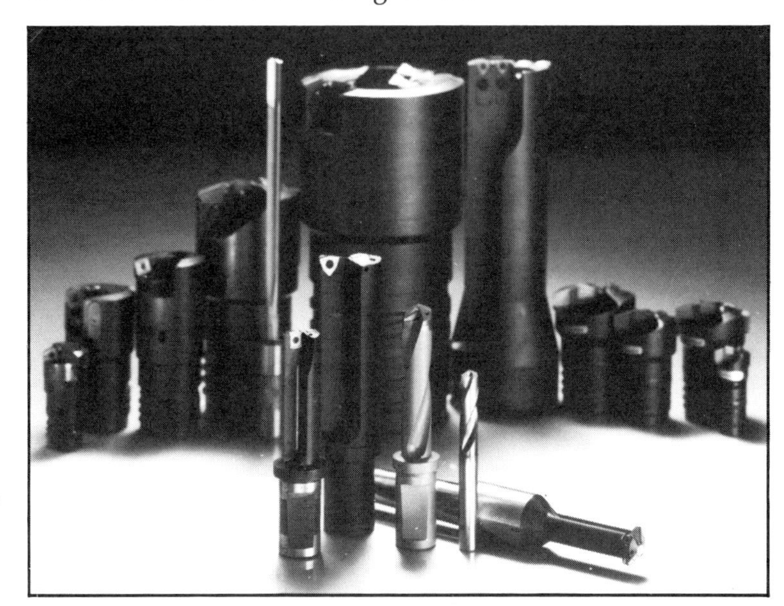

Fig. 1.33 A typical selection of solid and hollow drills, illustrating the varieties available and the range of sizes (Courtesy of Sandvik (UK) Ltd.)

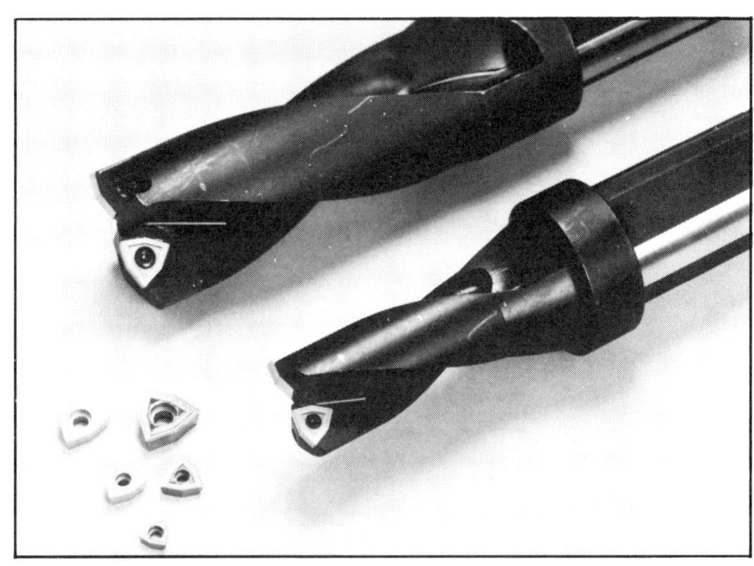

Fig. 1.34 Indexable-insert drills with positive-rake geometry for drilling materials such as aluminium and stainless steels (which work-harden whilst cutting) and high-temperature alloys (Courtesy of Kennametal (UK) Ltd.)

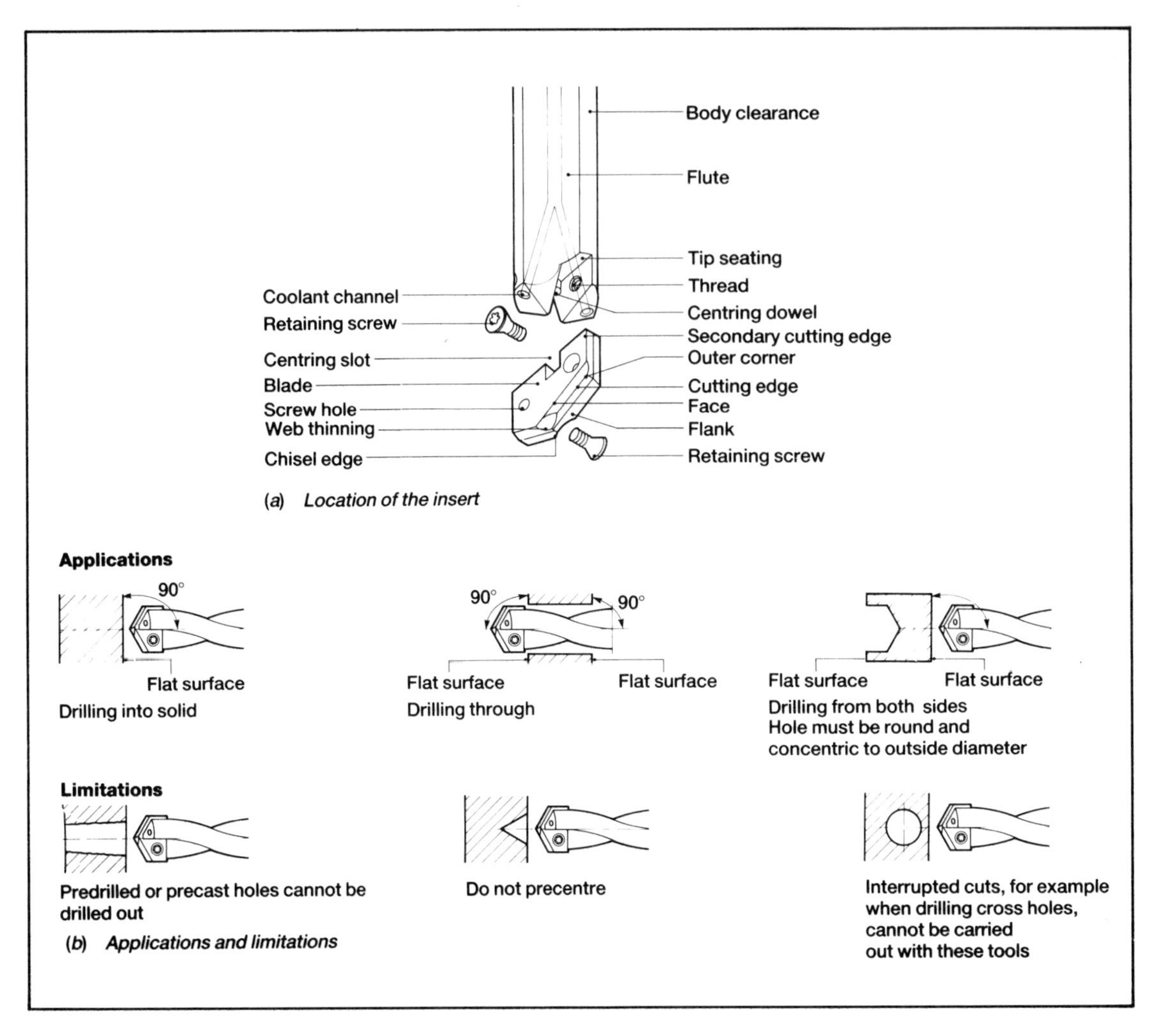

Fig. 1.35 A replaceable-insert drill (Courtesy of Stellram)

This review of drilling designs is not meant to be exhaustive, but simply a discussion of some of the types available. The effects of drilling wrought components, etc. will be discussed in Chapter 6, as will some of the problems associated with the drill's passage through the workpiece.

Drilling occurs for a variety of reasons: either so that shafts can be accommodated or bearings positioned, to introduce holes for lightness, or for many other applications. Very often, a thread is machined into the hole after drilling. There is a variety of processes for doing this, which are the subject of the next two sections.

1.9 Thread manufacture – an introduction

Screw threads have fulfilled fastening and translation requirements since Joseph Whitworth's thread of 1841 gained popularity; and many are still as popular now as they have always been. So it is useful to review the methods of thread production, but without going into a detailed discussion of thread elements and errors (which are presented satisfactorily elsewhere).

Threads can be produced in a variety of ways, involving one of two basic production methods, namely plastic working and metal cutting. The dominant method used throughout industry for producing threads commercially is plastic working. A typical example of this approach is thread rolling, which is employed for all conventional bolts and screws and which is also becoming increasingly popular for nut production. Thread rolling consists of pressing, or rolling, the thread's profile into the material's surface which is made to a slightly smaller diameter than the required major diameter of the thread (in the case of a bolt). When rolling begins, there is plastic deformation of the workpiece material, causing it to take on the required profile, reproduced on the rolls. Tangential or circular rolls are used in bolt production, depending upon the thread length; it is a very fast production method.

'Machine taps' are invariably used in the threading of holes; these are similar in appearance to conventional taps, but with a greater lead and with no flutes present. Once again, this method of thread production by plastic working is speedy, and it does not produce swarf.

Threads produced by plastic working are inherently stronger than those cut from the workpiece material. This is mainly because the grain structure is plastically deformed to the newly-formed thread profile, so the previous wrought directionality of the component is modified in this region to become similar to that of a forged structure. Threads can only be rolled on ductile materials, owing to the high forces required to produce plastic flow, which would be unacceptable or impossible using brittle materials.

A major problem associated with plastically-worked threads is that high accuracy and precision are required in many applications, and this method of production cannot achieve the necessary tolerance, profile, or pitch requirements demanded. In such cases, thread cutting is necessary. The production of threads by metal cutting can be subdivided into:

- Taps and dies – for manual or, most often, automatic threading, using opening, closing or solid versions for efficient thread production.
- Thread milling – this is principally used in the manufacture of large-diameter threads and for the production of worms and the like.
- Thread turning – this method automatically produces threads with perfect concentricity and of high precision.
- Thread grinding – a highly precise method of thread production.

It is not the intention to review the methods used for thread grinding, as this is generally considered to be outside the scope of most machining and turning-centre tooling. Similarly, the plastic deformation methods will not be discussed further, nor will the use of thread dies, as the speed and accuracy offered by turning and machining centres has to a major extent superseded their tooling application on these machines.

1.10 Threads produced by metal-cutting methods

1.10.1 Thread tapping

Tapping on machining centres almost invariably utilises automatic tapping devices, such as the one shown in Fig. 1.36. When using these reversible tapping units it is normal when programming the forward speed and rotation into the workpiece to reduce the tap's forwardly-driven speed to slightly less than the true pitch per revolution. This ensures that the tap is driven into the work by its own cutting action rather than by a preprogrammed feedrate. If this were not done, the thread might be damaged by incorrect pitching or have torn thread profiles, and in the worst case a broken tap could result.

Fig. 1.36 An automatic tapping holder (semi-floating design) cutting a thread on a component held in a fourth axis on a machining centre (Courtesy of Morris Tooling Ltd.)

Similarly, tapping units of the 'floating' type can be allowed to follow the previously-drilled hole, with the minimum of external influence to avoid thread errors. On some CNC machines, customised 'macros' can be developed for awkward tapping operations, or for difficult materials, so that once the problem has been initially overcome, the same 'macro' can be called up again as necessary.

1.10.2 Thread milling

Machining centres that have a helical interpolation facility can be used for thread-milling operations, in which the tool must be driven in a circular path whilst moving linearly at a feedrate which produces a movement of one pitch per revolution. Before milling a thread, the feature must be of the 'correct' diameter. In the case of internal threads a hole is required and for external threads a suitable boss is necessary. Holes are generally produced to the tapping size (or minor diameter) and bosses to the nominal thread diameter (or major diameter). If threads are required in blind holes, the normal procedure is to start the thread-milling operation at the bottom and move upwards. This ensures that the swarf can fall to the bottom of the hole where it will not interfere with the milling of the thread; whereas, if milling began at the top, trapped swarf could impede the cutter as it neared the hole's bottom and be a likely cause of errors.

It is highly desirable to obtain the smoothest entry into the workpiece by the cutter, as this avoids the likelihood of thread errors, therefore it is normal to 'arc' in and out of a cut. The point at which 'arcing' commences is chosen by considering the cycle time, which for a large internal thread feeding from the centre can be long. The smoothness of entry will also depend upon the arc radius: a small radius gives a more severe entry than a larger one. A general rule of thumb for thread milling is that the arc radius programmed is usually not less than the diameter of the cutter being used. To avoid 'thread thinning' when entering a cut, it is necessary to move the tool axis (Z) during the 'arc-in' block, so that the tool path reproduces the correct helix angle, before helically interpolating the thread.

External thread milling also requires smooth entry into the start of the cut in a similar manner to avoid 'thread thinning'. However, here the milling can be started tangentially to the boss rather than 'arced in', and at the finish of thread milling, it can be withdrawn in a similar way.

Cutting speeds used when helically interpolating threads can be increased to those of normal carbide milling, mainly owing to the short contact time between the cutter and the workpiece. The actual speeds and carbide grades used will depend upon various factors and specific applications, and cutting-tool companies should be able to furnish these details. Feedrates are normally in the range of 0.05 to 0.15mm per revolution; they are dictated by the requirements for surface finish and by the rigidity of the machine set-up.

It should be noted that when the tool is operating inside a circle (i.e. milling internally), the path that the centre-line of the tool follows is shorter than the path of the insert cutting the diameter. This is because the time taken to follow these two different paths is the same. Therefore the effective feedrate for the insert is considerably higher than that of the tool centre (which is the point at which the programmed feed is effective). Conversely, when the tool is operating outside a circle (i.e. in external thread milling), the centre-line path is longer than that of the insert, and the feed at the tool centre is faster than that of the insert. When using tool-radius compensation to position the tool, the control system automatically adjusts the feedrate to compensate for these differences, eliminating laborious manual calculations. When no radius compensation is used, or when it is used only to trim the diameter of the thread to size, the correct feedrate must be calculated to ensure that the right cutting parameters are maintained. For internal threads the feedrates must be reduced, and they must be increased for external threads.

Taper-thread milling involves programming a true involute. At present, only one control system in common use offers this feature, so it must usually be done using a 'high-level' language (i.e. parametrics), which some CNCs offer. This facility enables mathematical expressions to be developed in order to generate the involutes necessary for taper-thread milling. Alternatively, the involute can be split up into segments and programmed using normal interpolation; this gives acceptable results, which approximate to the 'true' involute. Owing to the taper on the thread, as the tool rises the radius at which it must cut increases according to the pitch. Having determined this figure the tool end-points can be calculated using increasing radius sizes. A difficulty

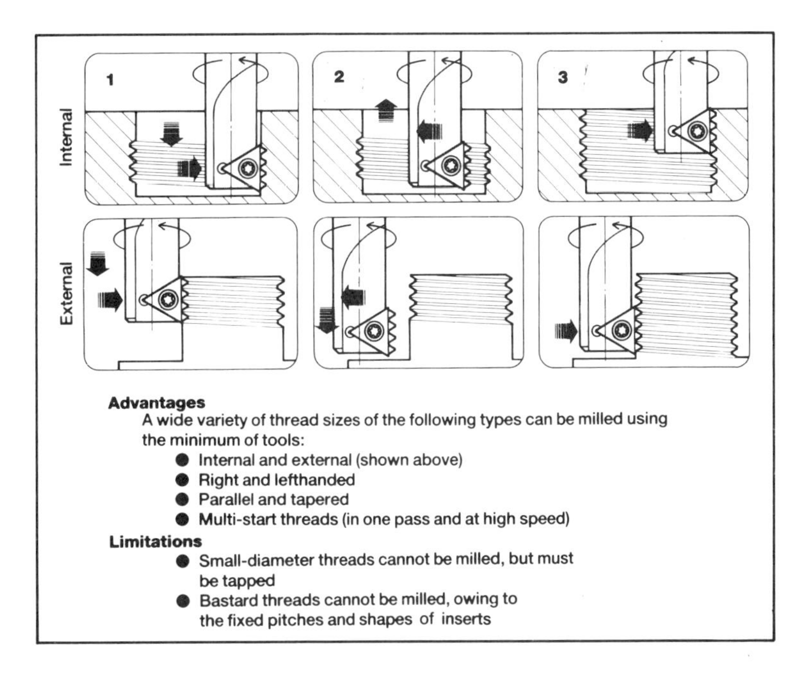

Fig. 1.37 Generating threads by thread milling: the machine must be equipped with a helical interpolation facility (Courtesy of Stellram)

occurs when calculating arc-centre coordinates, as the centres of each arc are in different positions, and are not at the thread centre. However, approximate positions can be calculated, which are accurate enough for most applications, by finding the mid-points between the start and finish coordinates of an arc, then projecting them back to the desired radius.

Typical thread-milling methods for internal and external threads are shown in Fig. 1.37. Here an inserted carbide milling cutter is being used; this will produce a precision machine thread that is suitable for most applications.

1.10.3 Thread turning

By far the greatest amount of thread production occurs on turning centres and lathes, using a range of single-toothed cutters such as those shown in Fig. 1.38. These can be used for both internal and external threading applications. The single-toothed cutters can be of the full-profile or V-profile design. The full-profile insert (shown in Fig. 1.41) offers benefits over the V-profile, whose main advantage is that it keeps tooling inventories smaller. There are a number of advantages of using full-profile inserts. Firstly, the workpiece diameter need not be preturned to the exact diameter but may be left slightly over-size so that it is skimmed during the final pass, and no final dressing is necessary. Secondly, compared with the V-profile insert, its nose radius corresponds to the thread specification, therefore the insert point is as strong as the relevant profile allows. Thirdly, as the full-profile insert is not unnecessarily pointed, the total feed depth will not be greater than that specified by the thread profile, so it may be possible to save one pass (or more), when cutting the thread. The fourth, and probably most important, point is that the full-profile insert assures that the correct profile is cut and that the thread is no deeper than specified. This results in a stronger thread whilst at the same time simplifying the threading operation.

Fig. 1.38 Internal and external indexable threading tools and inserts for use on turning centres, etc. (Courtesy of Sandvik (UK) Ltd.)

It is also possible to get inserts that are of the multi-toothed design. With this type of indexable insert, the second tooth cuts deeper than the first (leading) tooth and the third tooth, if employed, cuts deeper than the second. Only the last tooth has the full-profile form. When using these designs, the radial (plunge) infeed method must be employed. Use of these inserts is limited because:

- There must be a wide enough undercut at the end of the thread to accommodate the entire 'row of teeth'.
- Long cutting edges occur, giving rise to high cutting forces, so that stable machining conditions are essential.
- When internally threading holes of large diameter, a very sturdy boring-bar is necessary.

These are the multi-toothed insert's limitations; its major advantage is that when it can be employed it reduces the number of passes by two-thirds and increases the time between insert changes.

The ability to cut a thread on modern turning centres means that it is quickly and efficiently produced, in a time commensurate with other turning operations. There are various ways of turning a thread (Fig. 1.39). The workpiece can be rotated clockwise or anticlockwise, the tool can be fed towards or away from the chuck, and external or internal threads can be produced. In addition to this, the tool can be working in the normal position. or upside-down, depending upon what best suits the operation with regard to the machine, workpiece and chip clearance. The choice of method of screw cutting and the type of tool used depends upon prevailing conditions, such as accessibility of the start of the thread, and tailstock limitations of end space and limited run-out for the tool. When internally threading a workpiece, it is often advantageous to feed the tool away from the chuck. This simplifies chip clearance and allows the use of compressed air or a good coolant, which will further enhance chip evacuation. When externally threading tough materials, the chip clearance can be enhanced by using a tool in the upside-down position. The chips will then be able to fall straight down into the swarf disposal area without tangling around the tool or workpiece.

It is also worth mentioning the use of twin holders here. These are a combination of threading and turning inserts that are mounted so that a secondary machining operation can be done at the same time as cutting the thread. When a thread is cut on a tough material, a burr is inevitably formed at the start of the thread and is bent over the face where the thread began. This can be removed as the insert is taking its last pass along the thread, just prior to withdrawal. The turning insert is strategically positioned and adjusted so that it cuts the burr off as the final tooth pass occurs and withdraws as the insert exits the thread. The turning insert is of a modular nature and can be dimensionally adjusted until this effect is achieved.

The operation of threading is basically a turning operation in which the feed per revolution of the workpiece is equal to the thread's pitch. The tool tip (or insert) is ground to a profile which corresponds to the space between two adjacent flanks of the thread. The depth of cut will determine the chip thickness. A thread is never cut in a single pass, so the tool point must follow the exactly same path on every successive pass. To achieve this goal, the feed movement of the tool requires perfect synchronisation with the lathe spindle's rotation. The turning centre is designed to produce accurately synchronised linear and rotational motions, and the operator will program the correct data necessary for cutting the thread. The data fed into the controller will vary from machine to machine, but certain decisions must be made as to the thread pitch required, the position of the start of thread (to some extent this is controlled by whether the thread is left or right-hand), the thread profile, the amount of infeed of the tool per pass, the method used for generating the thread (more will be said on this shortly), and other associated information.

(*a*) *Threading insert geometry. For type A applications use positive helix; for type B applications use negative helix*

Righthand external threads

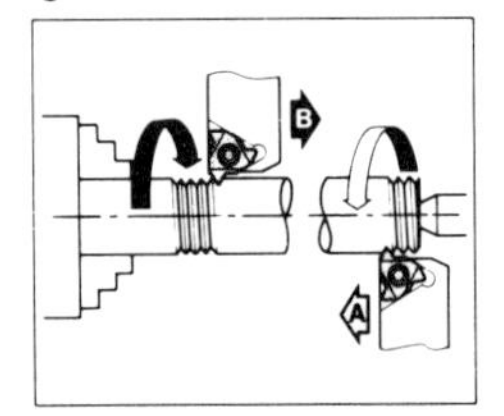

A Righthand external tool-holder with righthand insert
Component rotation: anticlockwise
Direction of cutting: from tailstock to headstock

B Lefthand external tool-holder with lefthand insert
Component rotation: clockwise
Direction of cutting: from headstock to tailstock

Lefthand external threads

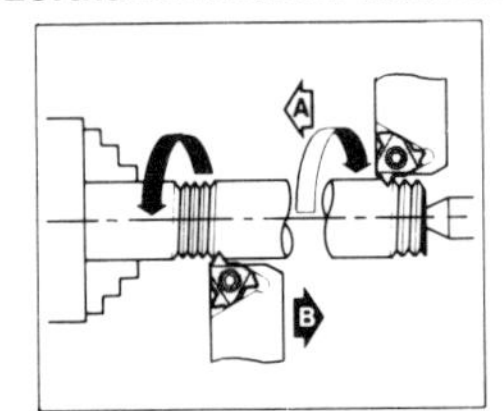

A Lefthand external tool-holder with lefthand insert
Component rotation: clockwise
Direction of cutting: from tailstock to headstock

B Righthand external tool-holder with righthand insert
Component rotation: anticlockwise
Direction of cutting: from headstock to tailstock

Righthand internal threads

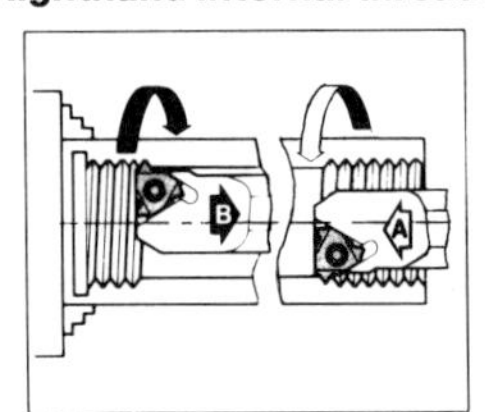

A Righthand internal tool-holder with righthand insert
Component rotation: anticlockwise
Direction of cutting: from tailstock to headstock

B Lefthand internal tool-holder with lefthand insert
Component rotation: clockwise
Direction of cutting: from headstock to tailstock

Lefthand internal threads

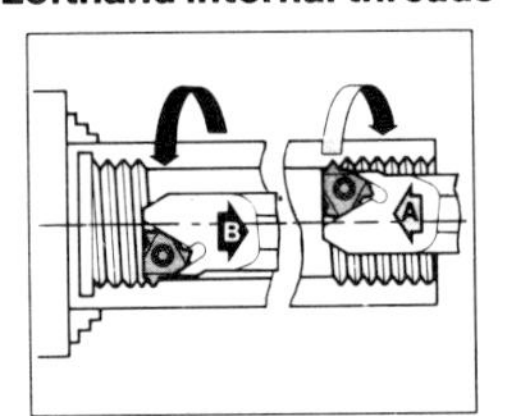

A Lefthand internal tool-holder with lefthand insert
Component rotation: clockwise
Direction of cutting: from tailstock to headstock

B Righthand internal tool-holder with righthand insert
Component rotation: anticlockwise
Direction of cutting: from headstock to tailstock

(*b*) *Helix angle calculations*

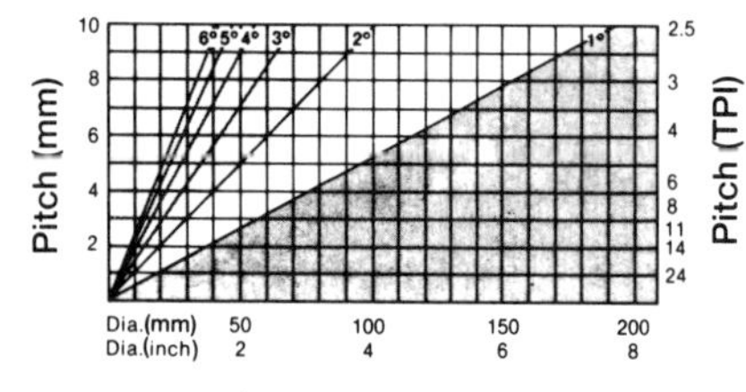

Helix $\tan\beta = \dfrac{L}{\pi \times D}$

where L is the lead and D is the effective diameter
Lead L = Pitch × Number of starts

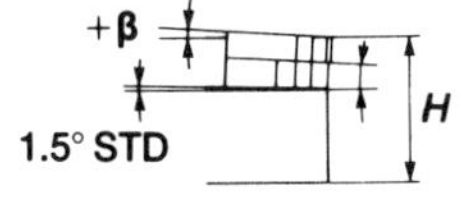

(*c*) *Recommended threading passes*

Threading pitch	0.5	1.0	1.25	1.5	2.0	3.0	4.0	5.0	6.0	8.0
Threading TPI	48	24	20	16	12	8	6	5	4	3
Recommended number of passes	4–6	4–8	5–9	6–10	7–12	9–16	11–18	12–20	12–20	15–24

These are general recommendations. These figures are for light and medium steels, iron and non ferrous material. For hard steels the depth of cut should be decreased and the number of passes increased

Fig. 1.39 Generating threads by thread turning (Courtesy of Stellram)

The screw-cutting insert's edge has to be ground by the manufacturer to provide the correct thread profile. It must operate with a radial cutting rake of 0° if the correct profile is actually to be imparted to the formed thread (Fig. 1.40). The lead angle of the flank surface varies at different points between the crest and the root of the thread, increasing towards the root of the thread. (The opposite is true on an internal thread.) Because of this effect, the actual cutting rake varies along the insert's cutting edge, becoming more positive on the 'leading' edge and more negative on the 'trailing' edge the closer it gets to the thread's root; this condition is particularly obvious in Fig. 1.40. It may be minimised if the insert is inclined, so that the top face is perpendicular to a line indicating the mean lead angle (measured at the pitch diameter), as shown in Fig. 1.40b. This inclination of the insert produces a symmetrical side clearance and is important in ensuring a uniform edge wear on both flanks; by so doing, it increases the insert's useful life. The fact that this small inclination of the insert has made one flank cut slightly below and the other slightly above the centre-line of the workpiece is of no practical significance at normal lead angles, for either the function of cutting or the thread's profile. Further, a small deviation from the exact symmetry required in the insert's inclination is also acceptable without too obvious a disadvantage. Thus, the inclined insert can be used to cut threads of between 0° and 2° with an inclination of 1°, and still produce a satisfactory thread. This is only true for the normal, symmetrical thread profile (i.e. V-forms); in the case of saw-toothed threads it should be borne in mind that the 'straight-flanked' ones (i.e. those with flank angles between 0° to 7°) in particular offer side clearances which are adequate.

Metric threading inserts are characterised by their thread profile and the pitch expressed in millimetres. The shape and the size of the insert will determine the completed thread form. One threading insert can be used to cut all threads of this shape and size, irrespective of their thread diameter, of whether they are right or lefthand, and of whether they are single or multi-start.

It was mentioned earlier in this section that a choice of infeed methods (Fig. 1.41) was possible; these are the radial-infeed and the flank-infeed methods. We shall now consider the effects on cutting a thread by each of these methods, starting with the radial-infeed technique. Here the tool is fed in radially before every pass and the cutting edges cut a V-shaped chip (Fig. 1.41a); owing to its stiffness as the cut progresses, the chip is difficult to deform to the desired shape. With radial infeeds, tests have shown that chip

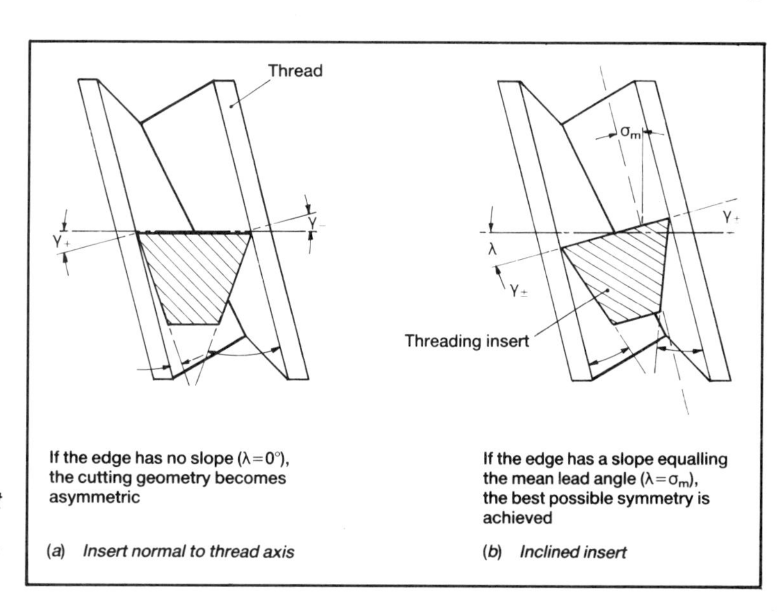

Fig. 1.40 Insert cutting angles when screw cutting (exaggerated). The lead angle is greater at the root of the thread (smaller diameter) than at its crest (larger diameter), so the cutting rake angle varies along the flanks of the thread (Courtesy of Seco)

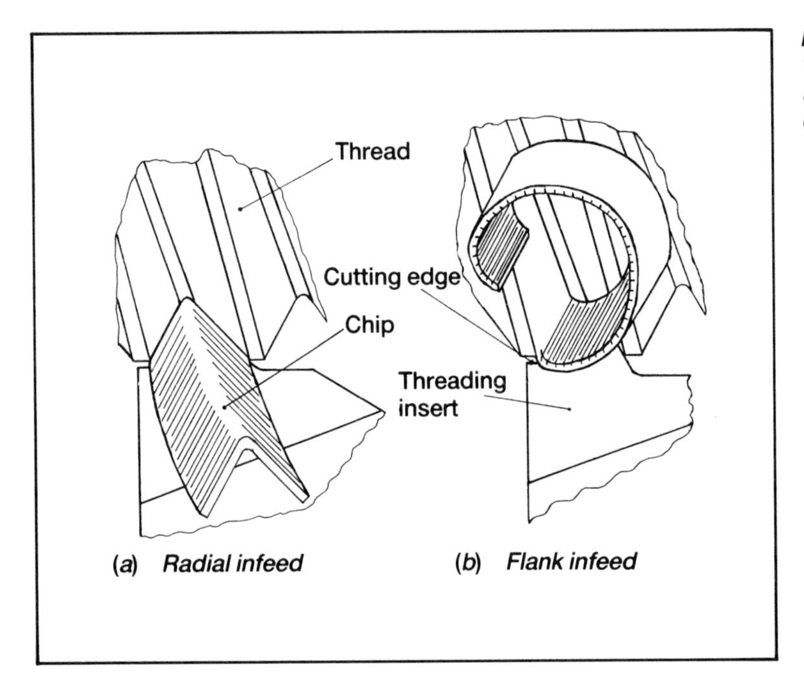

Fig. 1.41 Insert infeed methods: 'full-profile' inserts are shown, not at full depth of thread (Courtesy of Seco)

breaking is normally too soft in the first few passes and too hard in the last ones. This problem of hard chip breaking increases the cutting forces and heat input, and as a result tends to reduce tool life and increase the risk of tool failure. Therefore, it is normal to use the radial-infeed method when cutting threads on short-chipping materials and when using the multi-toothed insert types. Generally the use of chip-breakers is undesirable, and chip-formers have been developed instead by certain companies. These are placed on top of the threading insert with hard chip-breaking materials; they have a vertical edge that gives an easy-to-handle spiral (curled) shape to the chip. To reduce the problem of the chip breaking changing from soft to hard as successive passes occur, it is normal to reduce the depth of cut on each pass, until there is only the 'lightest' of cuts in the final pass along the thread.

A different chip form results from flank feeding the threading insert (Fig. 1.41b), which is similar to that found in conventional turning. Chips produced by this method are much easier to form, or to guide; and this fact, coupled to a more effective heat dissipation away from the tool point, allows heavier cuts to be taken. So the same number of passes result, but the single edge cuts chips roughly twice as thick as those from radial feeding.

There is a variety of flank-infeed methods (Fig. 1.42b to d). The modified flank-infeed method (Fig. 1.42c) was developed on some turning centres to overcome the fact that in the conventional flank-infeed method (Fig. 1.42b) the trailing flank carries out no cutting action and merely rubs along the flank previously generated. This rubbing causes edge wear and can produce a poor surface finish on the thread flank. In the modified method, therefore, the infeed direction is less than the flank angle (the particular difference in angle varies depending upon the CNC system utilised). By using this modified method, the negative effects of rubbing and surface finish can be avoided, without any loss of benefit of the flank-infeed principle. The profile is not affected by the method, as it is the last cut that forms the thread's profile. One drawback of the method is that the leading insert's flank is cutting the whole time, meaning that wear-rates are only slightly improved over those with the conventional flank-infeed method. The alternating flank-infeed method (Fig. 1.42d), which utilises both flanks alternately on successive passes to equalise the wear-rates and promote a longer useful life, has therefore become popular.

Fig. 1.42 The choice of infeed methods (Courtesy of Seco)

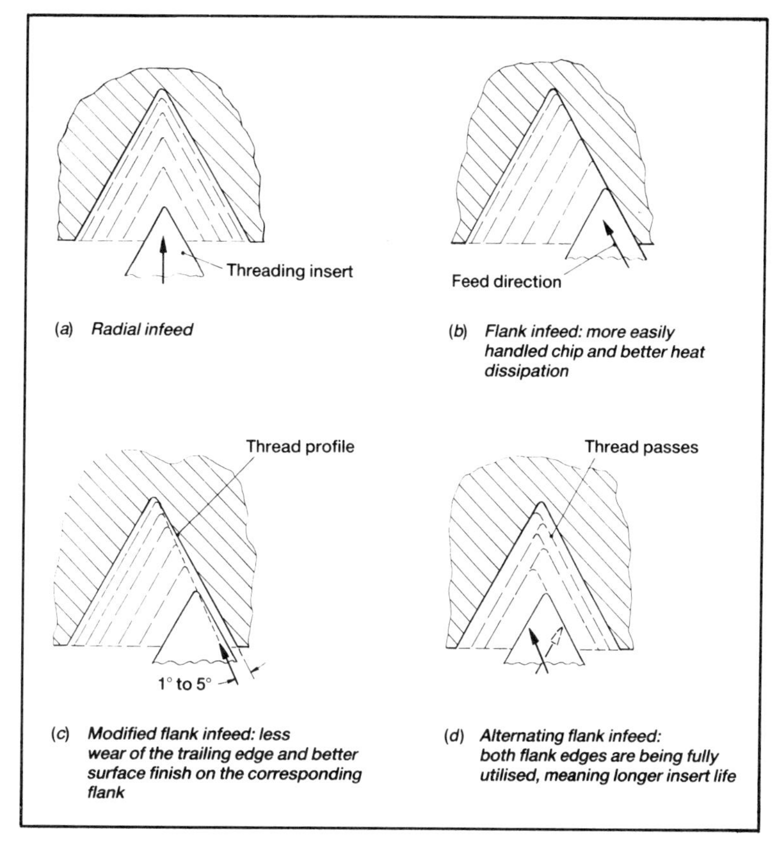

A variety of methods are used to produce the required profile on the thread, and another method (not shown) uses a variation on the radial-infeed method in which the tool point is programmed to increment within the gap between adjacent teeth in successive passes to 'nibble away' at the profile until completed. Radial-infeed methods can be used on short or long-chipping materials, but are most generally used with the former.

To ensure that good cutting performance occurs, excessively thin chips should be avoided: the tool needs to 'bite into something' to reduce the edge wear-rate. This is particularly true when cutting threads in work-hardening materials such as stainless and manganese-alloy grades of steel. The chip thickness must always be greater than the cemented carbide's edge radius, which is provided to improve the strength of the edge. Multi-coated threading inserts usually have larger edge radii than the uncoated varieties, with the sharpest of all edges occurring in the uncoated K-grades.

It is also worth mentioning that when heavy trapezoidal threads are to be cut, it may be advisable to premachine the thread by using a grooving tool with a width equal to the thread's root spacing dimension. Thus the threading tool need only be used to cut the final profile of the thread, in two or three passes. This reduction in use for the roughing operation has the added bonus of increasing the tool's life whilst at the same time improving the thread flanks' surface finish. This technique of premachining can also be employed when cutting coarse V-threads. In this case, the premachining is carried out using a more robust indexable turning insert during the rough-machining stage, which reduces the total number of passes required. Another bonus of using a suitable turning insert during the rough premachining of the thread is that a chip-breaker insert can be utilised with a 55° – 60° angle. It should be said that, in spite of using more robust tools for the roughing cuts, there is only a marginal saving of time, mainly because the tool must be changed. However, the 'real' bonus is the better utilisation of the threading insert, shown in the increased number of parts produced per insert edge.

Multi-start threads are used mainly for translation purposes. They have two or more starts present, which increases the thread's lead (the pitch multiplied by the number of starts equals the lead). They are often produced on turning centres, by one or two methods:

- Completing a single thread groove (start) in a number of passes, then repositioning for the next start and cutting this in a similar manner, and so on until all the required starts are completed.
- Adjusting the infeed, then machining one pass on the first start, then one pass on the second start, and so on until all the starts have begun. Then the infeed is increased and each start is again machined in turn, and so on until the final thread form occurs on all the starts.

Whatever the type of thread to be cut, whether it is V-form, trapezoidal or multi-start, it is not possible to vary such factors as the cutting speed, the feed and, to a lesser extent, the depth of cut, independently of one another, and certain limiting factors must be considered. The typical limitations imposed by the action of cutting a thread will now be mentioned. For threading it is normal to reduce the cutting speed by 25% compared with ordinary turning; this is partly because the insert's shape will limit heat dissipation. If a high chip load occurs through too high a cutting speed, the cutting temperature can approach the sintering temperature of the cemented carbide. As a result, the binder phase can soften, causing cutting-edge deformation. The answer to this problem seems easy: reduce the cutting speed. But this increases the risk of built-up edge being produced by chips being welded onto the cutting edge and later being broken off and taking a portion of the insert's edge along with them. This problem can be reduced if a tougher grade of carbide is used. However, the cutting speed should never be less than 40m/min when machining with carbides.

The feed (in millimetres per revolution) must coincide with the desired pitch or lead (when cutting multi-start threads), which means that the feedrate will also increase or decrease as the cutting speed is modified, in order to keep the feed per revolution constant. So, the critical factor is to have 'some' control over the depth of cut when threading. Each pass along the work causes an increasingly larger portion of the cutting edge to become involved in cutting, and the tool load will increase accordingly. If the depth of cut is kept constant during several passes, the chip-removal rate may increase by up to three times at each infeed. In order to minimise the stress on the cutting edge and keep it as uniform as possible, the depth of cut must be reduced with each pass along the workpiece.

To obtain a good surface finish, or close tolerance on the finished thread flanks, the machine can be programmed to make one or two finishing passes. These 'spring cuts', with a light tool pressure to eliminate the elastic deflections in the tool-machine-workpiece system, improve accuracy. 'Spring cuts' cannot be used on materials that work-harden (e.g. stainless steels, etc.), as they cause high tool wear. With work-hardening materials, cuts of less than 0.03mm should be avoided, as the material will deform elastically instead of being cut. This is even more of a problem with the austenitic steels, where it is recommended that the infeed should always be greater than 0.08mm.

These comments are only applicable to steels, and even for these the appropriate number of infeeds has to be found by trial and error. When an insert breaks, it is normally due to high stresses, so the number of infeeds should be increased. This is also true for the machining of threads in cast iron. It is obvious that the greater the number of passes the threading operation is divided into, the smaller the depth of cut and the lower the stresses on the cutting tip. However, if this philosophy is pursued too far and very many infeeds are programmed, the tool will not cut at all, owing to insufficient depth of cut, and elastic deformation of the workpiece material will result. Such a 'timid' approach to thread cutting will lead to a higher tool wear-rate, so it becomes necessary to reduce the number of passes!

Comparing the cutting forces for theading with those for normal turning, the power requirements are higher for thread cutting, especially when the chip thicknesses are small. If however, the chip thickness is increased, then the values for plain turning are approached. Therefore one should always try to employ higher chip thicknesses, as the benefits are twofold: an increase in the production rate and a decrease in the power consumption.

To conclude this discussion on thread cutting on turning centres, we shall now turn to the choice of carbide grades. Thread cutting demands an insert with a sharp edge, good resistance to wear and the ability to withstand fluctuations in temperature. A sharp edge and favourable insert geometry are demanded in order to obtain a good surface finish and at the same time reduce vibrations. High wear resistance is obviously essential, as otherwise the surface finish would be impaired and tolerance deviations on the thread would result. Temperature fluctuations must be withstood, as the very operation of threading produces fluctuations in the edge temperature: the fast machining pass along the thread causes heating, then the tool is withdrawn and returned to the start-point at speed, during which time the edge is cooled. This cyclic heating and cooling whilst the thread is being cut may occur from 10 to 20 times in quick succession. The heating and cooling cycle may produce fatigue cracks (known as 'comb cracks') as in milling, and could shorten the insert's useful life.

The choice of grade of the cemented carbide insert should be in accordance with prevailing conditions, but the following guidelines can be used to indicate starting-points for choosing the 'right' insert for the job. The K-grade has excellent wear resistance at low cutting temperatures and good toughness, providing the insert tip with good strength. This makes the K-grade the natural choice when cast iron must be threaded; it can also be used on other materials that have low cutting temperatures. The P-grades, on the other hand, are of benefit when machining long-chipping materials, particularly at higher speeds where their resistance to thermochemical attack (i.e. oxidation and the phenomenon of diffusion) increases the useful life beyond that of the K-grades. Finally, there are the coated grades; these types of insert offer the best resistance generally, but the coating process produces a less sharp edge than on the uncoated grades, with an edge radius of between 0.02mm and 0.05mm. If it is necessary to cut thin chips (i.e. below 0.05mm), coated threading inserts should not be used – but they often are, especially when a fine-pitched thread is to be cut! Apart from their resistance to high cutting temperatures, the advantage of using TiC-coated inserts is that they are less susceptible to pickup and built-up edge, which is of relevance at relatively low temperatures. Because of this, it is expected that coated inserts will gain importance in threading operations in which rounding of the cutting edge does not impair machining performance.

1.11 Parting-off and grooving

If a batch of components is to be bar-fed into the turning centre, then at some stage in its manufacture each component must be parted-off. Until recently, most parting-off tools, and indeed most grooving tools, were in the form of high-speed steel blades that were suitably ground for the application concerned. If a chip-former was ground onto these tools, its shape would depend on the operator's grinding experience and the technique used. Later, brazed cemented carbide tools became available and similar techniques were used to modify the tool shape to exploit the cemented carbide's properties. This was normally achieved by grinding in corner radii and sharpening the cutting edge, thus reducing the tendency for edge chipping and insert breakage.

The chip-forming requirements, cutting speed variations, tool-wear mechanisms and sensitivity to cutting height and component burrs, differ fundamentally for parting and grooving tools from those found in conventional metal-cutting techniques.

The chip-forming requirements of these tools can be divided into two parts: firstly, it is required to form or break the chip in such a manner that component production is not affected by long uncontrolled chips, and secondly, the chip must be made narrower than the actual groove being cut, so that the groove's side surfaces are not damaged by the work-hardened chip. The first exchangeable tool insert for parting and grooving with suitable chip-forming ability and geometry was developed in the early 1970s. Since then, there have been many developments in cutting geometries; more recently, the coated carbide grades have been introduced, which has increased productivity substantially.

During the parting-off of a bar (Fig. 1.43a), there will be a variation in the cutting speed as the tool approaches the bar's centre. This is true even with modern turning centres

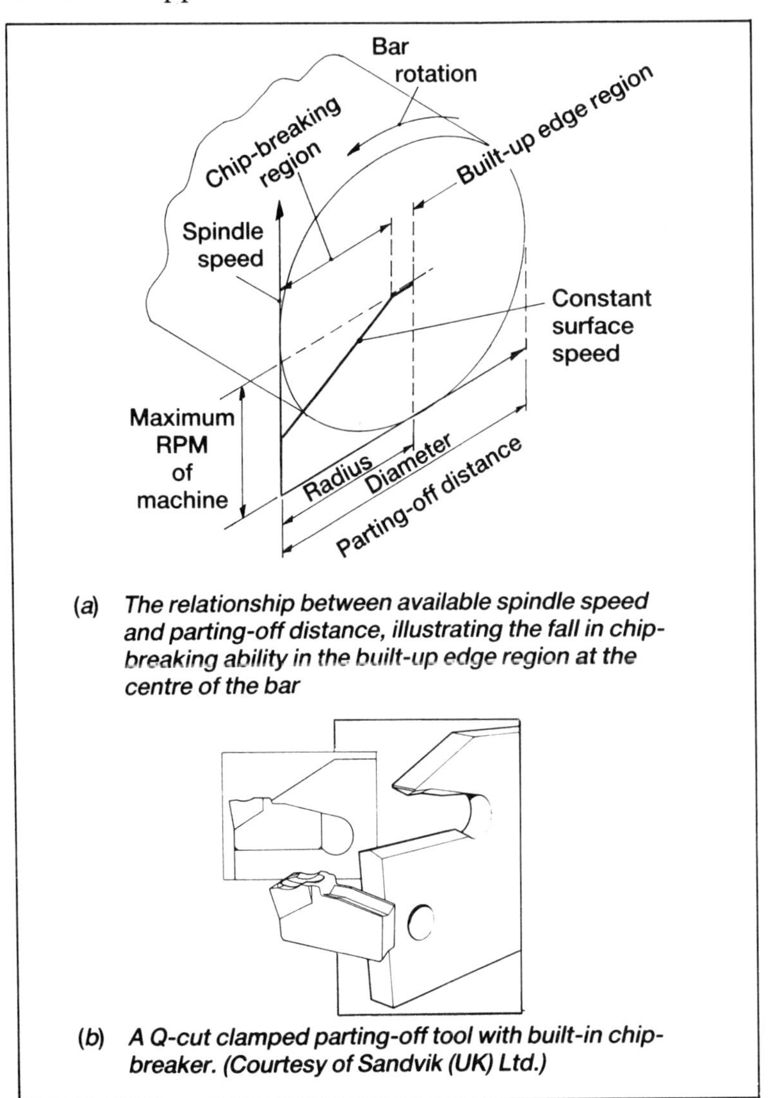

(a) The relationship between available spindle speed and parting-off distance, illustrating the fall in chip-breaking ability in the built-up edge region at the centre of the bar

(b) A Q-cut clamped parting-off tool with built-in chip-breaker. (Courtesy of Sandvik (UK) Ltd.)

Fig. 1.43 Parting-off: speed relationships and geometry

that can be programmed to achieve a constant surface speed, as the maximum spindle speed is attained just before the cutting edge reaches the centre of the bar. In the final stage, therefore, the cemented carbide tool must work in the built-up edge region. The demands this makes on the cemented carbide insert and its geometry require it to be extremely tough. Conversely, at the beginning of the parting-off operation the cutting speed is relatively high, which demands that the tool has a constant wear resistance together with a high ability to withstand plastic deformation. Therefore the demands of a parting-off operation, in particular, are stringent, but specialised cemented carbide grades and advanced tool geometries have made it possible to achieve successful parting-off consistently, if they are used correctly.

Unless the insert is firmly located in its holder, it cannot act as a reliable parting-off tool, so this is of paramount importance. Some manufacturers have designed their parting-off and grooving tools with a wedge-shaped notch in the holder. In these there is a tendency for the insert to move inwards into the pocket as the level of wear increases or the speed changes. Some designs have an insert pocket that is not parallel to the direction of feeding, and in these the cutting edge has a tendency to move downwards whilst cutting. This has a serious negative effect that increases tool wear, because the cutting edge is now below the workpiece's centre-height, owing to increased loads on the insert. A result of movement of the insert in its pocket is a loss of contact between the insert and its holder. This in turn causes poor clamping, leading to the insert splitting, followed by the holder being destroyed. Any movement of the insert within its pocket reduces the tool's ability to maintain close tolerances on the groove depth.

These problems have been considerably reduced, if not overcome, by the latest designs, a typical example of which is shown in Fig. 1.43b. This arrangement of the insert and pocket represents a completely new type of clamping, largely made possible by an extremely close tolerance on the insert. With this design, the base is obviously made parallel, so that any movement due to feeding does not alter the insert's centre-height and any sideways movement is restricted by the V-shaped location. The insert is locked into position against a radial stop, with a special key, so that consistency of dimensional accuracy is improved when inserts are renewed during a batch run. One problem with many inserted cemented carbide grooving tools is that the insert can become dislodged upon withdrawal from the workpiece; however, with this design the insert is positively restrained and is not afflicted by such a tendency.

Considering a typical insert geometry, with front and side clearances and a top rake, the important variable is the rake angle, which can be varied to suit the material to be cut and working conditions. In parting-off tools, as opposed to grooving inserts, the front edge is often ground to an angle of between 3° and 15°, in order to minimise the pip formed in the centre of the falling component. This is all very well, but it introduces an axial cutting force. If the angle is too large and is combined with a high feedrate, the force will cause the blade to bend. The blade's bending exerts pressure on the part to be parted-off and may cause an early break of the material at the centre, resulting in a larger pip than if a 'straight' tool had been used. A more serious problem than this is the fact that there may be a rounding of the parted-off workpiece, concave on the component just machined and convex on the next one (i.e. the part of the bar remaining in the machine).

If smaller front clearances are used, or the rake angle is small, a larger radial cutting force is produced. To overcome this problem, it is often necessary to use higher rake angles than would normally be recommended, particularly when parting-off small-diameter steel workpieces and other hard-to-machine materials. An increase in the rake angle causes a tendency for the tool to be pulled into the workpiece, or for the tool's wedge angle to be weakened; this is particularly true if the tool's centre-height is too low. Normally a larger rake angle is used when machining very ductile materials, such as aluminium, in order to obtain a free-cutting action and good chip flow.

A special chip-former is usually ground onto the insert to suit the machining conditions and workpiece material. These chip-formers are ground on the top surface of the cutting edge, producing a ridge over which the chip will have to climb when cutting. In this manner the chip-forming action is achieved more or less automatically. A typical chip cut on a ductile material curls into a 'watch-spring' shape as it is successively bent upwards by the ridge. The diameter of the 'watch-spring' is influenced by such factors as the width of the top land and the ridge height. Other influences on chip form are the programmed feedrate and the character of the workpiece material.

It has already been mentioned that it is important to set the parting-off or grooving tool on the centre-height. Setting the edges below or above the workpiece centre causes a pip to form as the work is parted-off from the solid bar. Any tolerances on the workpiece are normally set on their 'plus' side, because if the edge is set too low in relation to the workpiece's centre-height there is a likelihood that the edge will be damaged as the material remaining at the bar's centre starts to 'ride' on the edge before it breaks. Other problems occur when parting-off tubular material and bars with a pre-drilled hole in their centre. In these cases, setting the edge slightly below centre-height will produce no negative effects when parting-off, but will bring the risk that the tool could be pulled into the component. In the latest parting-off and grooving tools, considerable attention has been paid to the insert geometry to ensure that there is good chip control, and that they are more reliable with respect to insert breakage and wear than the old-fashioned solid HSS blades were. Parting-off tools like the one shown in Fig. 1.43b can be used to cut a wide range of materials, such as plain and alloyed steels, and cast irons, as well as stainless steels.

Parting-off and grooving operations are normally conducted at low feedrates, particularly when machining in the region of the 'breakthrough' phase; it is recommended that the 'constant cutting speed' facility is used. With the type of tool shown in Fig. 1.43b, the feeds recommended lie in the region of 0.07 to 0.30mm/rev, depending upon the programmer's requirements. These requirements may be for a pipless part-off, a good surface finish, tool consistency, high productivity, or component flatness to the workpiece. As mentioned above, the parting-off feed is normally at least halved during the 'breakthrough' phase, to not more than 0.03 to 0.05mm/rev. The objective of lowering the feedrate during this phase is to reduce the formation of parting burrs and at the same time prevent exceptional wear on the insert.

Having covered most of the conventional tools used on machining and turning centres, we shall now briefly consider a small sample of some of the 'specialised' tools that can be purchased whenever the occasion demands.

1.12 Specialised tooling

A typical 'numerically-controlled' or 'feed-out' facing and boring head is illustrated in Fig. 1.44. These programmable heads allow the machining of features such as large bores with intricate profiles (e.g. multiple diameters, grooves, tapers and threads) on prismatic parts. Until these heads became available for machining centres, such parts presented something of a challenge to the programmer! Traditionally, they would have been routed to multiple machines where a conventional jig-boring mill might have been used

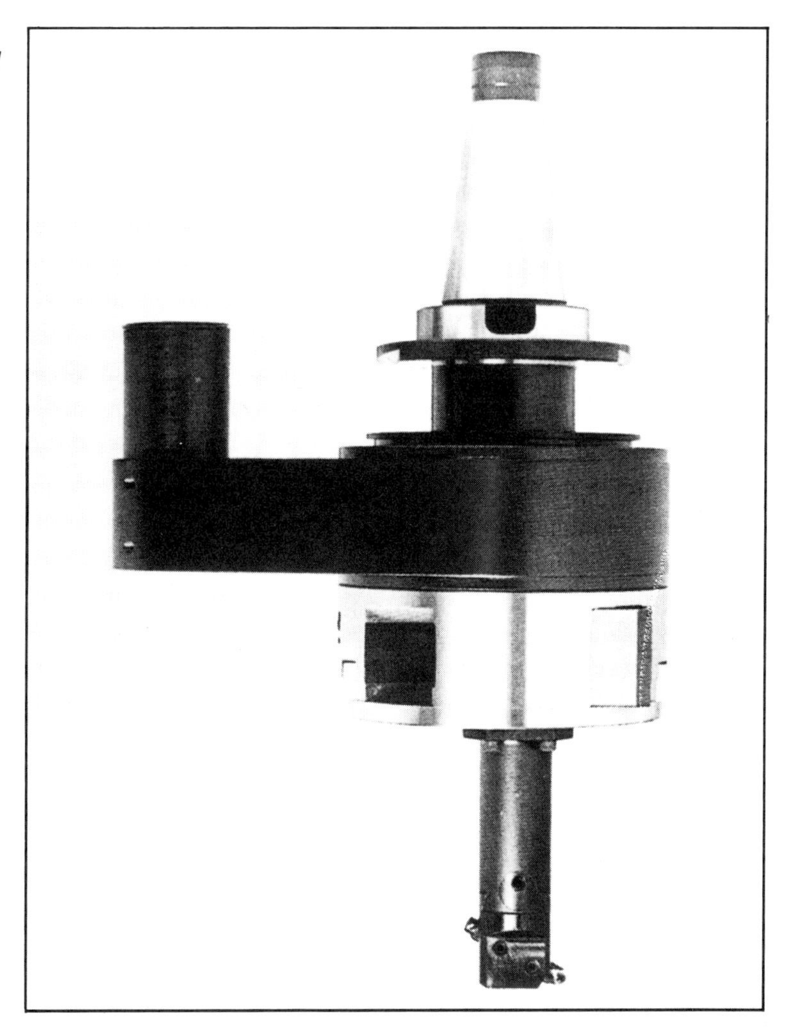

Fig. 1.44 U-centre numerically-controlled facing and boring head (Courtesy of Bristol Erickson)

to machine intricate bores, etc. Even these machines required considerable manual skill and many movements using numerous tools, which might have been reduced only somewhat if the purchase of custom-made multi-point boring heads could be justified by the high volume of work expected. In order to overcome these problems, boring heads were developed with an adjustable-diameter tool slide that was perpendicular to the centre-line of the spindle.

Such programmable heads can perform operations on the outside and inside diameters in one set-up of the workpiece. Their tool diameters can be set either automatically, using numerical control, or manually. If manual heads are used, the settings are usually made by vernier handwheel adjustment, or by the use of dogs. With these heads, manual radial feed occurs via the tool slide but on several types that have been available for some time, power feed is possible. The programmable heads are a more recent development, where it was felt that a broadening of their application would result in increased productivity if fitted to machining centres. These numerically-controlled heads have a fully programmable *U*-axis tool-slide feed that can be co-ordinated with that of the machine spindle's *Z*-axis, enabling it to produce tapers as well as contour bores. Such combinations of boring and facing operations permit a machining centre to machine complex prismatic parts in one setting, whilst at the same time reducing the tooling requirement and increasing productivity considerably.

Now consider the programmable boring head in Fig. 1.44 in a little more detail. The head uses a standard machining-centre tool-holder adaptor, so that it can be interchanged with any other tool in the carousel. Power to the head's tool slide is provided by means of a compact auxiliary d.c. servodrive motor that utilises a feedback unit and is mounted in a permanent position adjacent to the spindle. Once the head is located in the spindle, the drive unit is engaged, securing the head's main body in a non-rotating position. The radial motion of the tool slide is achieved by a *U*-axis command, using the CNC as for any other axis movement. On many CNC units it is not possible to provide a separate *U*-axis command for the radial motion of the head, and in these cases the head's tool-slide feed motion is shared with another axis. The problem of sharing feeds in this manner is that it usually limits the head's application to either straight boring or facing operations, which seems rather a waste of the programmable head's potential!

Where a difficult set-up of the workpiece, involving the machining of many faces, occurs on a machining centre, or where another axis control cannot be justified economically, then the right-angled head shown in Fig. 1.45 might be the answer. If a horizontal machining centre is required to machine slots, or drill holes in the top face of a component, for example, it may be achieved using the right-angled head without readjustment of the workpiece on the machine table. These angled heads can be either fixed, as shown in Fig. 1.45, or adjustable for one or two angles, allowing compound angled holes, etc. to be machined.

The two heads shown in Figs. 1.44 and 1.45 are available as listed options in cutting-tool manufacturers' catalogues; they are a compromise between allowing as many machining applications as possible and the higher productivity that can be obtained with custom-built tooling. If one requires tooling for a truly specialised high-volume production part, then custom-built, or specially-engineered, tooling (as it is sometimes called) is the answer. Typical examples of this custom-built tooling are shown in Fig. 1.46; here the holders and cutting tools are designed for a customer's production requirement for one job only. This approach allows the simultaneous machining of various workpiece features; it can only be justified if the pay-back is sufficient in terms of increased production or overcoming complex machining requirements, as this type of tooling is quite costly.

If some form of dimensional adjustment is required in custom-built tooling, then (within its limitations of movement) the Microbore insert shown in Fig. 1.47 might be what is needed. Microbore units placed strategically on the specialised tooling can be

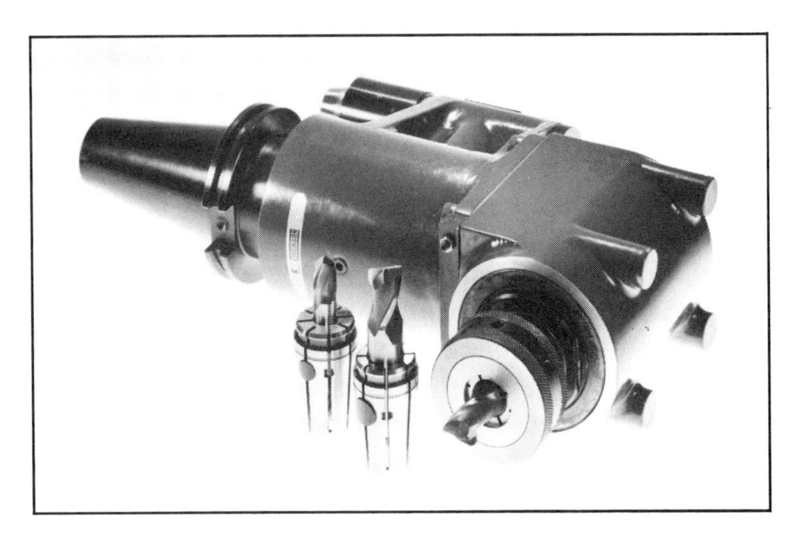

Fig. 1.45 A right-angled head for machining centres (Courtesy of Bristol Erickson)

Fig. 1.46 Specially-engineered tools, illustrating the complex arrangement of inserts (Courtesy of Kennametal (UK) Ltd.)

manually adjusted to preset dimensions in discrete steps of 0.002mm by inserting a key in the worm adjuster. The principle of operation of these units is simple yet effective: as the worm is turned by the key, this engages and turns the workwheel, and the threaded centre is moved radially inwards or outwards depending upon the direction of the key's rotation. This thread contains a cartridge containing the insert, which also moves radially according to the wormwheel's rotation. It is kept at the same attitude to cutting by the tags on the end of the threaded portions, so that the cutting edge always has the same attitude to cutting after radial adjustment. The cutting edge is spring-loaded through the cartridge-holder, which keeps it compressed whilst cutting; as soon as the cutting action is completed, the spring-loaded cartridge moves the cutting edge away from the diameter just machined so that it is clear of the hole and does not produce 'trail marks' on the machined surface.

Fig. 1.47 An adjustable Microbore insert (Courtesy of Devlieg)

These cutting units can be used on turning, as well as machining, centres. They are very popular because they are accurate and easy to adjust, which facilitates presetting and thereby reduces down-time considerably. At the same time, they are adaptable for automatic size control.

This completes this chapter's appreciation of conventional cutting-tool technology. In Chapter 2 we shall consider the recent growth in the influence of the modular tooling concept.

Chapter 2
Modular quick-change cutting-tool systems

Chapter 1 discussed how conventional cutting-tool technology, correctly applied, can make significant productivity savings, whether the emphasis is on increased production through longer tool life or on the reduction in the production time for each part. The trend in recent years has been to increase the productive cutting time of these expensive machine tools, and in order to achieve this objective it has been necessary to reduce tool-related down-time.

Cutting-tool manufacturers have not been slow in developing and producing quick-change tooling systems. Their initial steps towards automatic tool changing were made a decade or so ago. One early solution involved changing the indexable insert itself; the main drawback with this was that the changer was complex in design and could only change one type of insert. Therefore its use was limited to long-run turning applications, and even here it suffered with the advent of CNC.

Other approaches involved changing both the tool and the tool-holder, in a similar manner to current practice with CNC machining centres. This system also imposed restrictions owing to the relatively high weight and size of the tool-changer, which meant that its load-carrying capacity was limited. Even where a tool magazine is present (i.e. in the case of machining and turning centres) its capacity is rapidly exhausted, so that fully-automatic operation over a prolonged period is not possible. Further, the multitude of geometries and clamping systems necessary causes impossible demands on an automatic tool-changer, and the problem is exacerbated further by the fact that indexable inserts may not be suitable for all machining operations. A completely different approach is necessary for automatic tool-changing systems if these disadvantages are to be overcome.

Before we discuss some of the quick-change systems found today, it is worth mentioning that many machine-tool manufacturers can offer extra-capacity tool magazines holding more than 300 tools in certain instances. So one might rightly ask 'Who needs quick-change tooling when such machines have their own built-in storage and quick-change mechanisms?' This is a valid point, but a high financial outlay is required for these extra-large magazines, and even then only a finite amount of tooling can be accommodated whose variety is reduced considerably when the 'sister-tooling' approach* is adopted.

The machine-tool builders have spent much effort on reductions in the non-productive cutting time with conventional quick-change tools situated in magazines or carousels, by reducing their 'cut-to-cut' tool-change times. Another area where much benefit has been gained is in the initial tool set-up and maintenance of cutting tools, in the area of tool management, which will be discussed in Chapter 4.

So far, this chapter has dealt with early methods of quick-change tooling and the machine-tool builders' approaches in overcoming the problem. So why does one need modular quick-change tooling? The reason for using modular quick-change tooling systems on machining centres has been to standardise and thereby reduce tooling inventories, whilst at the same time making them more flexible to the cutting requirements that occur during a production run. Now that turning centres are used with a driven tooling facility more often than not, their requirements for modular tooling are similar to those of machining centres.

The well-documented survey commissioned in the early 1980s by the US Government, the *Machine Tool Task Force Study*, found that medium-volume manufacturing companies that used typical machine tools were doing productive cutting for only about 11% of the time. The non-productive time was taken up by such activities as loading and unloading (6% of the total time), changing tools (10%), set-up and gauging (10%), equipment failure (8%) and last, but by no means least, the incomplete use of shifts (55%). Thus the study illustrated that the times for activities related to cutting tools, namely the tool-changing and set-up and gauging times, amounted to at least 20% of the machine tool's available time. Therefore the cutting-tool companies advise users to focus their attention on reducing the times for these non-productive operations, as this will significantly improve the efficient utilisation of the machine tool over the working day.

It can now be seen that significant reductions in the machine tool's non-productive time can be made by eliminating, or at the very least minimising, the down-time associated with using cutting tools. If a company incorporates quick-change tooling systems on its machining and turning centres, great productivity benefits will accrue with relatively short pay-back periods. This is the basis for the discussions in the next sections, which will first consider the tooling requirements of turning centres, and then the applications of modular quick-change tooling on machining centres.

2.1 Tooling requirements of turning centres

Of all the machine tools that use single or multi-point cutters, the turning centre has undergone the greatest changes. The range extends from the earlier basic CNC lathes, with conventional square-shanked tool-holders and round-shanked boring-bars that are loaded manually by the operator, to the very latest turning centres, with features such as robot part loaders, flexible work holding, quick-change tooling, tool-wear probing, work-gauging systems and equipment to sense tool condition. The latest machines, where some or all of these features are fitted, cost a considerable amount of money. In order to recoup the financial outlay as fast as possible, they must increase the productive cutting time, whilst simultaneously reducing the direct labour costs. It is often this latter

*With the sister-tooling approach, there is at least a duplication of the most heavily-utilised tools within the magazine. Once the first of the pair of tools is near the end of its active cutting life, it is exchanged for its 'sister' and will not be called upon again during the production cycle.

aspect, of labour-cost reduction, which becomes the most attractive cost-saving item, as this is always a large component of the overall manufacturing costs in any factory.

If a company specifies a turning centre with a rotating tooling facility (sometimes called 'driven tooling'), which is held in the turret along with the usual tooling, the programmer will be able to program secondary operations such as light milling, tapping, drilling etc. to be carried out in a single set-up; this is known as 'one-hit machining'. These secondary machining operations may even eliminate the need for work to be carried out after turning – by a machining centre, for example; this gives a further saving in production time which is related to cost by reducing the work-in-progress and minimising the need to purchase or utilise a second machine tool.

The previous paragraphs have justified the need to use quick-change tooling and turning centres. There now follows a review of popular systems used extensively throughout the world, some of which can be categorised into two types: cutting-unit systems and tool-adaptor systems. The two systems vary in their basic approach to quick-change tooling and whether they are designed to be used on machining or turning centres separately, or for a more universal approach. The cutting-unit system is commonly known as the 'Block tool' system (Fig 2.1); it was the first to be developed by a leading cutting-tool manufacturer. This system is based on a replaceable clubhead for a square-shanked tool-holder; the coupling provides radial repeatability to within ±0.002mm. This high level of repeatability is necessary in order to minimise the coupling's effect on the diameter to be turned. To ensure that the forces generated whilst cutting do not deflect a Block tool, a clamping force of 25kN is used. The clamping may be done in a number of ways; manually, semi-automatically or automatically, as shown in Fig. 2.1b, c and d respectively. The clamping force is most commonly provided using a certain number of spring washers, which are preloaded to provide a reliable clamping force; the cutting units are released by compressing the washers so that the draw-bar can move forward. In the automatic clamping system, a small hydraulic cylinder mounted on the carriage behind the turret causes the draw-bar release, activated by the CNC.

So far we have mentioned accuracy, the clamping force and modes of releasing the Block tool; we now consider how the tool's precise location in its holder is achieved. The Block tool is located by the following process: the cutter unit slips in from above the coupling to rest firmly on a supporting face on the bottom of the clamping device; this supports the cutting unit tangentially during the cutting operation. Once the cutting unit is seated on the bottom face, the draw-bar is activated – either manually, using a key, or automatically by the hydraulic unit – to pull the cutting unit against a face; this makes a rigid and stable coupling that is easily able to support the loads produced during cutting. Both internal and external cutting units can be supported in this manner, as shown in Fig. 2.1e and f respectively.

A major advantage of all modular quick-change systems is the easier and quicker tool changing they offer, producing shorter cut-to-cut times. There is the added bonus of reduced operator fatigue, since tool handling – particularly of the heavy tools – is now a thing of the past with the manual and semi-automatic tool-changing methods. As a result of the smaller size of the modular tools, they can more readily be stored in a systematic manner, which allows them to be more efficiently located and retrieved from stores, as well as minimising the tool-stock space.

The advantages of the manual Block tool system over conventional tool-holders and their tool blocks, can be more fully appreciated through the following example. The numerical values in the table form the basis of the comparisons. The ones in the lefthand column

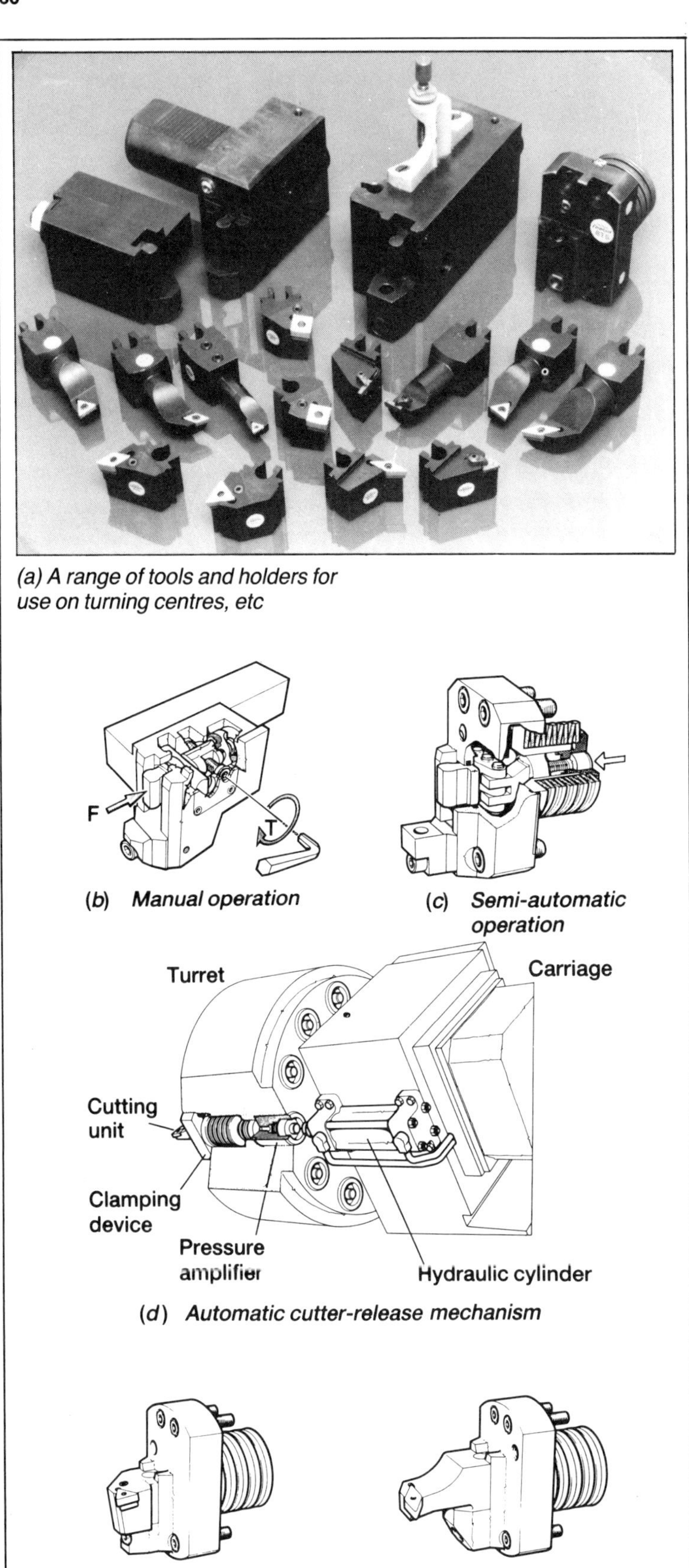

Fig. 2.1 The Block tool system (Courtesy of Sandvik (UK) Ltd.)

are typical of any turning centre where there is manual involvement in tool changing, securing and maintenance.

Operation	*Conventional tool-holder*	*Block tool system*
Setting-up time (minutes)	30	15
Tool-changing time (minutes)	3	1
Measuring-cut time (minutes)	5	0

This data can now be applied to the practical situation of a mixed production of small batches of turned components, where the actual cutting time is 15% of the total machine-shop time. Assume that an average of 30% of the tools needed measuring cuts (e.g. diameters measured, etc.), that 200 set-ups were required per year on the machine, and that some 1580 tool changes were needed during the year. So for these production parameters, the quantitive benefits of using the modular quick-change tooling system are as follows:

- The difference in the setting-up times is

  $15 \times 200 = 3000$ minutes per year

- The difference in the tool-changing times is

  $2 \times 1580 = 3160$ minutes per year

- The difference in the measuring-cut times is

  $\frac{1580}{3} \times 5 = 2630$ minutes per year

This means a total difference of 8790 minutes, or 146 hours – or, put another way, just over 18 working days. Hence the Block tool system allows for a significant increase in available productive time over this period. Alternatively, this saving in time can be multiplied by the machine's running cost per hour to further reinforce the correctness of the decision to purchase the quick-change tooling system, since it quickly builds up the pay-back on the investment.

The example given above clearly demonstrates the real benefits of using a manual-changing Block tool system. Incidentally, quick-change tooling can also be applied to conventional lathes with similar economies.

Fig. 2.2 shows a schematic representation of the whole quick-change tooling system and some of the positive features of its use. The process of changing tools can be further quickened if one semi-automates the process by using power assistance to clamp the cutting units, although this time saving must be set against the additional capital outlay necessary to add this feature.

So far the merits of using a quick-change tooling system have been praised, but one might ask 'What type of batch size can justify the financial expense of putting together such a system as the one shown in Fig. 2.2?' To answer this, we shall consider the two extremes, of large-batch production, and small-batch usage and one-offs.

Today, large batches, and even mass production, are increasingly performed in 'linked' turning centres with automated workpiece handling and process supervision. The aim is to limit operator involvement and for stoppages for tool changing and setting to occur according to an organised pattern, so that they usually happen in between shifts, or at the scheduled stops.

Using the Block tool system, say, tool changes can be organised and made very efficient, especially so when the tool changes are semi-automatic. The cutting units are small, light and easily organised for tool changing. They can be preset outside the machine environment, and their accuracy is assured by the precise coupling to the holder. Also, 'intelligent' tooling can be utilised (see Chapter 4), using coded or numbered cutting units, which allows the operator to change some or all of the tools in a turret in a very short time. The settings of these new preset tools can quickly be confirmed if touch-trigger probes are present (see Chapter 3), so they further reduce tool-changing down-time. These tooling aids also minimise the operator's activity and ensure that it is performed correctly, so eliminating the risk of mistakes being made during hectic machine stoppages. Another benefit of using quick-change tooling is that, as the time to change tools is very short, it may be possible to make unscheduled changes of critical tools if, for example, their wear-rate is unexpectedly high. This will raise the cutting performance, which in turn will lead to a more economical utilisation of each machine during large-batch production.

Where a company is involved in large-batch or mass-production runs, it is obvious that considerable savings can be gained by reducing the non-productive cutting times, and it might seem that this is the only production environment that can justify the extra expenditure for these tooling systems. This is not the case; companies producing small batches or one-offs can also gain, to a lesser degree, from an efficient quick-change tooling system. This is particularly true when machining families of similar components, or where there is a complicated machining requirement on one-offs.

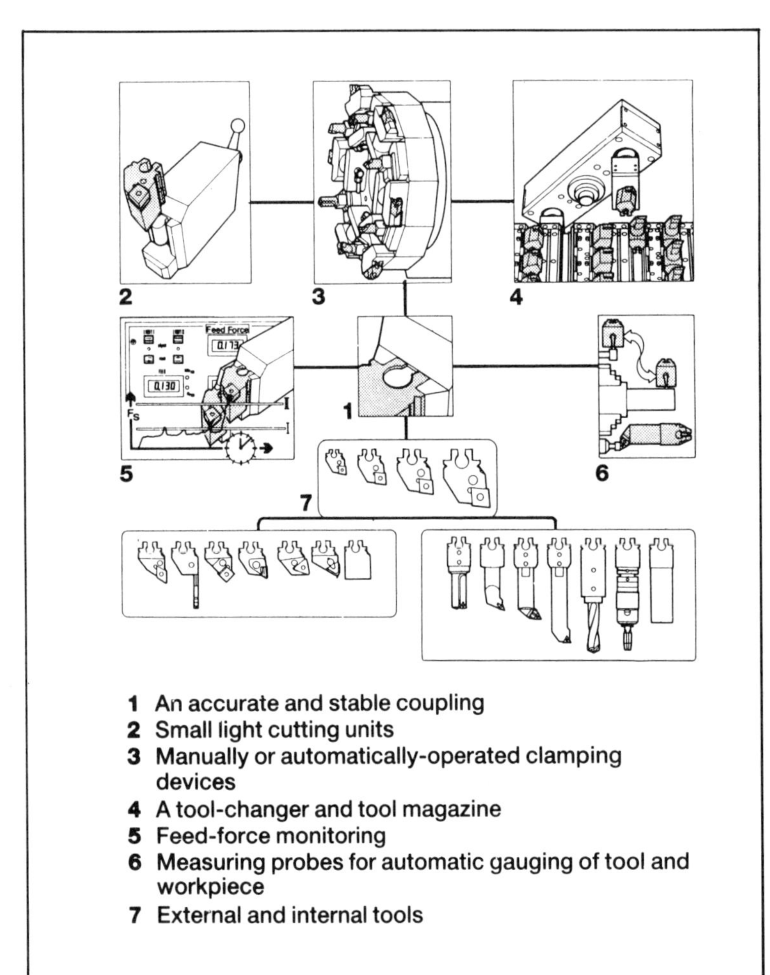

Fig. 2.2 Features of the Block tool system for turning centres (Courtesy of Sandvik (UK) Ltd.)

A frequent problem is that of insufficient tool storage on the machine tool, particularly on single-turret turning centres, and quick-change tooling is often the answer in these circumstances. Using the semi-automatic Block tool system, say, extends the turret capacity with minimal loss of productive cutting. Replacing a new cutting unit simply requires the operator to lift out the old unit and push in another; optional stops can be programmed into the CNC. Presetting the tooling, in conjunction with the repeatability of the coupling between the cutting unit and the holder, ensures that the cutting edge is correctly positioned in relation to the workpiece. Therefore, the operator no longer needs to adjust the machine when changing the workpiece for different configurations.

Some figures to reaffirm the company's decision to use quick-change tooling on conventional machines, for the particular case where heavy or large tooling was previously used (e.g. on turret or capstan lathes), are that the average time to change the old-style tools would have been about five minutes using square or round-shanked tooling, whereas with quick-change manual-clamping tooling it is less than 50 seconds. One company that used this method saved up to 84 minutes a day in reduced down-time, and this cost, when related to the machine's running cost, worked out at a saving of £38 per day. The total savings per year more than paid for the investment required in purchasing the quick-change tooling system, which could be used on other work without extra cost, ensuring even greater profits later!

In spite of all this convincing evidence in favour of quick-change modular tooling, some pessimistic production engineers may still remain sceptical as to the advantages to be gained from such expenditure. Another factor, which might be of even greater importance, could be that the company simply cannot afford the luxury of purchasing a complete tooling system. In either of these cases some caution may be advisable before a company becomes fully committed to purchasing such a system. In such cases, the solution might be to purchase just a few quick-change units, for a specific medium-sized batch, and later appraise the situation in terms of the likely productivity increases and the operators' experiences. In this manner only a relatively small financial outlay will be necessary and the company will not become too disenchanted if the results are unfavourable owing to some extraneous circumstances beyond its control.

The discussion so far has basically concerned just one system – the cutting-unit method, typified by the Block tool system. The tool-adaptor systems tend to be more diverse in their approaches to modular quick-change tooling. Typical of this design philosophy is the KM system, which was developed out of the experience gained using the original KV system. The system is diagrammatically represented in Fig. 2.3. The next section discusses how this system was designed and how it operates, highlighting the benefits of its application in the metal-cutting industry.

2.1.1 Design and development of quick-change tooling

Prior to designing a new quick-change tooling system, a number of key decisions had to be made. The basic criterion of the system's configuration for use with turning, boring and rotating tooling required that it be round and on the centre-line. Also, for ease of tool changing and accuracy in the radial direction (*X*-axis), a tapered shank was necessary. So that an equal accuracy occurred in the axial direction (*Z*-axis), there was a face contact requirement (Figs. 2.3b and 2.4c). The cutting-edge height was deemed to be a less critical dimension and a reasonable tolerance was allowed here, which would give good results for the great majority of metal cutting operations using modular quick-change tooling. Together, these design criteria gave the following requirements for repeatability:

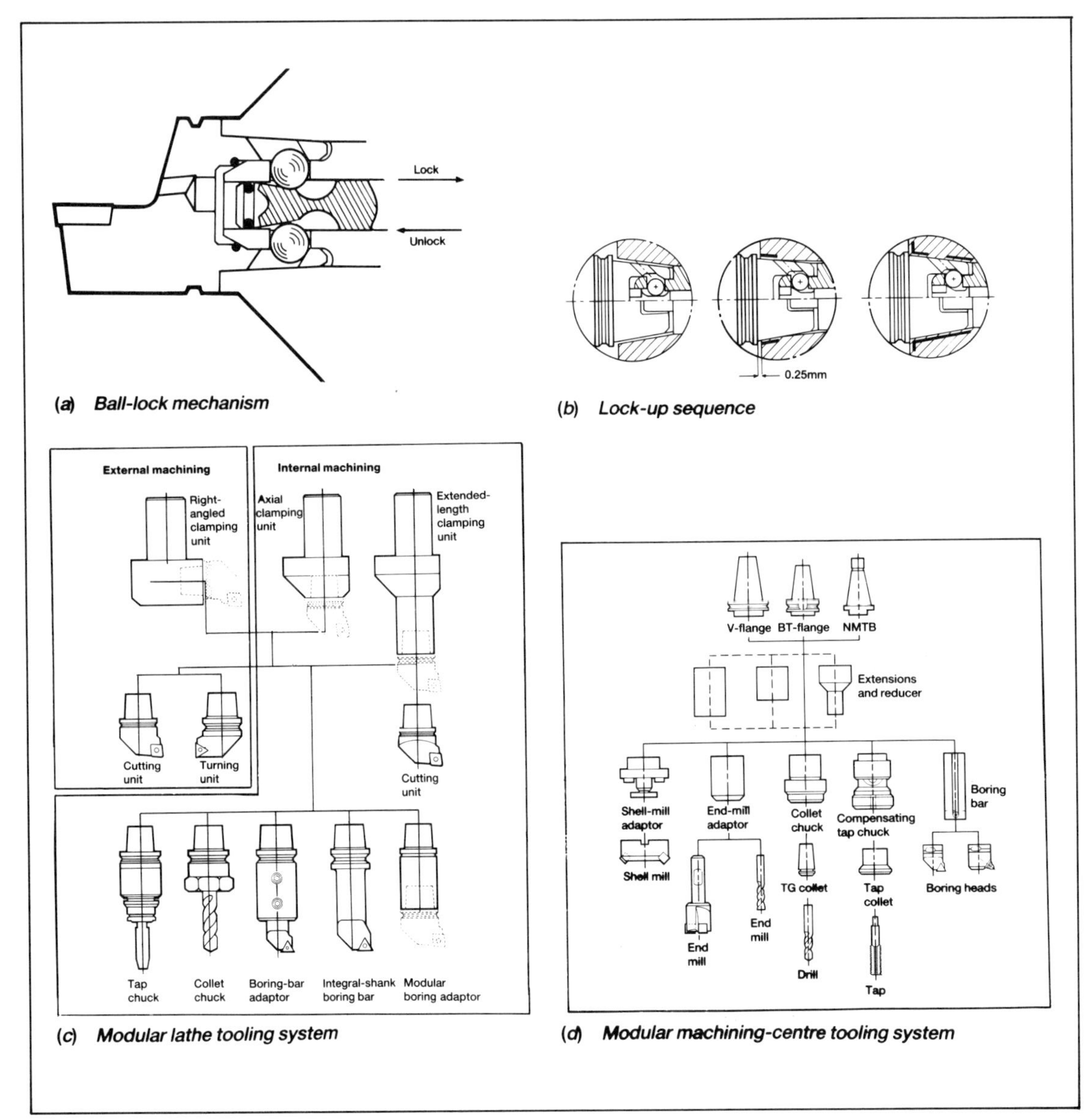

(*a*) *Ball-lock mechanism*

(*b*) *Lock-up sequence*

(*c*) *Modular lathe tooling system*

(*d*) *Modular machining-centre tooling system*

Fig. 2.3 The KM tooling system (Courtesy of Kennametal (UK) Ltd.)

- Axial tolerance = ±0.0025mm.
- Radial tolerance = ±0.0025mm.
- Cutting-edge height = ±0.025mm.

Quick-change turning-centre tooling installations are in the 'intermediate' size range, and are capable of handling tangential cutting loads of 12 000N. With this load requirement, the cutting unit closely approximates to a 32mm square-shanked tool-holder, and this is the KM's smallest size, as shown in Fig. 2.4d. However, when a review of the dimensional envelope of machines of this size was made, it was found that a 40mm round-shanked system was the maximum that could be accommodated. This size was chosen for the coupling, with adaptors for sizes from 25mm to 80mm for use on turning as well as machining centres.

Once the basic configuration to meet the dimensional and repeatability criteria had been established, the coupling shape could be considered. It was quickly decided to use the male part of the joint on the cutting-tool unit, as its overhang was the smallest and it was less influenced by deflections produced by high tangential loads. Another advantage of using a male cutting unit, is that it provides more protection for the taper and the locking mechanism once the cutting tool is removed.

With the taper's configuration determined, it was then necessary to decide on the method of achieving contact between the taper and the face. There are two methods of providing this contact: by providing metal-to-metal contact by holding very close tolerances on both halves of the coupling, or by designing a small amount of elastic deformation into the assembly. As the male half of the joint was located on the cutting tool, any such deformation would take the form of an expansion of the female taper in the clamping unit. In testing, an optimum performance occurred with a combination of the pull-back force coupled with the elastic deformation. This resulted in better static and dynamic stiffness, and was less costly to manufacture than a metal-to-metal configuration.

Then taper size was considered; it was determined that the gauge-line diameter had to be as large as possible, in order to promote high stiffness. As the wall thickness would also be affected, a compromise of 30mm was decided upon. The final decisions to be made about the joint concerned its length and taper angle. If a steep taper was utilised, the greater angle would cause an increase in the force required to produce a given amount of elastic deformation in the female half of the coupling. However, a slow taper, of smaller angle, has the effect of increasing the force required to separate the male and female tapers (an effect used in 'self-holding' tapers). After evaluating both types of tapers, the latter method was chosen as it gave the optimal taper, 1 : 10 by 25mm long. This gave the best combination of stiffness and forces for locking and unlocking. This taper equates to the well-known Morse taper, and has the added bonus that limit gauges are commonly available for checking tolerances during production.

Once the coupling shape had been established, the locking mechanism could be considered. This was fully investigated using the latest computer-aided design (CAD) techniques, which allows information to be quickly transmitted to the manufacturing process without errors. Techniques such as finite element analysis were used on the key components to ensure the correct strength and durability levels (Fig. 2.4a and b). Extensive life testing was also conducted, to avoid unexpected failures of the tools in use, which would have been costly.

The locking mechanism used precision-hardened balls to produce a system which has high mechanical advantage coupled to low frictional losses and is of low cost. The newly-designed coupling required a mechanism that produced up to 31 000N locking force, which would fit inside a taper with a gauge-line of only 30mm. The ball-lock mechanism used two balls that locked into holes machined through the tapered shank of the cutting unit, as illustrated in Fig. 2.3a. This configuration allows a 9mm draw-rod to be used to apply the pull-back force. The holes in the tapered shank, which the balls are seated in, have a machined 55° angle, resulting in a mechanical advantage of 3.5 : 1. The resulting coupling allows a high clamping force of 31 000N to be produced by a draw-rod pulling force of 8 900N. As the draw-rod is pulled back it forces the two balls radially outwards, until they lock into the tapered machined holes, as shown in the lock-up sequence in Fig. 2.3b. Applying a force and pushing the draw-rod releases the balls and at the same time 'bumps' the cutting tool to release it from the self-holding taper.

Referring to the lock-up sequence illustrated in Fig. 2.3b again, once the cutting unit is inserted in the female taper it makes contact at a stand-off distance of 0.25mm from the face. As the locking force is applied, a small amount of elastic deformation occurs at the front of the female taper. Once the cutting tool is locked, there is a three-point contact, at the face, the gauge-line and the rear of the taper, as shown in Fig. 2.4c.

If one compares the coupling's stiffness with that of a solid-piece unit which has been machined to similar external dimensions (for example, cutting unit and clamping assembly), then when a 12 000N load is applied to simulate tangential cutting loads to the tool point, the difference in deflection between them would be of the order of only 0.005mm. Hence, this new modular coupling and its assembly closely approximate to the ultimate rigidity of a solid-piece cutting tool.

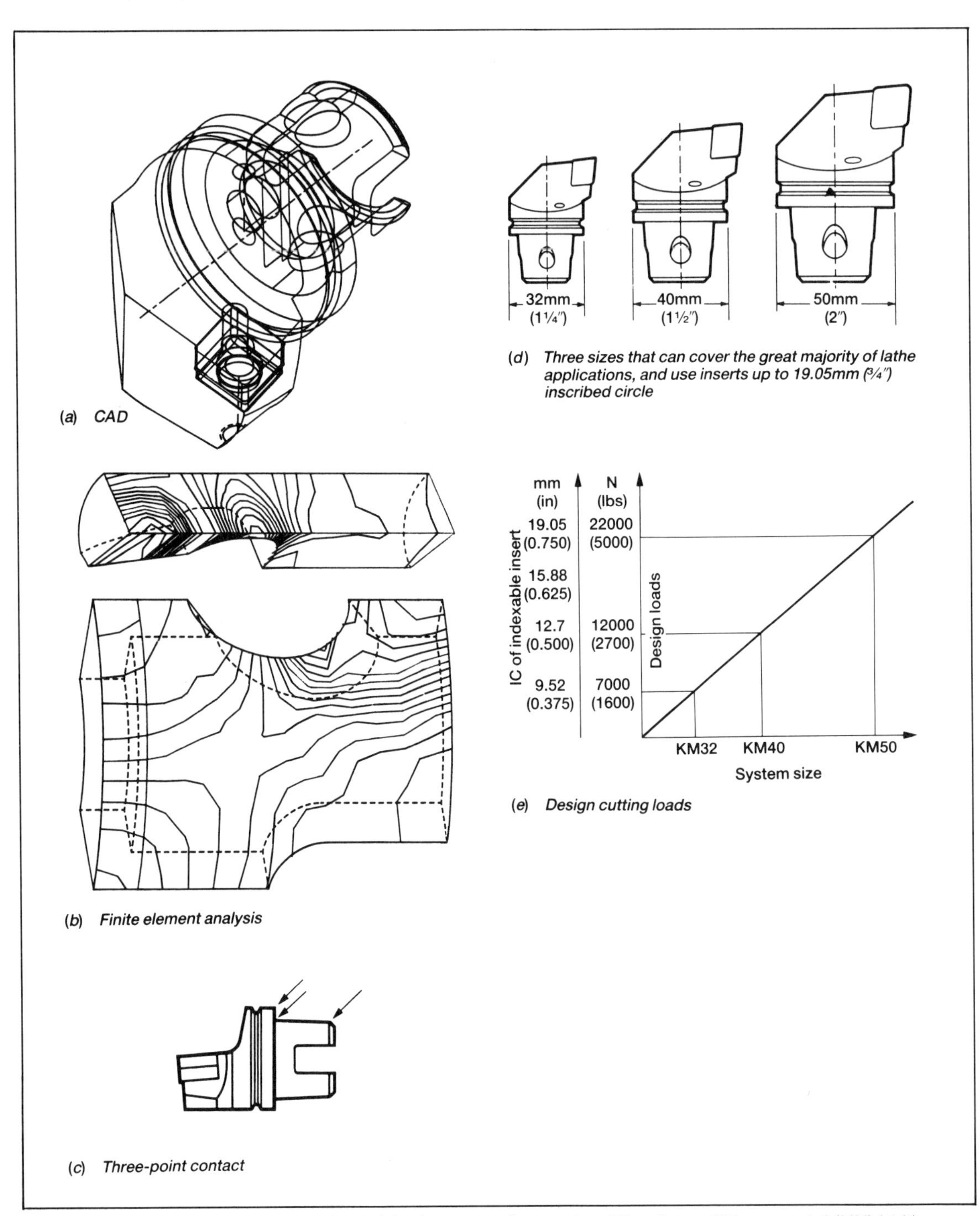

Fig. 2.4 The design of the KM modular quick-change tooling system (Courtesy of Kennametal (UK) Ltd.)

The holders for this new modular tooling system, easily fit into the VDI (DIN) and VDMA standard envelopes as shown in Fig. 2.5b. This torque-nut clamping method has a threaded draw-rod which is engaged by the drive nut. Rotating the drive nut in a clockwise direction will lock the tool; similarly, counterclockwise rotation will release it. Five turns of the drive nut are required to lock or unlock the tool, with a torque of 20Nm. The torque-nut designed holders, or 'axial units' as they are often known, are primarily used for internal machining; right-angled units are used for external machining, with a wedge mechanism to redirect the lock-and unlock force. These right-angled units are carefully designed so that the mechanical advantage needed for 'torquing' the nuts remains the same as for the axial-clamping types. The torque-nut design (i.e. the axial-clamping unit) can be utilised in a semi-automatic or a fully-automatic installation.

The disc-spring designed clamping method illustrated in Fig. 2.5c also allows for either semi or fully-automatic application. In this design the disc springs apply the clamping force through an end-cap which is threaded onto the back of the draw-rod. Pushing on the end-cap releases the cutting tool. A hydraulic cylinder mounted on the machine-tool carriage, behind the turret, provides the release force. A force of the order of 20 000N with a stroke length of 6.5mm is sufficient to release the cutting unit. The disc-spring clamping unit is designed to fit into the same dimensional envelope as the torque-nut type discussed earlier. Designing clamping systems in this manner allows the machine-tool builders, who also design turrets, to offer these new modular tooling systems with only minor modifications to the turret top-plate and the release cylinder.

The last clamping method to be considered is the cheapest to buy and install on turning centres. This is the manual clamping unit, which can be easily retrofitted to current CNC machines. On most existing machines, there is no access to the turret's rear, and in such cases a front-activating mechanism is necessary. This manual right-angled clamping unit is shown in Fig. 2.5a. It is applicable on either turning or machining centres, and is activated by a key at a right angle to the holder centre-line. Rather than orienting the

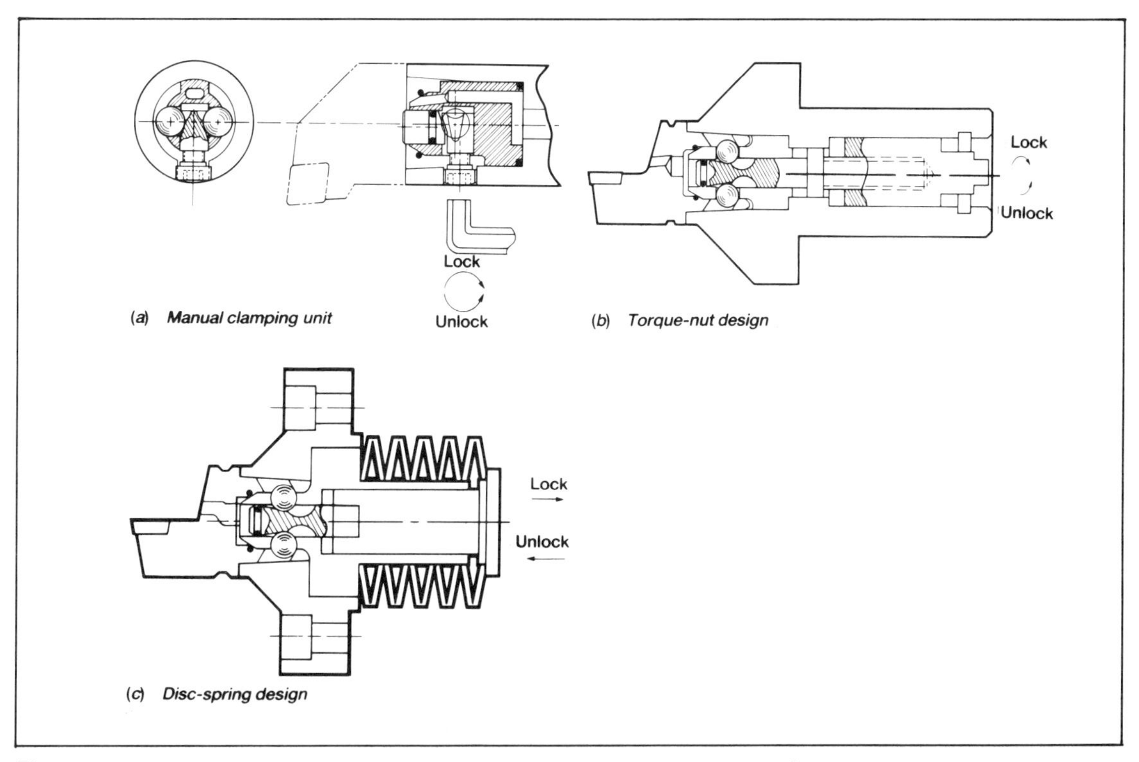

Fig. 2.5 Clamping methods for KM tooling (Courtesy of Kennametal (UK) Ltd.)

draw-rod axially, as in the other two designs, in this design the draw-rod is turned through 90° to the centre-line of the coupling. It is pushed, not pulled, to lock the coupling, using a differential screw connected to a threaded draw-rod. As this draw-rod cannot 'bump' the cutting tool to release it from its taper, a wedge, which is activated when the draw-rod's differential screw is reversed, performs this function. Four turns are needed for the differential screw to lock and unlock the tool, with a torque of 15Nm.

The problems a cutting-tool company has to overcome in designing, testing and developing a new modular quick-change tooling system have been discussed. Before considering the requirements of machining centres, it is worth looking briefly at the finished product as designed for a turning centre, which is illustrated in Fig 2.6. The photograph shows the compact nature of the cutting units for external and internal machining. The short self-holding taper and the ball-locking holes can also be seen, the function of which was discussed earlier in this section. Fig. 2.7 shows the actual cutting unit during a

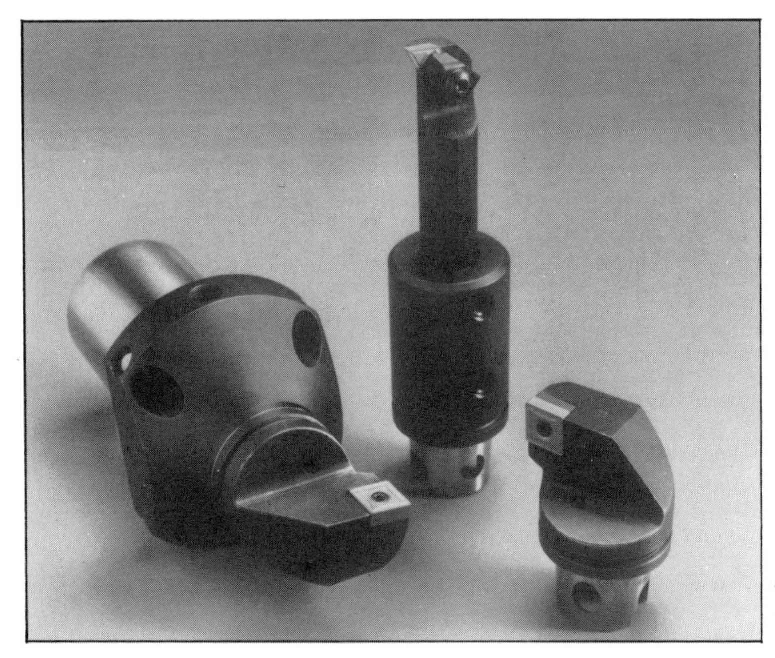

Fig. 2.6 KM modular quick-change tooling (Courtesy of Kennametal (UK) Ltd.)

Fig. 2.7 KM modular tooling in action (Courtesy of Kennametal (UK) Ltd.)

cutting cycle; it can be appreciated how sturdy the assembly is and that the cantilever effect (i.e. overhang) is reduced to a minimum to give greater rigidity and improve the dynamic cutting stability.

Fig. 2.8 shows the earlier KV modular quick-change tooling system held in the turret of a turning centre. The picture highlights how compact tooling is; this is even more true on the latest KM system, and enables a high density of tooling to be situated on a turret. Fig. 2.8 also shows the diversity of the cutting tools that may be accommodated without fouling each other. This is often a problem with solid tooling: sometimes tool pockets either side of a large tool have to be left empty, which clearly limits the turret's tooling capacity.

Yet another tool-adaptor system – the 'FTS' – uses a Hirth gear-tooth coupling, as shown in Fig. 2.9b. The coupling between the tool cutting unit and the clamping/adaptor unit is by means of a Hirth gear-tooth system with a collet, which guarantees a high positioning accuracy with a near-perfect transmission of torque produced whilst cutting. The clamping method consists of a collet with several clamping elements which open wide, so that the tool cutting unit does not require precise positioning to be inserted. Once the cutting unit is inserted, the clamping is carried out by axial movement of the draw-bar, either by manual means or using a torque motor. The same type of clamping arrangement can be used for cutting units with internal or external, right or lefthand tooling, as well as tools held at 90° to the axis. Just some of the range of tooling that makes up the FTS system are illustrated in Fig. 2.9a.

Fig. 2.8 A turning centre with KV quick-change tooling and automatic clamping units (Courtesy of Kennametal (UK) Ltd.)

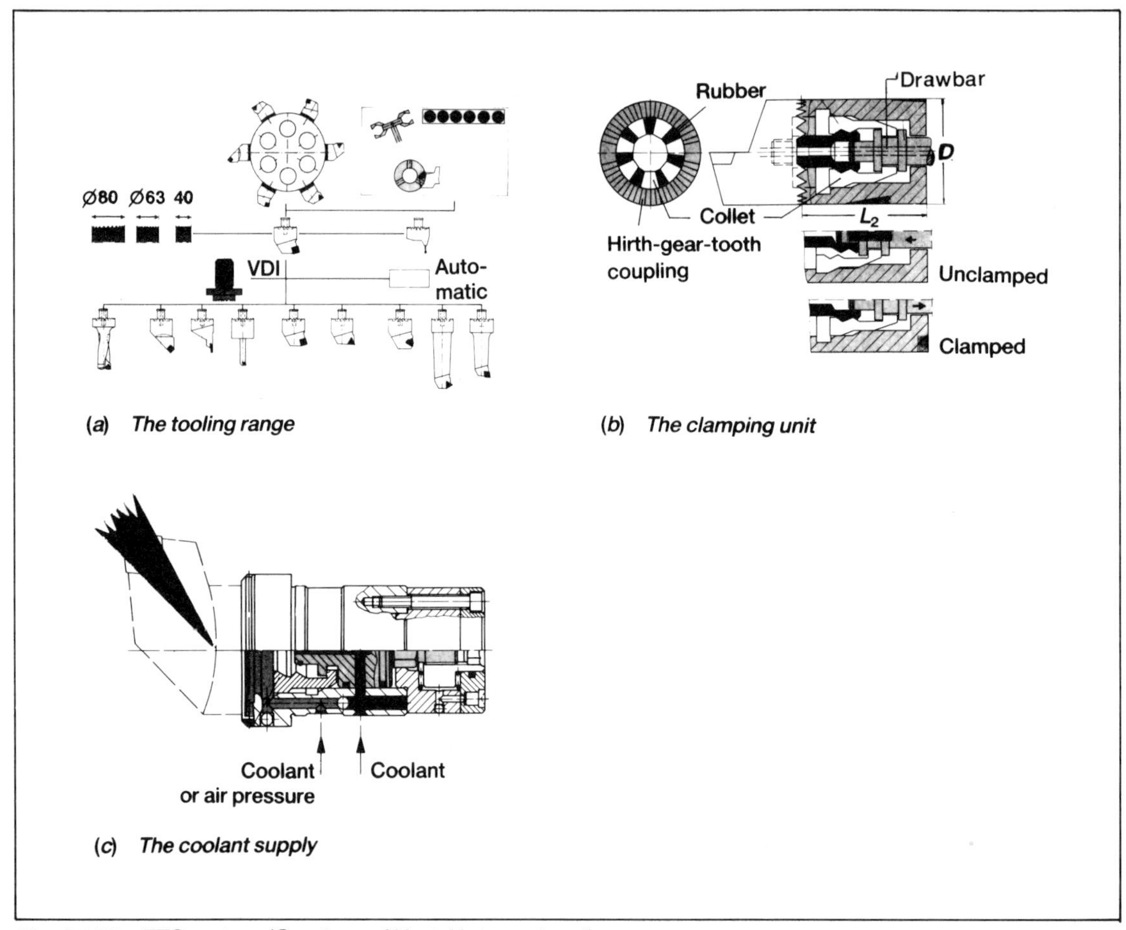

Fig. 2.9 The FTS system (Courtesy of Hertel International)

As with most of the modular quick-change tooling systems, the delivery of coolant is vital if the cutter is to perform satisfactorily throughout the insert's working life. Therefore coolant can be delivered, when called for by the material to be cut, through the adaptor body and coupling right onto the insert, without obstruction, as illustrated in Fig. 2.9c. So, by having a coupling that clamps the cutting unit directly onto the adaptor, it can be internally flushed with coolant, and the Hirth gear-tooth system does not become polluted during its lifetime.

The FTS quick-change modular tooling has a repeatability of coupling of less than ±0.002mm. This tool adaptor system is a standard component, with identical installation dimensions for the 40mm and 63mm diameters. It may be mounted in a disc, drum, row, flat or chain magazine.

This completes the review of quick-change modular tooling systems for turning centres. It is not meant to represent an exhaustive survey of all those available, but simply to highlight some of the popular systems made available to the metal-cutting user by the major cutting-tool specialist companies.

2.2 Tooling requirements of machining centres

Since machining centres were developed, their tooling requirements have undergone relatively few changes, compared with those of turning centres, in the search for more productive cutting time. Their development is typified by the basic configuration of a machining centre with an automatic tool-changer and a tool-storage magazine, or carousel. In recent times some manufacturers have reduced tool-changing times by including a load-and-unload facility on the tool-change arm. That is, whilst the 'old' tool is still in-cut, the 'new' one is selected from the carousel by the CNC; then the 'old' cutter is removed from the spindle and replaced by the 'new' one, in a double tool-holder design. This allows for a faster response to the next cutting requirements of the CNC program.

When one looks at the tool-storage capacity of machining centres, it can be seen that for several reasons most of them have a less-than-total capacity. This may be due to one or more of the following reasons:

- Heavy tooling may be required in the tool-storage system, and because of the system's configuration (such as with the chain type of carousel), tools must be widely spaced to keep the magazine correctly balanced.
- Large tools might require the pockets adjacent to them to be left empty to avoid the likelihood of them fouling each other.
- 'Sister-tooling' requirements might be necessary, i.e. a duplication of all, or most, of the most commonly-used tools, or those that might be susceptible to breakage or wear.

In order to increase the capacity of a tool-storage system whilst simultaneously expanding the range of tools that can be called upon during a production run, modular tooling has been developed which further extends the machine's capability and versatility.

With some of the advanced systems of modular tooling, the tools can be automatically loaded from a centralised preparation and storage facility. This requires a complex tool-presetting and tool-management ability for the retrieval of old tools and their replacement by new ones whilst the cutting cycle continues unhindered. More will be said about this in Chapter 4, and some of the in-cycle tool-monitoring systems necessary to protect the tools, and more importantly the workpiece, will be discussed in Chapter 3.

The use of modular tooling systems on machining centres is further justified by the other predictable advantages of their use. For example, they provide lighter units for the operator to load and unload, thus reducing fatigue. The tools are smaller (although in a machining centre's case, not drastically so), which decreases the tool-storage space required and the inventory. Lastly, but possibly most importantly, tool-changing times and the initial set-up times are reduced considerably. The arguments that were pursued in the discussions of the advantages of quick-change modular tooling on turning centres are equally true for machining centres; they are also valid for CNC milling machines that do not have the benefit of an automatic tool-changing facility.

So far the merits of using modular tooling on machining centres have been praised, but what does a typical system look like, and how adaptable is it to most shop-floor production requirements? The diagram in Fig. 2.3d answers the first of these questions, showing a typical modular tooling system for a machining centre. All such systems have been based upon popular rotating adaptors, such as the V-flange, BT-flange, NMTB and others not specified here. They are available as modular tooling in six sizes from 25mm to 80mm in diameter in this tool manufacturer's range, and have a manual right-angled coupling device.

The rotating modular quick-change tool-holders and cutting units illustrated in Fig. 2.3d will fit any of the rotating adaptors with the same coupling size. To further expand their versatility , a range of extensions and reducers can be used; these have the same manual right-angled coupling as the adaptors and a tapered shank to accommodate a rotating adaptor. These 'mechanical interfaces' (i.e. reducers and extensions) can be used to assemble cutting tools of extended length or to attach a small rotating tool-holder to larger-sized adaptors. These extensions and reducers have also been used in turning-centre applications with modular tooling, specifically where there is a need for internal machining or rotating tooling applications. So by using them in a modular machining-centre tooling system, a wide variety of cutting configurations can be assembled from a limited tooling inventory.

Recently, two cutting-tool companies pooled their resources to develop the quick-change modular tooling concept a stage further, whilst at the same time expanding modular tooling's versatility. The system they developed could be termed a 'universal' tooling system; it is schematically represented in Fig. 2.10. This system of modular tooling allows cutting tools to be shared equally between machining and turning centres, such that turning, boring, drilling, reaming, spot-facing, fly-cutting, face milling, etc. can be done on either machine, provided that the machine tool is configured accordingly. This has the extra advantage of reducing the tooling inventory still further on top of all the other benefits that accrue from this type of tooling in terms of enhancements in productive cycle-time. The system is designed around the KM system mentioned previously. It may be mounted in the machine spindle of a machining centre, or in the turret of a turning centre or its derivatives, by using the well-known VDI or ISO, taper-shanked tool-holders, or it may be mounted directly into adaptors in the spindle or turret.

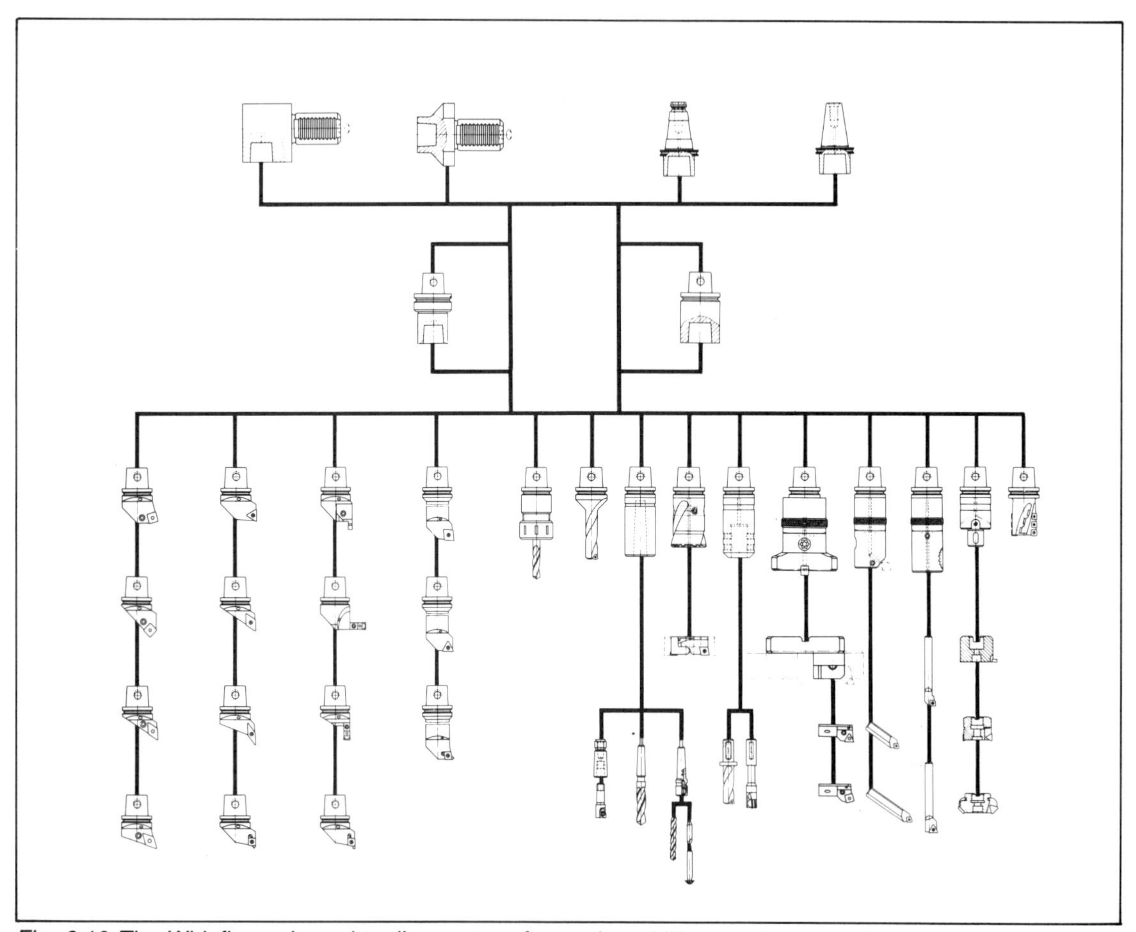

Fig. 2.10 The Widaflex universal tooling system for turning, drilling, boring and milling (Courtesy of Krupp Widia)

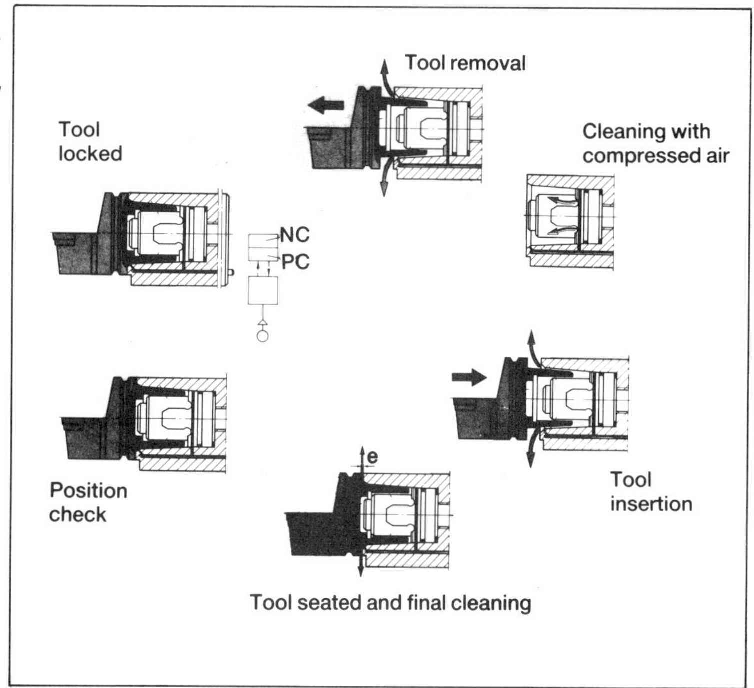

Fig. 2.11 The tool-locking cycle for the Widaflex quick-change modular tooling using contract-pressure monitoring of the position of the tool head (Courtesy of Krupp Widia)

To enhance the universal tooling system even further and ensure that the tool is positively located in its mating taper, an electronically-activated back-pressure device coupled to the CNC can be used. The tool-locking cycle is shown in Fig. 2.11; its operation is as follows:

- The 'old' tool is removed by either the tool-change arm (on a machining centre), or a tool-transfer mechanism (on a turning centre).
- Compressed air purges the female taper, cleaning out debris from the previous tool's cutting operation.
- A 'new' tool is inserted into the holder. Its male taper is cleaned, and it begins to seat itself in the female taper.
- As it is pushed firmly home to register with its opposing taper, the back pressure is monitored electronically, and a signal to indicate that the seating has occurred is sent to the CNC, confirming that the coupling is firmly locked.
- The tool is then ready to begin the next turning or machining operation.

Quick-change modular tooling of this level of sophistication needs to be coupled to some form of tool-transfer mechanism to gain the full benefits of its range of machining applications and speed of operation, and to minimise the pay-back period on its purchase. This will be the next topic to be considered.

2.3 Tool changing using an automatic tool-changing mechanism and tool-holders

To improve the tool-carrying capacity of conventional CNC machine tools still further, automated tool-storage devices with tool-handling equipment can be supplied. These units provide the following benefits:

- They increase the machine tool's productivity times.
- They significantly reduce the times for changing worn tools.

- New cutting units are automatically delivered to the machine by the tool-changing system.
- The rotating magazine holder can provide tool-storage capacities ranging from 60 to 240 cutting units, with up to 24 different types of tool geometries.
- They provide storage of the cutting units, which are returned automatically from the machine tool.
- The integration of these mechanisms onto machine tools can be easily accomplished.
- These systems can cover a wide range of sizes and options to suit most machining applications.

So these are the advantages of installing such systems, but how do they achieve such attractive production advantages, and how do they operate? This will be the subject of the next section.

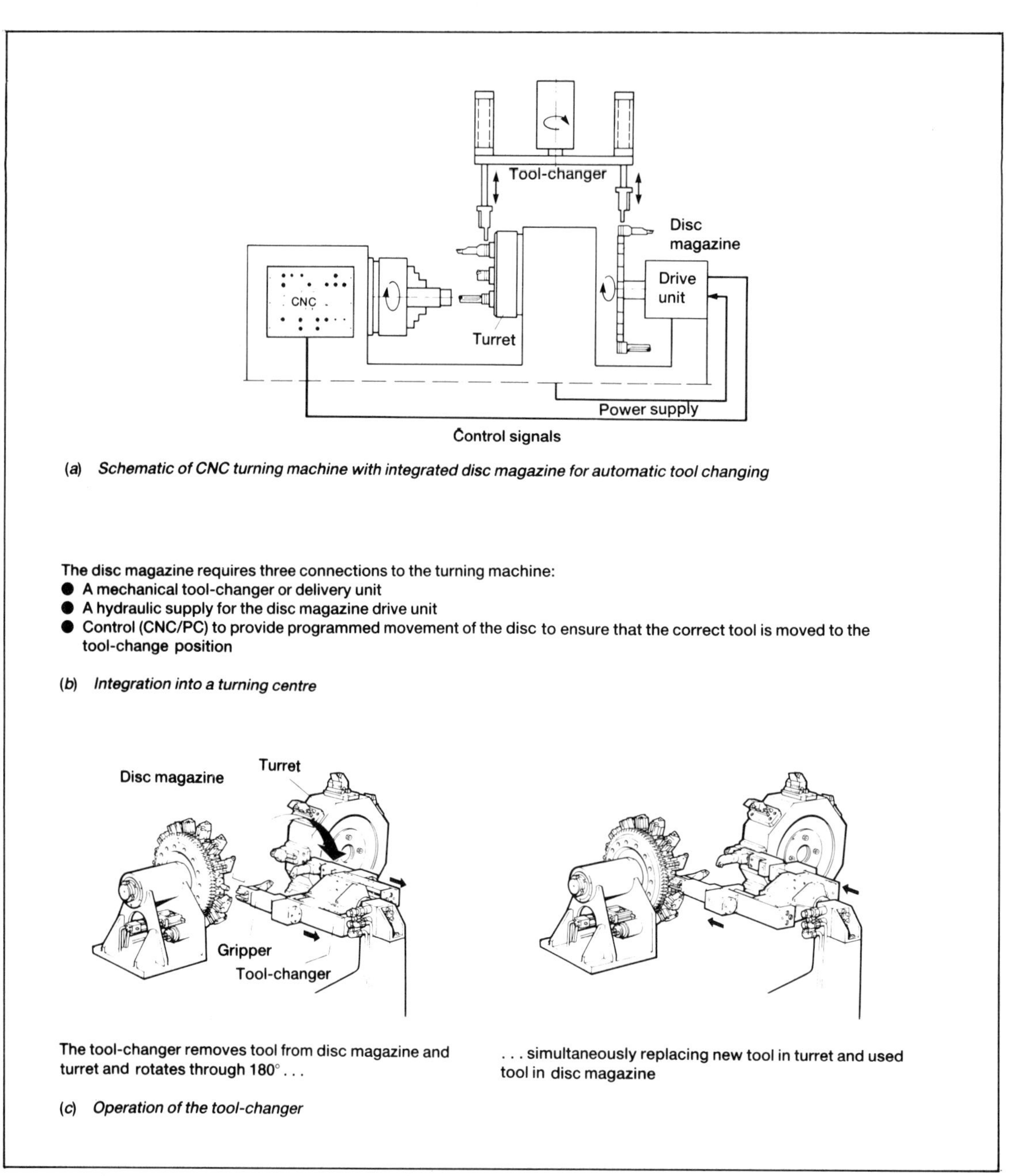

(*a*) *Schematic of CNC turning machine with integrated disc magazine for automatic tool changing*

(*b*) *Integration into a turning centre*

(*c*) *Operation of the tool-changer*

Fig. 2.12 The random-access automatic tool-changer for Block tools (Courtesy of Sandvik (UK) Ltd.)

2.3.1 Automatic tool-changing mechanisms on turning centres

Cutter units can be stored in either a drum or a disc type of storage device. A typical example of the storage facilities and their automatic tool-changers is shown in Fig. 2.12. If one looks at the schematic representation of an installation (Fig. 2.12a) in conjunction with Fig. 2.12c, the method of operation can be simply discerned:

- The rotary tool-changer swivels between the tool magazine and the machine's turret. Each gripper can rotate through 90°, to deliver tools to the front face of the turret (Fig. 2.12a).
- The tool-changer simultaneously removes the cutting units from the disc magazine and the turret, then it rotates through 180° (Fig. 2.12c *left).*
- The new tool is delivered to the turret and accurately positioned in the correct orientation, whilst the used tool is replaced in its correct turret position, for either re-use or replacement (Fig. 2.12d *right).*
- The tool-change arm is withdrawn from the working area of the machine, and the cutting cycle commences.

Where space in the vicinity of the machine is limited, a gantry-type of tool-changer using a twin-gripper assembly can be used for high-speed tool-changing. In this design, the gripper assembly also rotates through 90° to deliver tools to the turret's periphery, as shown in Fig. 2.13a. The major advantages of using this type of tool-changer are that it can be fitted to a machine tool with very little modification, and that it keeps the working area free and unencumbered from any mechanisms that might get fouled by swarf, or hit by the machine tool's moving parts.

The disc-type turret magazines mentioned above have 'random-access' capability. This means that the tool-changer has completely free access to any tool in the store, providing an almost instant delivery to the turret of any tool required. If a company requires even more versatility from the automation of a Block tooling system, it is also possible to change the disc turrets automatically; by doing this the tool-storage capacity approaches the infinite!

The advantage of using any of these universal tool-changing mechanisms is that they are designed on the modular principle and can be fitted to whatever configuration a company requires, allowing for a degree of customisation.

Consider the drum-type tool-storage facility shown in Fig. 2.14, which is being used in conjunction with a gantry-type loading configuration. The rotary drum-type holder permits a continuous supply of cutting units over long periods of untended machining, in the same manner as with the disc turret. When first setting up the machine for a new family of workpieces, the magazine racks can easily be substituted with new ones (Fig. 2.13c) holding other types of tools. The cast iron drum (Fig 2.14a) consists, for the most part, of the rotating magazine holder assembly, and it encloses the interior working parts. This drum is mounted centrally on a column and runs freely on two bearings. A locking mechanism is attached to the drum and hinge plates for mounting of either five or ten cutting-unit magazines, depending on the magazine holder's capacity.

The drum can be removed easily (Fig. 2.14b) allowing for access to the interior working parts. The tool-actuator mechanism is stationary inside the rotating magazine holder; this transports the cutting units out of, and back into, the magazines and is the main interior working part of the drum assembly. When a particular tool cutting unit is called for, the drum magazine is indexed into position before the tool actuator is energised. Then the actuator will load, or unload, the gripper, as shown in Fig. 2.14c. The gripper is

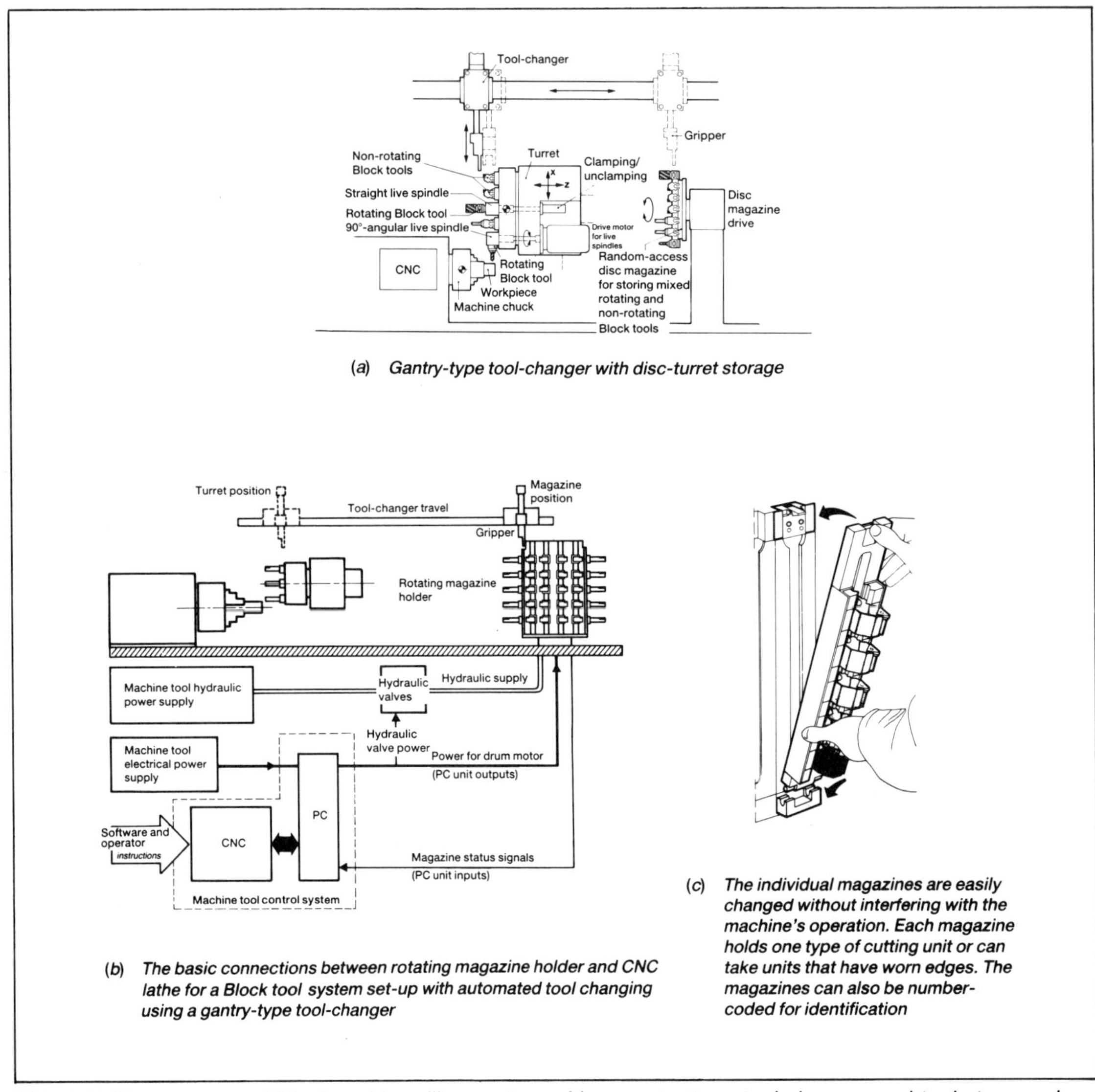

(*a*) *Gantry-type tool-changer with disc-turret storage*

(*b*) *The basic connections between rotating magazine holder and CNC lathe for a Block tool system set-up with automated tool changing using a gantry-type tool-changer*

(*c*) *The individual magazines are easily changed without interfering with the machine's operation. Each magazine holds one type of cutting unit or can take units that have worn edges. The magazines can also be number-coded for identification*

Fig. 2.13 The automated Block tool-handling system with a gantry-type tool-changer and tool-storage drum (Courtesy of Sandvik (UK) Ltd.)

attached to a transportation mechanism – typically a gantry robot – and is taken to the machine-tool magazine, as shown in Fig. 2.14d to f. Gantry systems such as the one in Fig. 2.14d can cater for a whole range of machining conditions on many complex parts. 'Sister tooling' can be used for heavily utilised cutting units, so that the systems have a large and versatile cutting capacity on the shop floor, particularly if powered (driven) tooling is also incorporated into the system as mentioned earlier. Their versatility can be further enhanced if the gantry robot is designed to include such features as touch-trigger probing, of which more will be said in Chapter 3, and the capacity to load and unload the workpieces as shown in Fig. 2.14f. This level of sophistication, with tool and work loading coupled to a 'probing' ability, is almost the 'state of the art', but not quite so: completely automated machine tools, in which chuck-jaws for different work-holding needs can be offered, or a complete chuck-changing facility using a gantry robot is possible, are listed by some machine-tool builders. When machines of this high level of specification are used, either as 'stand-alone' machines or as part of a flexible manufacturing system, the throughput of work and the variety of applications are considerable. When coupled to a high utilisation, with many shifts, the pay-back period, namely the time for the return on the investment, is much shorter.

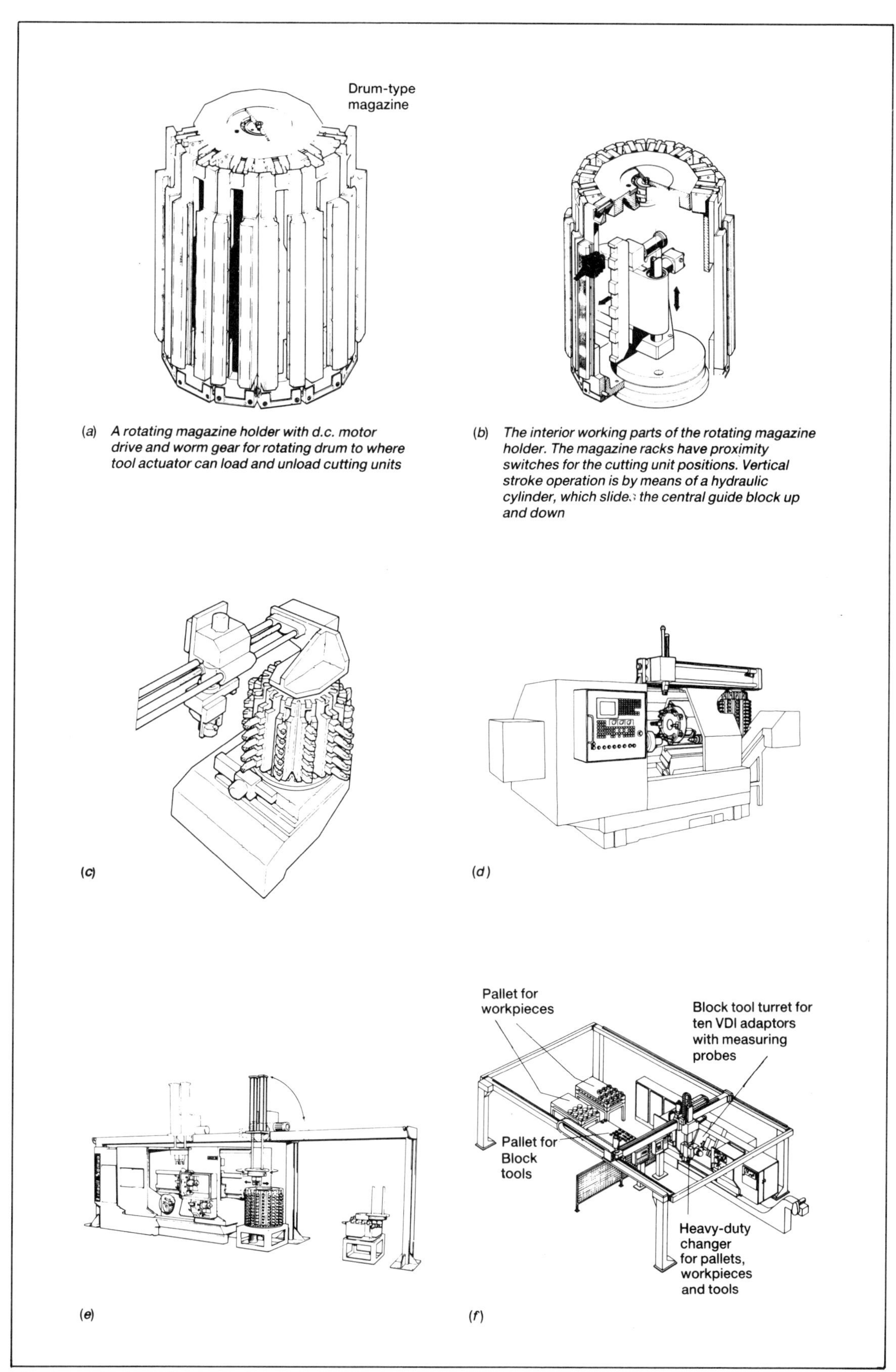

(*a*) *A rotating magazine holder with d.c. motor drive and worm gear for rotating drum to where tool actuator can load and unload cutting units*

(*b*) *The interior working parts of the rotating magazine holder. The magazine racks have proximity switches for the cutting unit positions. Vertical stroke operation is by means of a hydraulic cylinder, which slides the central guide block up and down*

(*c*)

(*d*)

(*e*)

(*f*)

Fig. 2.14 Block tooling mechanisms and transportation devices for turning centres (Courtesy of Sandvik (UK) Ltd.)

Fig. 2.15 KV quick-change tooling carousel and turret with automatic clamping units on a turning centre (Courtesy of Kennametal (UK) Ltd.)

The level of investment required to incorporate all of the features described above is considerable, and is only possible when a company has an expected volume of work which justifies the cost. An alternative approach – which might be suitable for medium to large-batch situations in which the product line is diverse and calls for a certain amount of flexibility in coping with the production demands – is illustrated in Fig. 2.15. Here, the conventional turning-centre turret has quick-change tooling supplied by a chain-type tooling carousel, which can carry approximately 20 additional tools that may be utilised. This tooling configuration, with automatic tool clamping in the turret, is within the reach of most budgets and considerably expands the machine's capabilities for coping with a more diverse range of jobs.

To reiterate earlier statements, whichever the configuration and whatever level of sophistication one requires from a machine tool, it should be apparent that quick-change tooling expands the tooling potential vastly, whilst reducing non-productive cycle times. It should be enthusiastically considered as an alternative to conventional tooling, bearing in mind the productivity savings previously discussed. Once bought, it may be utilised on other machines, for example CNC variants or even conventional machines, simply by changing the adaptor. If a company cannot afford the whole system, it should experiment with just a few units to gain experience in their use – bearing in mind, of course, that the full production savings will only result when a whole system is utilised in an effective manner.

No matter how sophisticated the quick-change tooling system is, or how well-tried and tested the conventional tooling might be, or how complex the feasibility studies, once the tool begins to degenerate (possibly through extraneous circumstances), the productive capability is lost. At the very least there is a need to protect the tool from the vagaries of chance failure by using some form of tool-condition monitoring system, and this in turn will also protect the workpiece. This type of monitoring system, and others, will be discussed in the next chapter.

Chapter 3
Workpiece and tool-monitoring systems for CNC machine tools

A basic requirement for increasing the productivity of CNC machines is the ability to operate them in the unattended, or at the very least, minimally-attended condition, whether they are 'stand-alone' machines or part of a flexible manufacturing system (FMS). If unattended machining is desirable, the absence of an operator will create a considerable number of problems which must be overcome if the machining system is to operate satisfactorily. These problems arise in performing the monitoring and service functions which would normally be seen to by the operator, who would usually do such things as monitoring the cutting tool's condition and performance, replacing worn or defective tools by interrupting the cutting cycle, assessing the workpiece quality during machining, changing speeds and feeds if required, and responding to unusual conditions that are seen or heard during the cutting operation. In unmanned situations, the monitoring systems must provide the artificial intelligence necessary to mirror the experience gained by fully-skilled operators and their instinctive reactions, and to provide the type of expertise normally associated with human involvement. A considerable number of monitoring systems are available from both machine-tool builders and cutting-tool companies. As some of these systems have been introduced only recently, many machine-tool users may not be fully aware of them and of their potential benefits.

Monitoring systems can be classified as process-monitoring, workpiece-monitoring, and machine and tool-monitoring systems. Typical applications of these monitoring systems for untended machines are as follows:

- Monitoring the correct loading of the workpiece, correcting any set-up misalignments or datum offsets, and checking the quality of the workpiece.
- Checking for use of the correct tools, identifying the tools and setting tool offsets, monitoring for tool wear and breakage, and initiating tool replacements as necessary.
- Adjusting speed and feed where necessary, and compensating for such effects as tool wear, thermal deformation and congestion of chips.
- Monitoring of machine elements including the controller, and taking any necessary action in response to program failure, diagnostic error messages, etc.

Whatever the function that is to be monitored, there is a need for some form of sensor to be included in the system to detect any problems, so that action can be taken if necessary. The sensor's output (triggered by an error message, for example), must be processed to obtain the correct information to allow a decision to be made. The machine's control unit receives the 'sensed' results and initiates controlled actions to correct or recover the situation.

Various types of monitoring and sensing systems are currently available for turning and machining centres. This chapter will briefly discuss just some of the systems that are available for workpiece-recognition and monitoring, and then spend more time discussing some of the various methods of cutting-tool monitoring.

3.1 Workpiece-monitoring systems

The workpiece monitoring functions needed for unattended machining are as follows: identification of the workpiece, automatic set-up of the workpiece, and gauging of the workpiece. Approaches to each of these will now be discussed briefly.

3.1.1 Identification of the workpiece

Whenever an automated cell or system is required, the first requirement before machining can commence is that the different parts must be successfully selected from a random mixture. In order for the individual parts to be identified from the diverse range available, a sensing device is required. Once the part has been selected correctly, the required CNC program is read from the memory of the controller. Various methods are available for workpiece identification, particularly in the field of pallet-recognition systems, using such features as bar coding and binary coding, together with the touch-trigger probing method.

Usually, the components are delivered to machining centres on pallets with appropriate fixtures, and they are recognised by the machine tool through an indirect identification system. These systems can use discrete coding which uniquely identifies each pallet, ranging from sophisticated vision systems and electronic radio-frequency transponders to optical character-recognition devices and binary-coded pins. Alternatively, the touch-trigger probing method may be used. In principle, this system consists of a very sensitive omnidirectional switch which is capable of detecting deflections in any direction. If it is deflected by an obstruction, e.g. the workpiece, a signal is sent to the controller indicating the actual coordinates at the point of contact, which are stored in the memory for further processing. Depending upon the machine tool configuration, the signal transmission may be either hard-wired or inductive, or it may use telemetry techniques. Figs. 3.1a and b show the probing configurations for turning and machining centres, respectively.

These probing systems can fulfil a number of functions; in the case of workpiece recognition, the touch-trigger probe is programmed to identify various predetermined features on either the pallet, its fixture or the component. A range of workpieces can be accommodated, identified and defined, based upon the results of probing using conditional part programming.

However, pallets are not normally used for loading rotational components onto turning centres, so workpiece identification is rather more difficult in this case, since indirect

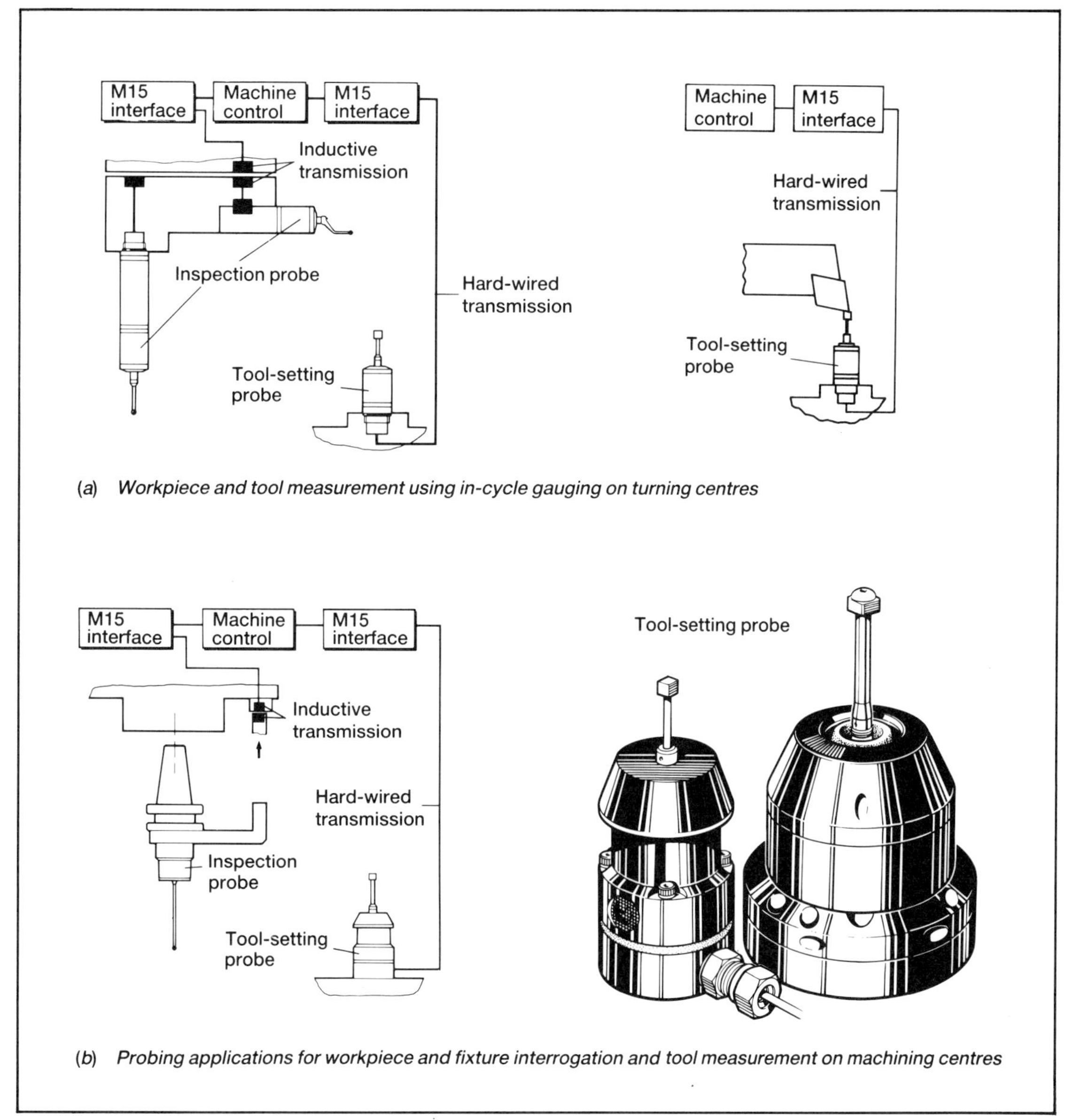

Fig. 3.1 Touch-trigger probing applications for turning and machining centres (Courtesy of Renishaw Metrology Ltd.)

methods cannot be used. Instead, the workpiece itself must be identified; this is most commonly done by using either a vision-based system or the touch-trigger probing system.

By using an image processor with the configurations of the various workpieces stored in the memory, a machine-vision based system will be able to recognise a workpiece by using the image-comparison technique. A real advantage of this workpiece-recognition system is that no cycle time will be lost, as identification may be undertaken whilst the workpiece is waiting to be loaded beside the machine. If the alternative workpiece-recognition system, namely the touch-trigger probe, is utilised, the probe, which is mounted on the turret, is programmed to confirm that the desired component is in position once the work is held firmly by the chuck, mandrel or fixture. The probe will assess the unique dimensions of the workpiece to check that the correct part is being loaded from a random mix of a family of components; when these features are affirmed or identified, the CNC will call up the correct tool data and program.

3.1.2 Automatic set-up of the workpiece

Once the workpiece is positioned on the machine tool it must be correctly aligned with respect to the machine axes, and the datum point of the part program has to be adjusted to correspond with the new position of the component. Using touch-trigger probes accelerates this otherwise time-consuming activity which is a crucial function on unattended automated machining operations. The ability of touch-trigger probes to interrogate the component is almost invariably one of the main reasons for their incorporation on turning and machining centres. These probes are either automatically loaded in the spindle of the machining centre (Fig. 3.2a), or permanently mounted on the turret of the turning centre (Fig. 3.2c). Datum points or surfaces can be found by the probes, and the axes can be automatically zeroed at these locations by the controller. If the workpiece is misaligned on the pallet when it is positioned on the machine its misalignment can be measured and compensated for by the software. When a casting requires machining, as in Fig. 3.2a, positions on the casting's contour can be sensed, and the depth of cut can be optimised by an automatic calculation.

On turning centres, the set-up of the workpiece is much simpler to achieve owing to the symmetrical nature of most rotational parts. The techniques of identifying part features and zeroing axes are similar to those already described for machining centres; they are illustrated in Fig 3.2c.

Automatic workpiece set-up requires a considerable programming knowledge, and many machine-tool builders have developed standardised software packages which reduce the effort required from the end-user and allow for a degree of customisation. Such automatic systems reduce set-up times dramatically, and the use of touch-trigger probes cuts work-in-progress and idle times considerably, so they are gaining widespread acceptance by industry.

3.1.3 Gauging of the workpiece

Once the workpiece has been machined, certain of its features need to be checked in order to confirm any further similar components automatically generated by the CNC will be dimensionally correct. In recent years, a variety of new gauging concepts have been developed and new technologies evolved, that allow for consistency and speed in checking dimensions and then using the data for any corrective action necessary. The techniques for monitoring the quality performance of a machine tool may be classified into four groups: prediction and correction of errors, in-process gauging, in-cycle gauging, and post-process gauging; the last of these groups may be sub-divided into near-machine gauging and off-machine gauging. We shall consider each of these techniques in turn and then go on to discuss further methods of measuring workpiece features.

Prediction and correction of errors

The basic idea of the method involving error prediction and corrective action is to try and anticipate machining errors in real time, and then correct them. As with any machining process, the object is to cut a good part and so eliminate the need for further inspection. A detailed mathematical model must be used in which the error-producing parameters, namely the effects and the interactions, are accurately described. This technique, known as deterministic metrology, has yet to be applied to the production environment, but it may have a considerable impact on work measurement in the near future.

(a) Using a touch-trigger probe to interrogate a casting on a machining centre (Courtesy of Cincinnati Milacron.)

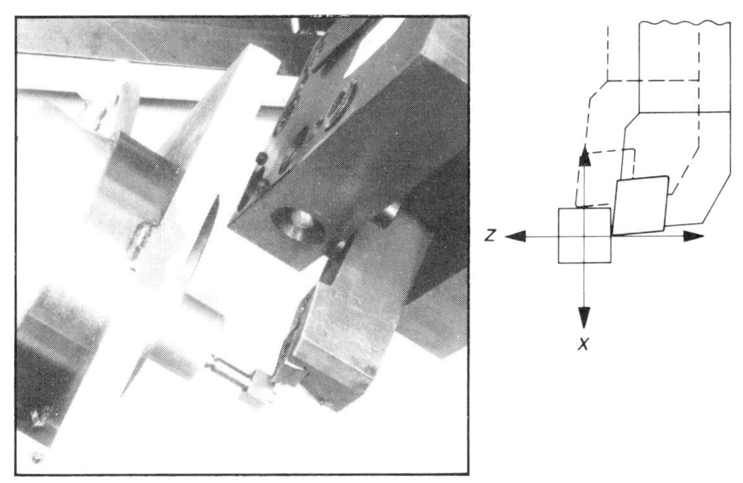

(b) Automatic tool offsets preset on a turning centre (Courtesy of Gildemeister (UK) Ltd.)

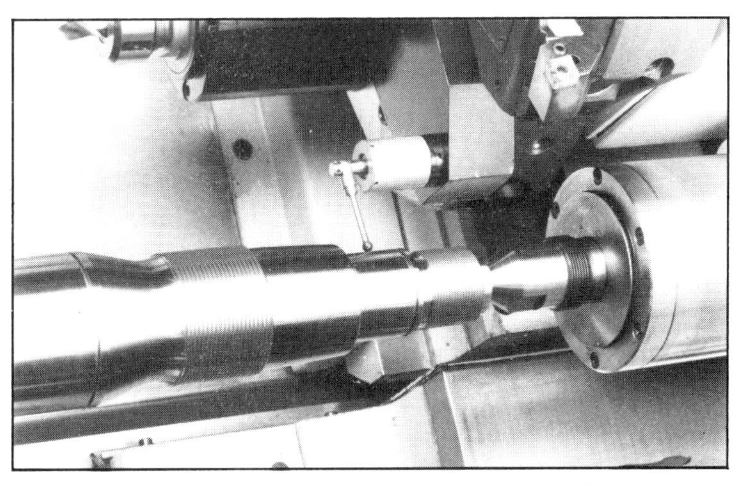

(c) In-cycle gauging of job using probing (Courtesy of Gildemeister (UK) Ltd.)

Fig. 3.2 Touch-trigger probing applications on turning and machining centres

In-process gauging

The technique of in-process gauging allows the information produced by measurement during machining to be used to control the process. The information is provided continuously, and the workpiece quality is controlled by the data generated. This technique is well-known in grinding applications and it may be developed for use in turning operations, offering practical solutions for real-time measurement and control. It is less adaptable for machining centres, and more developmental work is required to gain the full benefits from this system.

In-cycle gauging

Most machining and turning centres use the in-cycle gauging technique, which is also referred to as on-machine gauging. With this measuring method, the quality of the workpiece can be assessed either between cuts or after completion, prior to removal from the set-up, as illustrated in Fig. 3.2c. The driving force for the introduction of in-cycle gauging is almost always the need to gather information for quick quality control so that the waiting time of the machine tool can be reduced. Measurement is achieved by utilising the axes' feedback systems; a disadvantage of this is that the machine's systematic errors and other process-related inaccuracies can go undetected by in-cycle gauging. The other problem is that it appears to increase the non-productive cycle time. However, this increase does not actually occur, because the set-ups and inspection required without probing also increase the true costs and cycle times, and amount to the same thing, as shown by Fig. 3.3a. In fact, a large saving occurs, as the gaps between production cycles are dramatically shortened by using probing, compared with the conventional set-up and inspection methods.

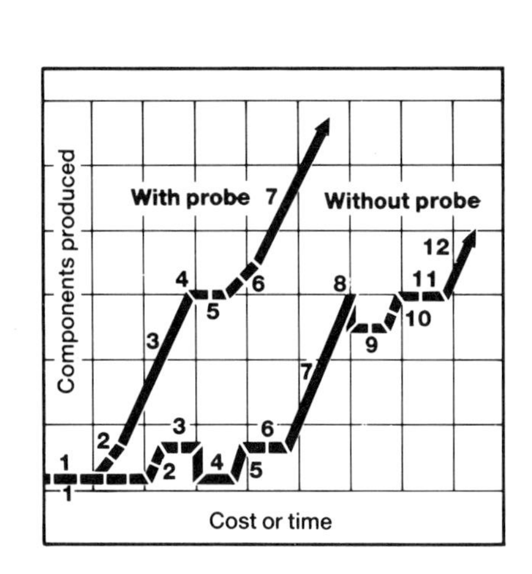

WITH PROBE
1. Set-up 2. First component produced Auto inspect and update offsets 3. **Production** 4. Parts out of tolerance 5. Reset tool 6. Reset offsets using probe 7. **Production**
WITHOUT PROBE
1. Set-up 2. First component produced 3. Inspect 4. Reset 5. Rework 6. Inspect 7. **Production** 8. Tool breakage 9. Reset 10. Rework 11. Inspect 12. **Production**

(*a*) *The production increases due to in-cycle probing*

(*b*) *Automatic tool measurement and tool offset-updating*

Fig. 3.3 Touch-trigger probing on CNC machines (Courtesy of Renishaw Metrology Ltd.)

A useful technique to compensate for errors in the machine is the so-called 'footprint' method. With this technique, the finished workpiece is first measured using in-cycle gauging and then inspected under full metrological conditions by an independent measuring system. If, as expected, there is a dimensional difference between the measurements, then this is introduced as a tool offset in the machine control unit for that type of workpiece. This offset will compensate for any dimensional errors caused by in-cycle gauging of further workpieces produced on that machine.

Near-machine post-process gauging

Post-process gauging is a widely-used method of assessing workpiece quality and at the same time correcting for tool wear. Two popular methods are used, namely near and off-machine gauging. Near-machine gauging has been developed to elimate the problems associated with in-cycle gauging, but with the benefits of performing a closed-loop machining function. Workpieces are transported by a suitable handling device to an independent gauging station situated on or near the machine tool, where they are inspected by a multiple or single feature receiver gauge similar to that shown in Fig. 3.4.

Fig. 3.4 Near-gauging of workpieces using sophisticated robot loading from the machine tool (Courtesy of Gildemeister (UK) Ltd.)

(a) Robot loading of machine and near-gauging fixture with workpieces

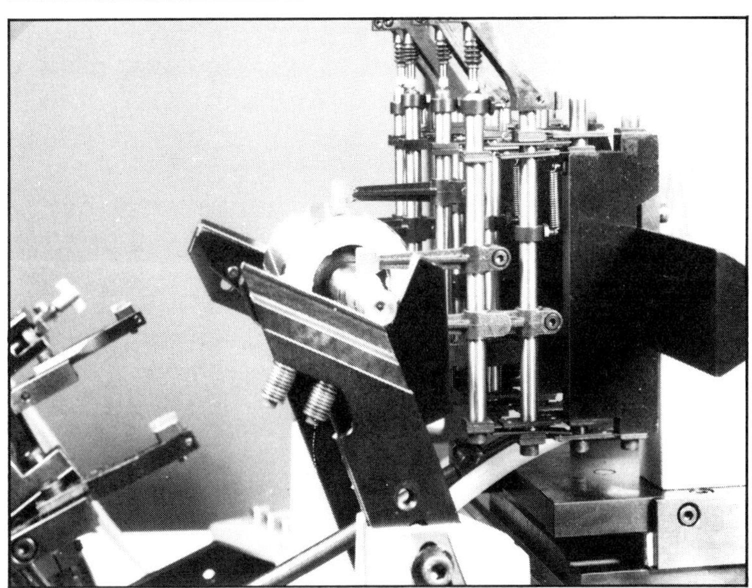

(b) Detail of a typical transducer-based receiver guage for workpieces

The results of the gauging are usually fed back to the machine control unit, where tool offsets are updated within specified limits, thus maintaining the desired quality of workpieces by closing the dimensional control loop. The use of transducer, caliper or non-contact receiver gauges gives reliable inspection results, coupled with the capability to compensate for errors whilst not increasing the machining cycle time. Near-machine gauging systems are normally applied on turning centres, as rotational parts require simpler and less expensive gauging than the prismatic parts normally associated with machining centres.

Off-machine post-process gauging

In the off-machine metrology techniques, the parts are inspected in a separate quality-control area, using dedicated systems specifically designed for the purpose. Gauging the part in this manner is usually time-consuming, but it gives the most accurate results. However, in a flexible manufacturing system (FMS), these off-machine inspection systems must themselves be flexible and allow for automatic measurement of features to increase their versatility. Machines such as CNC coordinate-measuring machines are normally employed as part of an overall system as they are highly adaptable; they are easily utilised for automated inspection of parts and for sending updating information to the machine controller using closed-loop feedback. A typical coordinate-measuring machine, in which a touch-trigger probe can be used for inspecting parts as frequently or infrequently as required, is shown in Fig. 3.5a. If high-volume inspection is required because the short cycle times for each component necessitate very frequent tool-compensation updating, then it might be advantageous to use non-contact gauging, utilising such features as laser inspection devices. A typical system is shown in Fig. 3.5b; here the amount of dimensional data that can be processed can approach 50 discrete values per second which is more than adequate for any CNC machine tool.

So far, the main emphasis in this chapter has been on workpiece monitoring, and on how the information gathered may be used to update tool offsets. Many companies have expressed more interest in monitoring the condition of the tool itself, as this directly affects workpiece quality. There are a number of ways in which tool monitoring can be achieved; some of these will be discussed in the next section, which explains how and why each system is installed and their principles of operation.

3.2 Tool-condition monitoring on turning centres

If operators are present, one of their major functions is to monitor the tools whilst the machine tool is cutting. They continuously assure themselves that the tools are performing productively. The tool-related monitoring functions performed by an operator during manufacture may be classified into four groups: identification of the correct tools, measurement of the tool offset, monitoring tool life and detecting breakage. In the case of untended machining, where no operator is present, these functions must be automatically executed by suitable control and monitoring systems. Just some of the methods of achieving these tasks will be described below.

3.2.1 Tool identification

Some method of identifying the correct tool to be used for a specific operation can be considered to be an essential requirement: if the wrong tool is selected, at the very least the workpiece will be scrapped and, worst still, the machine tool itself may be seriously

Fig. 3.5 Off-machine measuring techniques

(a) A CNC coordinate-measuring machine using touch-trigger probes for workpiece inspection. It may be used in a 'stand-alone' mode (as illustrated), or can be incorporated into an FMS with full statistical data generation for post-process control of machine tools (Courtesy of LK Tool Co. Ltd.)

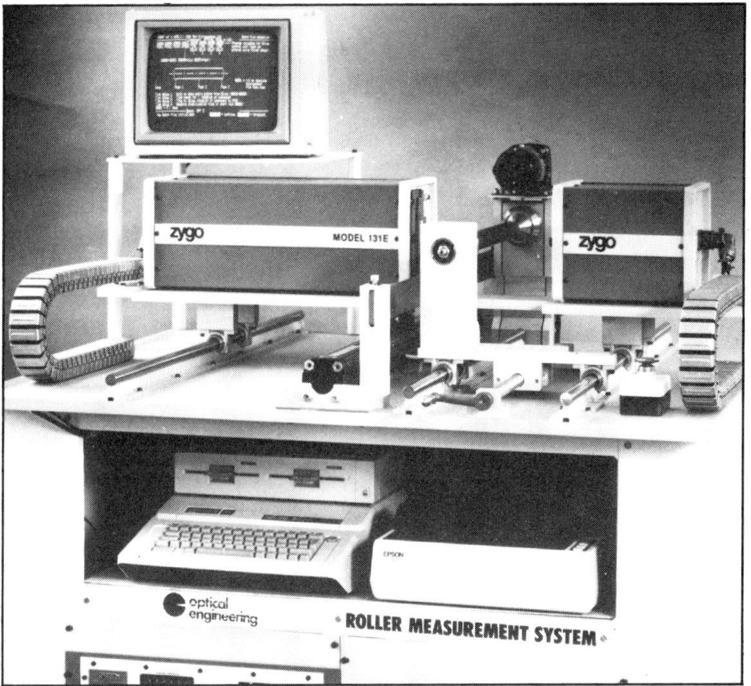

(b) Laser measuring systems for high-speed data gathering and control of workpiece dimensions and tool offsets in a near-machine or post-process receiver gauging station (Courtesy of Scanatron)

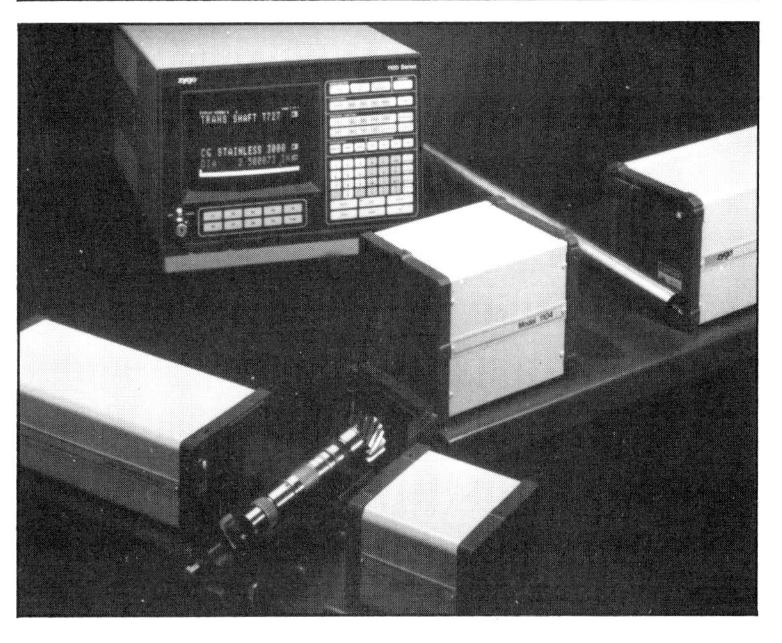

damaged. An efficient automatic tool-handling system, in which the tools may be accessed at random as required, will depend upon each tool being identified correctly at every stage of the part's manufacture. Two techniques can be used to identify a tool successfully: geometrical checking of the tool, and coding of the tool.

In the first technique, the tools are identified by using touch-trigger probes, etc. to establish their key geometrical features, as illustrated in Figs. 3.2b and 3.3b. Enabling a machine controller to recognise the 'correct' tool with this system requires extensive computer storage, whereas with the tool-coding system each tool's code contains enough data for it to be successfully identified. A variety of coding systems are available, and some include a ring, or bar, code for holders and tools. Lately, a more adaptable system of tool identification has been developed which stores tool data on silicon chips known as EEPROMs. These EEPROMs (electrically-erasable programmable read-only memory chips), or their derivatives, are built directly onto tool-holders. Data such as tool insert features and offsets, holder type, tool life, etc are stored on the chip, which truly identifies the tool to the machine control unit. The information concerning the tool selected from the storage facility is transferred to the controller by either a contact or a non-contact read/write sensor. The data can be input onto the chip at the presetting stage. More will be said on this topic in the section on the use of 'intelligent' tooling in Chapter 4.

3.2.2 Tool-offset measurement

As mentioned above, knowledge of the offset position of the tool's cutting edge and its relationship to the part's datum point is an important requirement when CNC machining. This data can be established off the machine tool by using tool presetting devices (which will be dealt with in more detail in Chapter 4) to determine the tool's position. When a replacement or a new tool is required for a particular job, the position of its cutting edge must be determined, and this information must be fed into the machine control unit. Even if the tool itself is not replaced, errors are introduced whenever the insert is indexed in the holder, owing to the insert tolerances, and these factors must be allowed for and the machine adjusted accordingly. Where the tool-holders are replaced, there will be variations between similar holders which must also be considered, requiring further tool presetting. To achieve accurate and speedy tool setting on turning and machining centres, a variety of on-machine tool-sensing devices are available. The two main types of on-machine tool-setting devices in use are the touch-trigger probe (already mentioned) and the vision-based sensor.

The hard-wired omnidirectional touch-trigger probe shown in Fig. 3.1a and b is the most popular method of setting tool offsets; the arrangement of the probe varies according to the machine-tool builder's tooling philosophy. On turning centres the probe is often mounted on a swing-arm as shown in Fig. 3.3b, so that it can be withdrawn to the protection of its casing once it has been used in order to avoid damage from swarf or tool collisions. The tool-offset probe is usually cube-shaped (Fig. 3.2b), with the faces of the cube aligned with the machine axes. Once the probe has touched the tool tip, its position is known relative to a datum point: this allows the machine controller to set the tool offsets in the tool table automatically, allowing the 'first-off' part to be machined accurately.

Vision-based tool-offset sensors have recently become available on turning centres. In this system, the tool is driven to a position in the field of view of a video camera, and the image produced by the tool tip is analysed to obtain the required X and Z-offsets in under three seconds. Accuracies of ±0.025mm on the diameter and ±0.010mm on the length are claimed for this system, and it is also stated that the sensor can be utilised for insert identification. Whether the system is sufficiently 'workshop-hardened' to withstand the rigours of the cutting-tool environment has yet to be established: the lens system, in particular, may need extra protection from the coolant spray associated with metal removal at the high speeds typically utilised on turning centres.

3.2.3 Tool-life monitoring

Owing to the interaction with the workpiece and with the chips produced, cutting tools will wear. The worn tools must be replaced prior to tool failure. Where unmanned machining occurs, a sensing system must be used to determine the extent of wear and when the tools need to be changed, in order to maximise their useful life.

A variety of sensing techniques have been developed to sense tool wear; some are more generally available than others. These sensing devices can be classified into two main groups, direct and indirect methods:

- Direct sensing methods include radioactive techniques, measurement of electrical resistance, optical observation of the wear zone, and measurement of workpiece dimensional changes or of the distance between the workpiece and the tool-post.
- Indirect sensing methods include techniques based on temperature, sound, vibration, acoustic emission and force (either measured directly, or through measurements of power, current or torque).

The monitoring performances of most of the techniques listed above have proved satisfactory in the laboratory, but only a few have been adopted for tool-life monitoring of unmanned machines. These tool-life monitoring systems fall into three groups: those that record the machining time, those involving touch-trigger probing, and the extension of tool-breakage detectors to indicate tool wear. A brief appraisal of each of these systems is given below; later in the chapter, there is further, more in-depth, discussion of several of the popular systems now available.

Regarding the recording of machining time, a number of the controllers now available provide software that allows permitted tool life to be specified for individual tools. During the cutting operation, the controller monitors and accumulates the actual tool time used. Once this time reaches the specified value, a 'worn tool' alarm is initiated, and the tool is changed when it is safe to do so.

Using touch-trigger probes it is possible to measure changes in the tool's offset due to wear, or to measure the effects of wear on the change in workpiece size. If allowable values for these measurements of wear are correctly specified, the machine control unit can be programmed to initiate tool changes once these levels are reached.

A popular system for indicating tool wear-rates is by measuring the force increase as the tool progressively wears. A variety of systems are available; some of the methods of measuring the forces are mentioned in the 'case studies' in section 3.3.

3.2.4 Tool-breakage detection

It is essential that the tool has a good cutting edge before it enters a cut; once it has started cutting it is necessary to continuously monitor the cutting edge so that there can be a fast response if catastrophic tool failure occurs. If a broken tool is undetected during unmanned machining considerable damage may be caused to the workpiece, the toolholder or the machine. When tool failures occur, it is normal to act quickly in order to minimise damage, and the appropriate action might be an emergency stop of the feed action or spindle rotation and a fast retrieval of the cutting tool. It is therefore obvious that an in-process tool-breakage sensing system is essential for any unmanned machining operation. It is possible to divide tool-breakage detection systems into two broad groups: post-process or between-process sensing, and in-process sensing. With post-process sensing, the cutting tools are checked either between cuts or prior to taking a cut. This method is appropriate for cases where tool failure is not too critical, or when inexpensive workpieces and tools are used. The post-process sensing method for the detection of broken tools can be either direct or indirect. In the direct sensing method, the existence or position of the tool edge is inspected using a suitable sensor; the normal method is to use a touch-trigger probe, a proximity sensor, or an optical sensor for this purpose. This method is particularly popular for detecting breakages of small tools, the main reason being that most in-process techniques are not sufficiently sensitive to detect this type of breakage. In the indirect post-process sensing technique, broken tools are detected by measurements of the workpiece's dimensions and roughness. Alternatively, touch-trigger probing methods or other in-cycle workpiece-gauging sensors may be utilised to estimate the workpiece's condition after machining. For continuous monitoring of the tool whilst it is cutting (i.e. in-process tool-breakage detection method), an indirect sensing technique is normally used. There are a variety of signals which have been proposed for detecting tool breakages; these include cutting-force monitoring, acoustic emission, vibration analysis, sound analysis, cutting-temperature monitoring, electrical-resistance monitoring, and surface-roughness analysis. Out of these techniques, the commercial systems tend to be either force-related or based on the acoustic emission method. The commercially-available systems based on force-related signals may be sub-divided as follows:

- Those that use a dynamometer, either on the tool block or below the turret.
- Those with thrust force bearings.
- Spindle-bearing instrumented.
- Those with a sensor for assessing spindle deflection.
- Power monitoring.
- Torque monitoring, sometimes known as 'torque-controlled machining'.
- Motor-current monitoring.

The cutting-force monitoring systems are based on the principle that an increase in the cutting force gives a good indication of the tool's condition as it approaches failure. From the list above it can be appreciated that there are many ways of sensing forces on the tool. Cutting forces may be sensed at any point in the force flow path of the machine tool, either on the tool side or on the workpiece side. The real difference between the various sensing approaches is the level of sensitivity that may be expected, in terms of the change in their response as the tool's condition progressively deteriorates during its cutting life.

When sensors are integrated onto the machine tool, they should be positioned in such a manner that they cannot influence the machining process, or the dynamic and static behaviour of the machine. Turning and machining centres can commonly use similar types of force sensors. However, the dynamometer-type sensors are usually favoured on turning centres, whereas spindle-deflection sensors are normally associated with machining centres.

The development of suitable processing algorithms and microprocessor-based signal-processing units can be attributed to the successful use of force-related signals for detecting tool breakage. The algorithms, and the logic developed for detecting breakage differ from system to system, but a typical approach is to set a threshold limit for the sensed signal, and to initiate the appropriate action once this limit has been exceeded. Most systems provide a 'learn' mode, in which the nominal signal levels are defined. The limits are then set as a percentage of the nominal values, with this percentage being user-programmable in most cases. In the more advanced systems, dynamic or floating-threshold techniques may be incorporated; this reduces user effort and improves their operational reliability.

The other commonly-available system for detecting tool breakage utilises acoustic emission (AE), which is now emerging as a popular low-cost application. More will be said about the system in section 3.3.3, but the principle behind its operation is that high-frequency elastic stress waves are generated during cutting, owing to the rapid release of strain energy. This is known as 'acoustic emission', and it is caused by such factors as material deformation, fracture and phase changes. The amplitude of the acoustic emission, which is normally monitored by a piezoelectric sensor, will increase by a considerable amount just as the tool is about to break. Owing to the very high frequency of the acoustic emission signal, it can easily be separated out from the noise signals generated during cutting a workpiece. Development of acoustic emission sensing systems continues and many of the early problems are now resolved, but more work remains to be done to fulfil the true potental of this monitoring system. The principle of acoustic emission is schematically represented in Fig. 3.6, which shows how one company has applied this technique in tool monitoring.

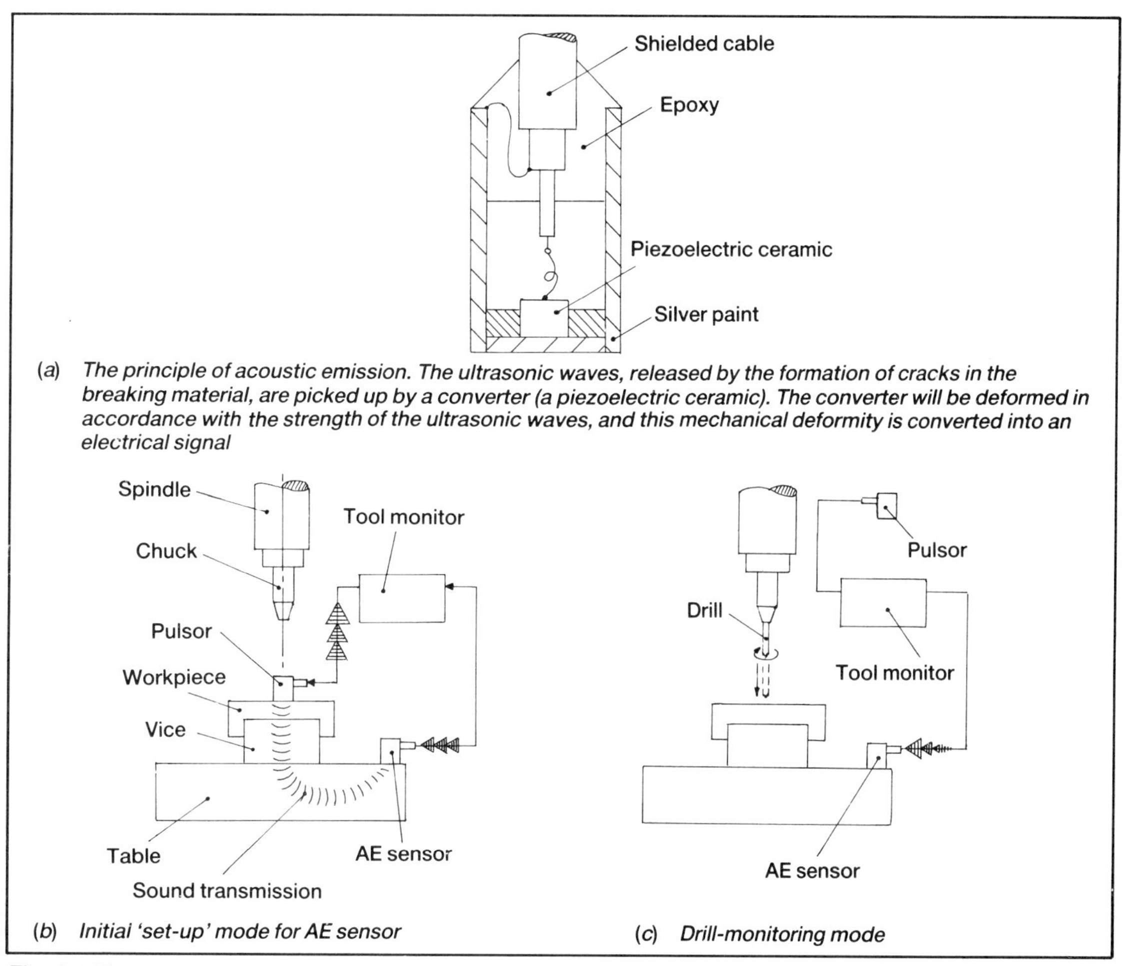

(*a*) *The principle of acoustic emission. The ultrasonic waves, released by the formation of cracks in the breaking material, are picked up by a converter (a piezoelectric ceramic). The converter will be deformed in accordance with the strength of the ultrasonic waves, and this mechanical deformity is converted into an electrical signal*

(*b*) *Initial 'set-up' mode for AE sensor*

(*c*) *Drill-monitoring mode*

Fig. 3.6 The principle and operation of tool-life monitoring using acoustic emission (Courtesy of Kennametal (UK) Ltd.)

After this brief résumé of tool-monitoring systems, we perhaps need to look at some commercially-available systems in more depth in order more fully to understand the principles behind their operation and the benefits a company obtains with their use. The next section will therefore consider why, how and where they may be incorporated on machine tools.

3.3 Case studies on tool-condition monitoring on turning and machining centres

Rather than giving a general description of how tool-condition monitoring is achieved, it is more profitable to look at how several different companies have designed their systems, and at how these systems monitor the cutting tool's action throughout its unmanned life.

3.3.1 Tool monitoring using feed-force sensors – strain gauges

Modern microprocessor-designed tool-monitoring systems can be used for a number of reasons, for example to monitor the tools's condition or to cut costs.

The advantages of using monitoring to detect the tool's condition are:

- Tool wear is monitored and tool changes initiated when necessary, so avoiding damage to the machine or the workpiece
- If there is a breakage, a signal will be be produced to stop the machine tool within milliseconds.
- The system will detect if a tool or the workpiece is missing, thus eliminating wasted machine time and the likelihood of unpredictable crashes.

The cost advantages of using tool monitoring are:

- Tool life can be optimised, which means that tools need only be changed when they are worn and so reduces tool costs.
- Down-time (which is normally associated with crashes) is lessened, and this increases the machine tool's output.
- Repairs to the machine tool and cutting tools may be reduced to a minimum so the maintenance costs are lower.
- The metal-cutting operation is monitored automatically, limiting operator involvement.

The lists above are quite an impressive recommendation for using a tool-monitoring system, but how does it achieve consistent and accurate tool monitoring, whilst at the same time controlling the cutting process? These questions will now be considered, dealing first with how the system monitors the tool's performance whilst cutting.

It is well known that a tool will produce relatively higher loads as it begins to wear during the cutting operation. For effective process monitoring it is important that the signal used should vary progressively as the tool wears, and not just at the point when it breaks. It has been shown (Fig. 3.7) that during a drilling operation the axial force component (F_A) provides a better indication of the cutting edge's condition as a function of tool wear than the torque value (M): the increase in the axial force for the worn tool is more clearly defined (Fig. 3.7a *right*). This change in the force generated whilst cutting is

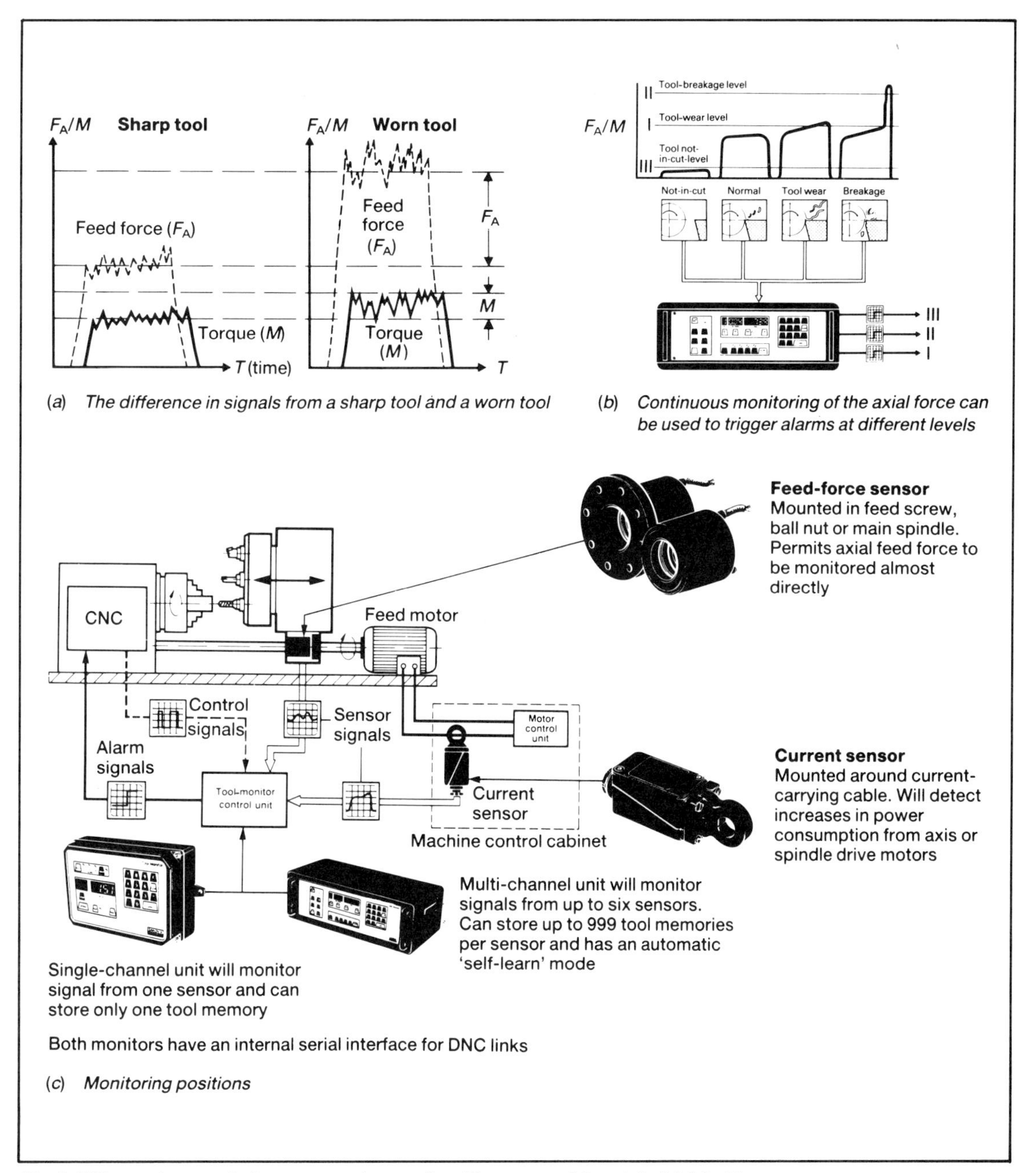

Fig. 3.7 How, when and where to monitor tooling (Courtesy of Sandvik (UK) Ltd.)

instantly detected by the feed-force sensor, via the spindle and the bearing. The sensor transforms the force change into an electrical signal which is transmitted to the signal-processing device. Once the signal is received, the processing device can immediately initiate action by the machine controller if the tool is worn, broken or not-in-cut.

This is all very well, but when should monitoring take place, and what action should result? Fig 3.7b shows how continuous monitoring of the axial force can be used to trigger four alarms:

- Level I can be used to detect tool wear. The alarm signal can be used to initiate a tool change on completion of the operation.
- Level II can be utilised for tool breakage. This signal should be used to stop the machine immediately the breakage is detected.

- Level III may be used to monitor whether the tool is in-cut or not-in-cut, as the case may be, meaning that either the tool or the component is missing.
- A further level can be used for crash protection. Like level II, this alarm signal should be used to stop the machine immediately and, in so doing, protect the machine tool.

For any particular workpiece operations, the forces that will trigger the alarm signals are programmed into the instrument in the 'set-up' mode. It is possible to vary the tool's reaction time: normally it is between 0.1 and 1 second, but for tool breakage a shorter reaction time is desirable, typically from 1 to 10 milliseconds. Fig. 3.7c illustrates the monitoring positions for a turning centre, showing the possible positions for the sensor, such as in the ball-screw nut or main spindle, in conjunction with a current sensor. These signals are continuously monitored by a single or multi-channel control unit, as are the control signals from the CNC machine control unit; any alarm signals are passed back to the machine's control unit for the appropriate action to be taken.

The multi-channel unit (Fig. 3.8a) processes analogue signals from the sensing devices; it can be integrated into most machine tools to measure and compare the tool-condition signals generated by the sensors. The unit is self-contained and can be fitted into any machining environment, requiring only basic power and signal connections to be made. It performs the following functions:

- It *senses* and processes tool-cutting information from up to six separate signal sources.
- It *learns*, by automatically memorising the signal values obtained from sharp cutting tools, whilst in the 'learn' mode.
- It *stores data* for up to 999 different cutting operations per channel in its memory, for each cutting operation that occurs, it automatically sets the appropriate levels for each of the alarm signals from its memory.
- It *reacts* by sending alarms to the machine's control unit, informing it if the tool is worn, broken or not-in-cut.
- It automatically *coordinates* machining and monitoring on commands from the machine's control unit.
- It *adapts* to the particular machine and cutting environment; once installed and programmed to suit the machine tool, the set-up parameters can be modified to adapt to further machining requirements.
- It *communicates* between the operator and the machine via the control panel, informing operating personnel about cutting-tool conditions and providing an interface for control of all functions.

Fig. 3.8b shows a typical turning-centre application of tool-condition monitoring. The machine is controlled on two axes, with sensors on the feed-drive bearings of both the *X* and *Z*-axes. A typical nominal force for these sensors is 40kN, but this rating will depend upon the end-user's requirements. The sensors can be designed for tapered or angular contact, or for combined axial and radial bearing applications, to suit the machine tool.

If tool monitoring is required on a four-axis turning centre, two tool-monitoring units will be required, since the turrets can be regarded as two separate machines, operating more or less independently.

The key elements in any tool-condition monitoring situation are the sensor's position and its design. For universal installation on a variety of machine-tool configurations, the positioning of the sensing devices is usually on the recirculating ball-screw nut assembly. Therefore the feed-force sensors are designed to be incorporated in this area; a

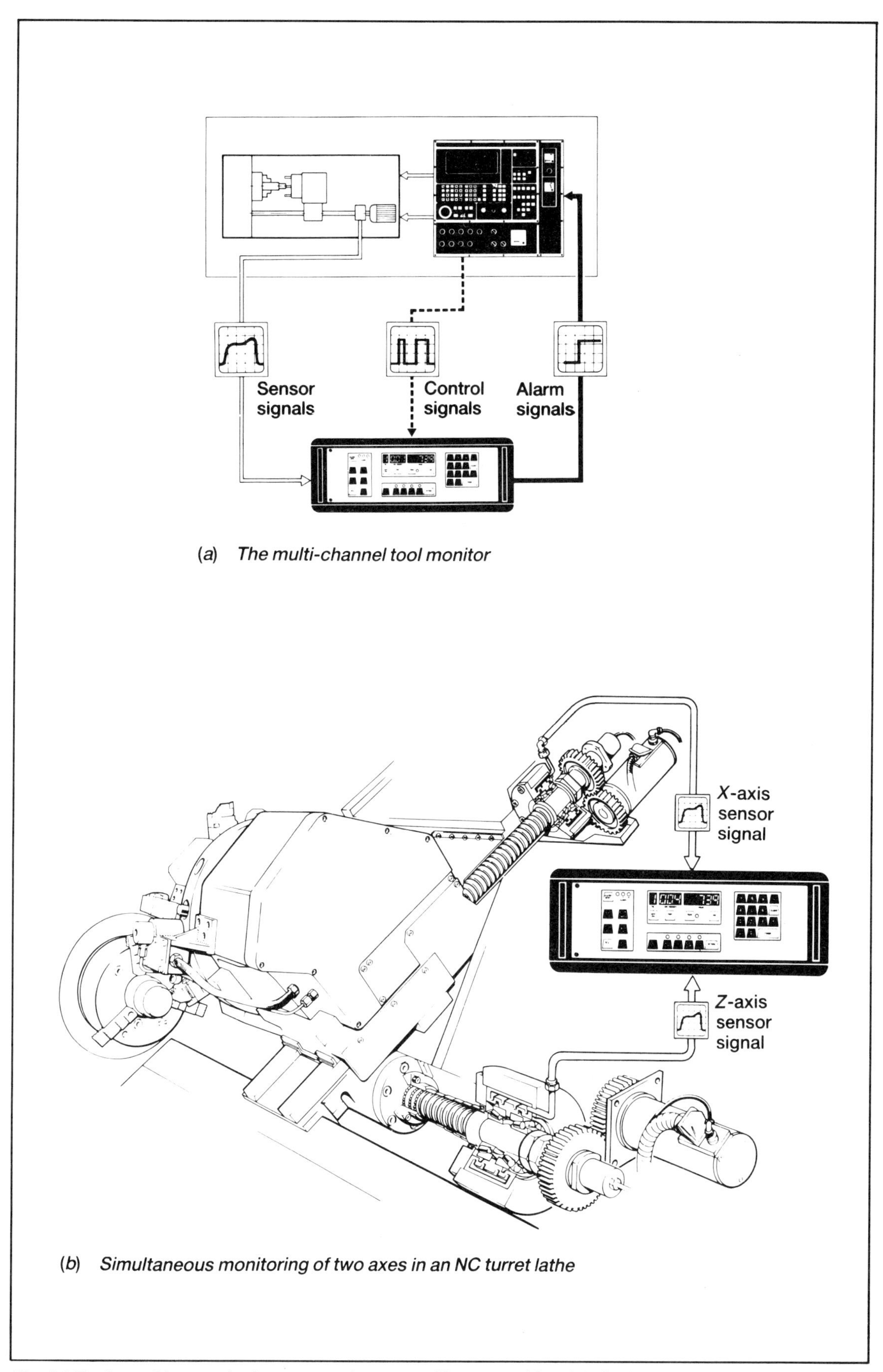

Fig. 3.8 Tool-condition monitoring on a turning centre (Courtesy of Sandvik (UK) Ltd.)

typical example is shown in Fig. 3.9a. The sensor design consists of two concentric rings which form a precision housing for the bearings of the rotating spindle. The profile of the inner ring has a special form, providing two force-sensing zones, into which are mounted symmetrically-positioned strain gauges which are part of a full Wheatstone bridge circuit (Fig.3.9b). This design ensures that changes in temperature or the preloading of the bearings will not influence the sensor signal. The strain gauges are protected from dust, coolant and lubricating oil by the outer ring, which is electron-beam welded to the inner one. The distribution of a number of strain gauges around the edge of the force-sensing zones minimises any fluctuation in the measured signal during normal working conditions. The cross-section of a typical assembly of the Z-axis feed-force sensor can be seen in Fig. 3.9c.

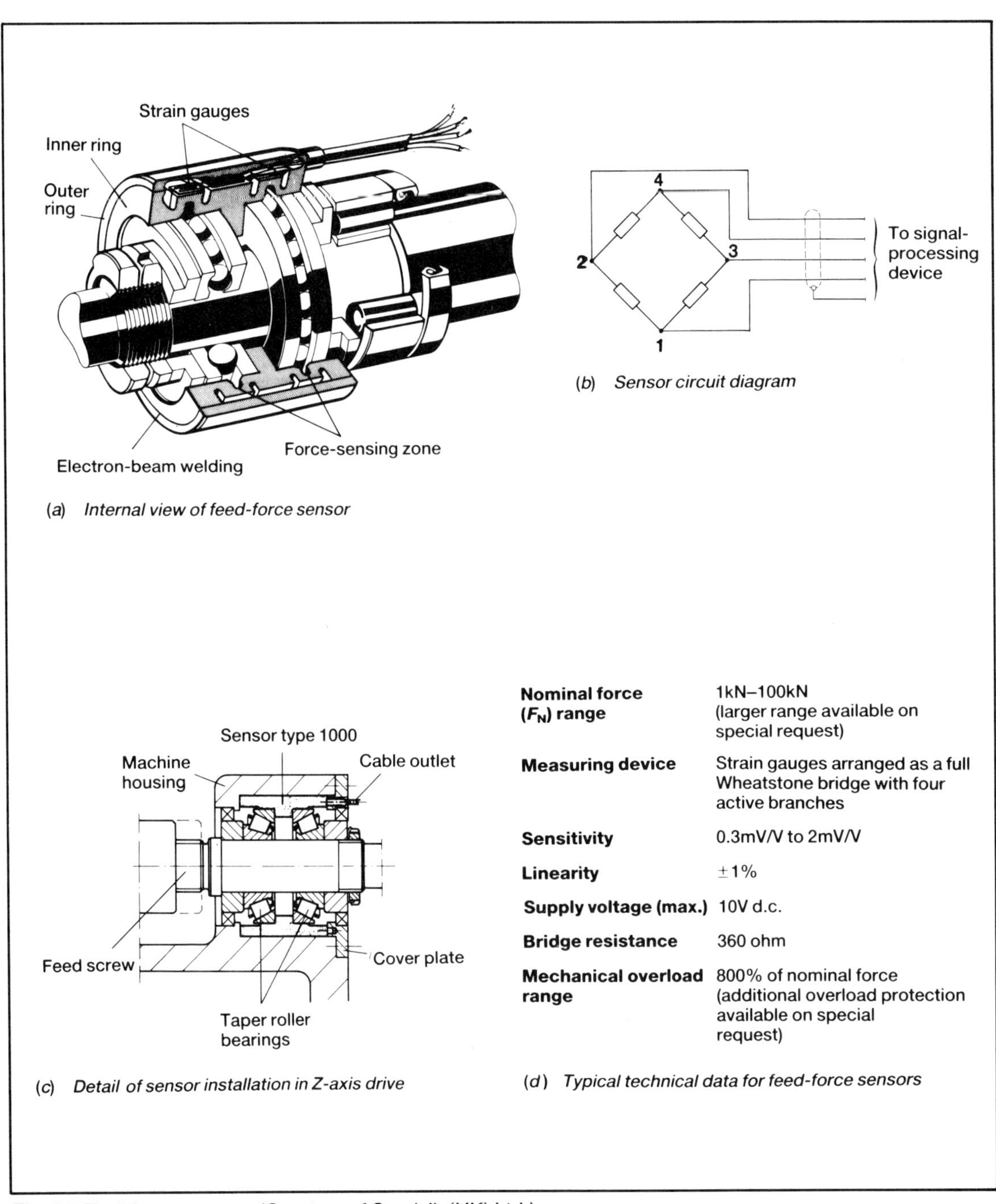

(a) *Internal view of feed-force sensor*

(b) *Sensor circuit diagram*

(c) *Detail of sensor installation in Z-axis drive*

Nominal force (F_N) range	1kN–100kN (larger range available on special request)
Measuring device	Strain gauges arranged as a full Wheatstone bridge with four active branches
Sensitivity	0.3mV/V to 2mV/V
Linearity	±1%
Supply voltage (max.)	10V d.c.
Bridge resistance	360 ohm
Mechanical overload range	800% of nominal force (additional overload protection available on special request)

(d) *Typical technical data for feed-force sensors*

Fig. 3.9 Tool-force sensors (Courtesy of Sandvik (UK) Ltd.)

The multi-channel tool monitor can process signals generated by the machining process from up to six different sources as shown in Fig. 3.10a. Each input channel samples the on-going cutting operation 2500 times per second. It is important to be able to monitor more than one channel simultaneously – for the contouring of a workpiece's profile where the tool is moving in two or more planes, for example. A single tool-monitoring unit can easily cope with the required number of monitors, and the instrument can receive signals from a diverse range of signal-generating devices. Signals from most kinds of analogue sensors can be used to serve as inputs to the tool monitor without any external adaptation. The unit can also be adapted for different signal types, by programming it for signals such as filter frequencies and input sensitivities during the 'set-up' mode. These inputs may be in the form of signals from feed-force sensors, motor-current sensors, power transducers, etc.

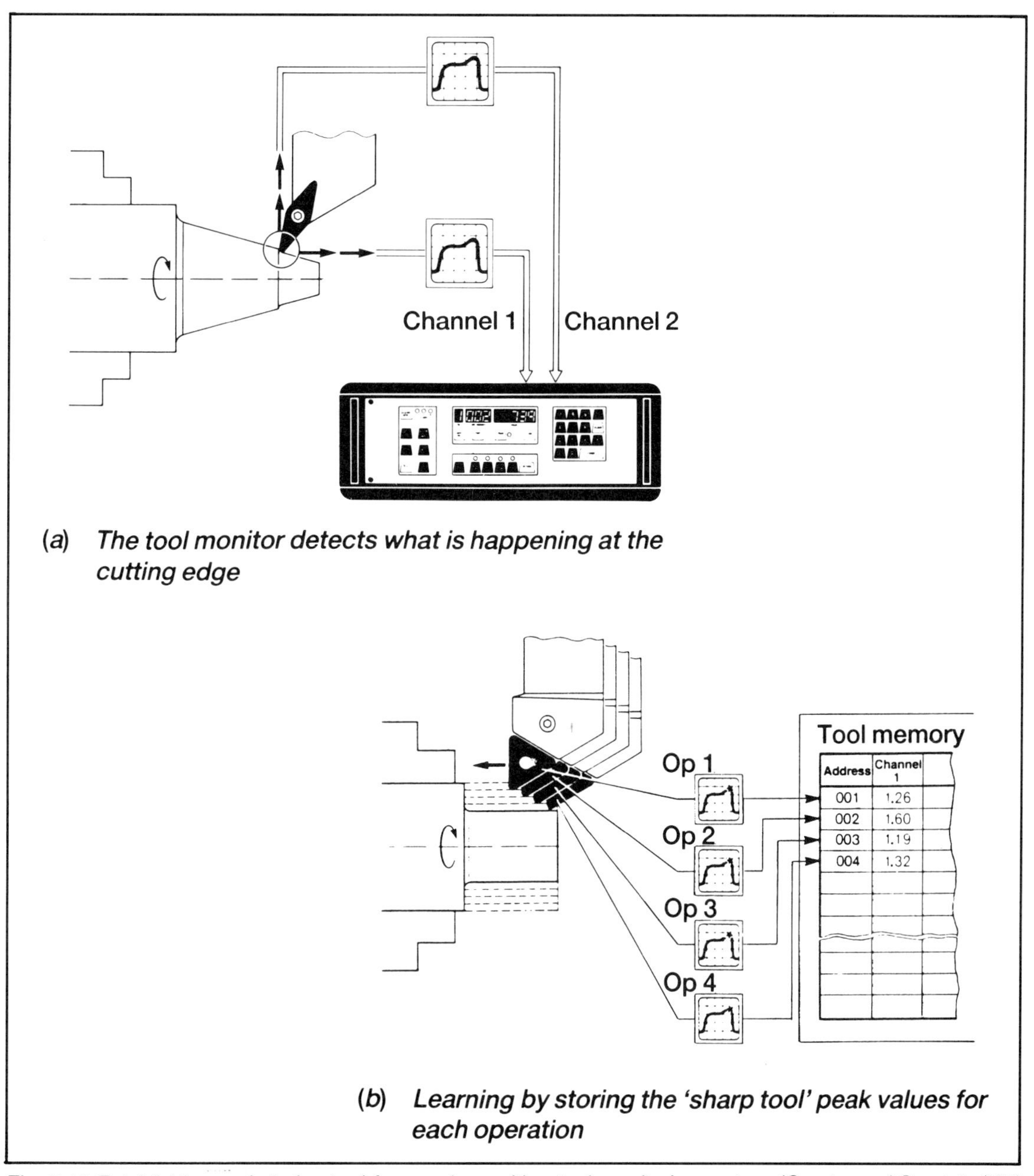

(*a*) *The tool monitor detects what is happening at the cutting edge*

(*b*) *Learning by storing the 'sharp tool' peak values for each operation*

Fig. 3.10 Establishing and storing tool-force values with a tool-monitoring system (Courtesy of Sandvik (UK) Ltd.)

With any sensor device it is necessary to be able to eliminate the influence of unwanted interference from sources such as the friction forces in the machine, so the instrument contains a 'tare' function for zeroing the signal. The tare function's contribution to zeroing the machining parameters means that close repeatability is achieved. During the machining of the first workpiece, with the tool monitor in the 'learn' mode and using sharp tools, the monitor automatically stores the peak value of the sensor signal for each tool or operation, as shown in Fig. 3.10b. These values are stored in the tool memory until recalled for use in the 'monitoring' mode for identical machining operation. If new values are required, they can be manually entered into the tool memory or previous values can be changed from the front control panel, as necessary, thus increasing the monitor's versatility.

The application of tool-force monitoring on machining centres is very similar to that described for turning centres. Once again, the sensor signals could be supplied by feed-force sensors on the ball-screws of the *X*, *Y* and *Z*-axes, as shown in Fig. 3.11. However, a combination of a feed-force sensor in the main spindle and a current sensor in the main spindle's drive motor could be used as a reasonably reliable alternative that would cover applications of drilling and rough milling.

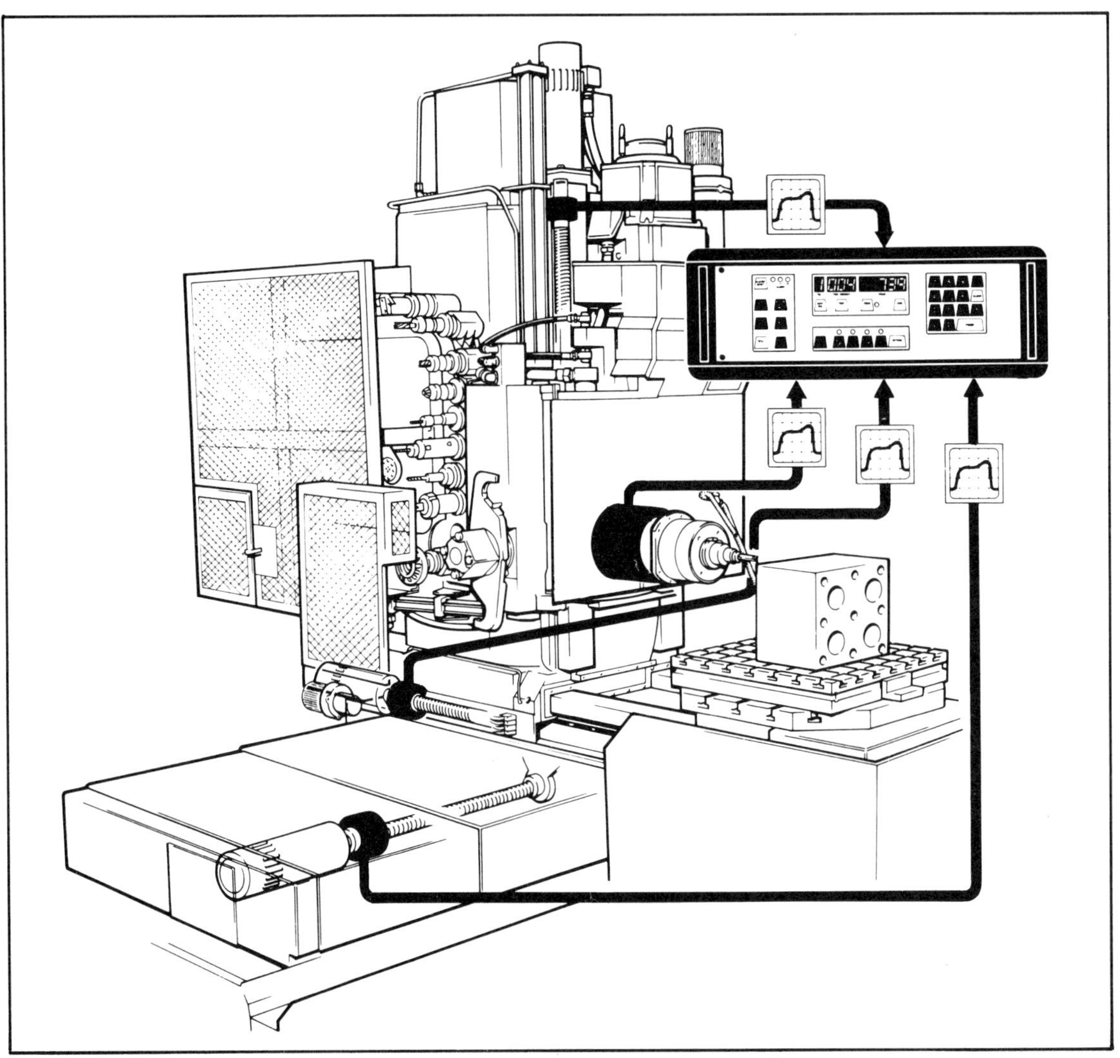

Fig. 3.11 Tool-force monitoring on a machining centre. Signals can be obtained from a feed-force sensor in the main spindle and/or feed-force sensors mounted on the feed screw bearings. In some cases, current signals could be used from the feed servo drive or the main spindle drive (Courtesy of Sandvik (UK) Ltd.)

Fig. 3.12a illustrates the manner in which this monitoring system is interfaced to the CNC by a simple but effective connection principle. The coordination is achieved by two digital control signals from the machine control unit which controls four basic functions in the tool monitor. The two signals are:

1 A control signal, which defines that an operation is taking place. This may be initiated either by a 'programmed feed signal' which is activated when any of the machine's

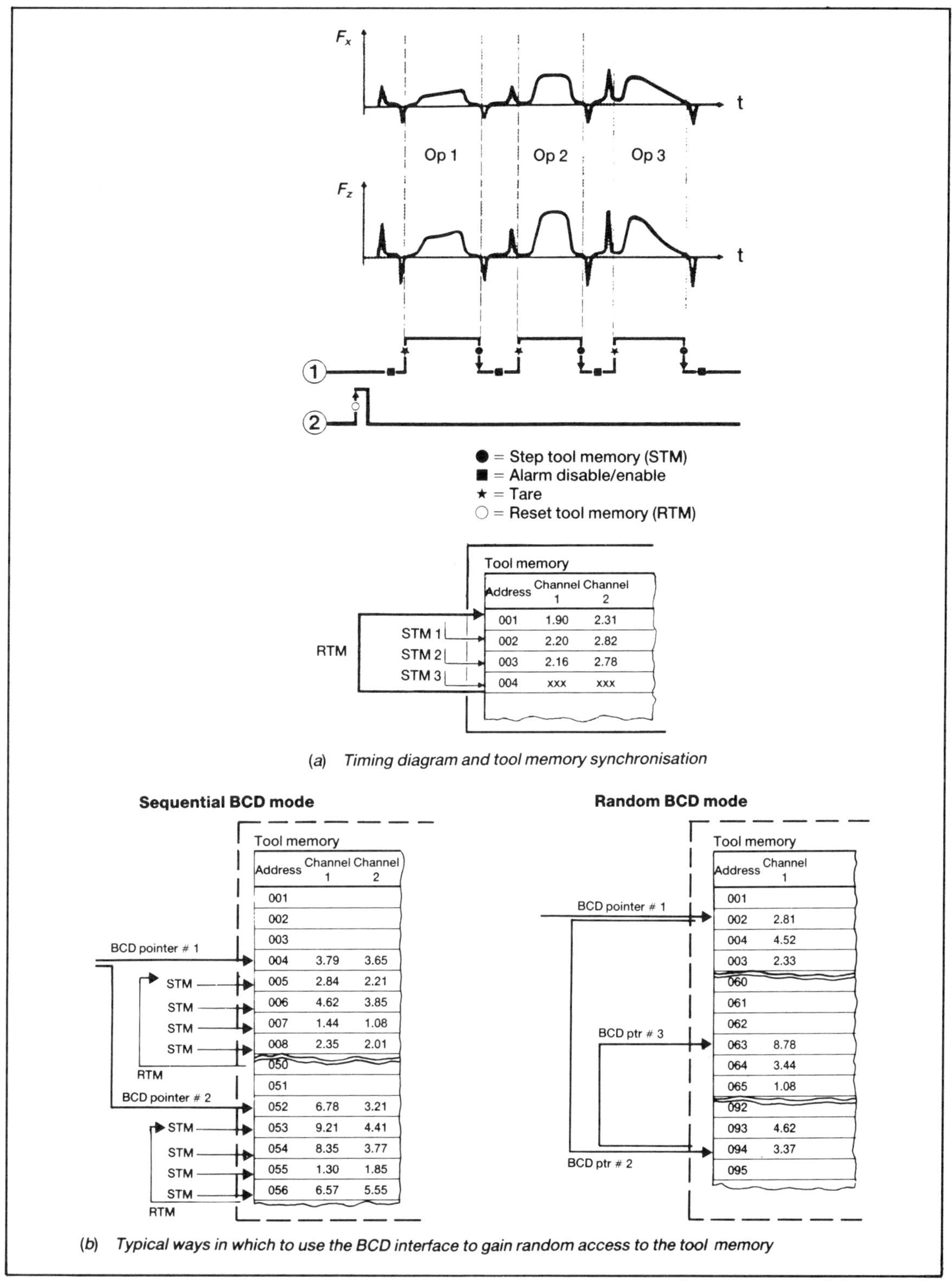

(*a*) *Timing diagram and tool memory synchronisation*

(*b*) *Typical ways in which to use the BCD interface to gain random access to the tool memory*

Fig. 3.12 Interfacing to CNC (Courtesy of Sandvik (UK) Ltd.)

slides is transversing in the working feed, or by a 'bi-stable M-function', which is part of the CNC program. Once this control signal is received, it initiates the following functions in the tool monitor:

- The 'tare' fuction, which is activated on the rising edge of the signal, that is, after each operation.
- The 'step tool memory', which is activated on the falling edge of the signal, that is, after each operation.
- The 'alarm disable/enable' which is in the deactivated condition when the signal is at its lowest level, that is, during rapid transverse, tool changes, etc.

2 A control signal at the start of the program initiated by an M-function, which activates the 'reset tool memory' on its rising edge.

The limitation with this example, which is the basic version of the multi-channel tool-monitoring unit, is that it is restricted to following one CNC program from start to finish, sequentially stepping from address 001 in the tool memory. However, random access to any desired tool-memory address is possible using an optional binary-coded-decimal (BCD) interface, as shown in Fig. 3.12b. This interface may be used in either the 'sequential' or the 'random' modes. The 'sequential' mode allows different sections of the tool memory to be selected, corresponding to different CNC programs, whilst the 'random' mode permits fully-randomised selection of any desired tool-memory address, which is useful in machining centres when identical operations are repeated many times, such as when drilling hole patterns.

3.3.2 Monitoring for tool breakage, wear and collisions using piezoelectric sensors

Tool breakage, wear and collisions are frequently the causes of stoppages in machine shops. A survey conducted by the Munich Technical University in 1983 showed that when machines were run for 24 hours per day for 364 days of the year (i.e approximately 100% utilisation), only 8% of that time was converted into actual cutting performance, (i.e. 700 to 800 hours per annum). With such a low utilisation of capital, any increase that can be gained through tool-condition monitoring can be warmly greeted! Furthermore, if the degree of automation is high so that there is little visual control by the operator, any stoppages will produce a rapid decline in operating efficiency, and the number of parts produced falls considerably.

The system described below differs markedly from the one mentioned in section 3.3.1, in that piezoelectric strain sensors are attached to specific points – on the turret housing, say, on turning centres, or at any convenient position for a strong signal on machining centres. Each sensor measures the strains produced during the cutting action, and the signals emitted from these sensors are evaluated by monitoring systems for tool breakage, wear and collision, as depicted schematically in Fig. 3.13.

We shall consider the tool-breakage detection device first, and then go on to consider the others, in order to build up a reasonable picture of their designs and operating principles. At the moment a tool breaks, there is a significant rise in the cutting force, followed by a sharp drop, as illustrated in the graph in Fig. 3.13a. This pattern is a telltale sign of tool breakage and forms the basis on which the tool-breakage system works. The digital breakage-monitoring system continuously monitors the upper and lower force values. If these values deviate from the norm because of tool breakage, the sensor reacts immediately. Within two milliseconds of the breakage, a command is issued to stop the feed motor, thus reliably preventing any continued operation with a damaged tool. The

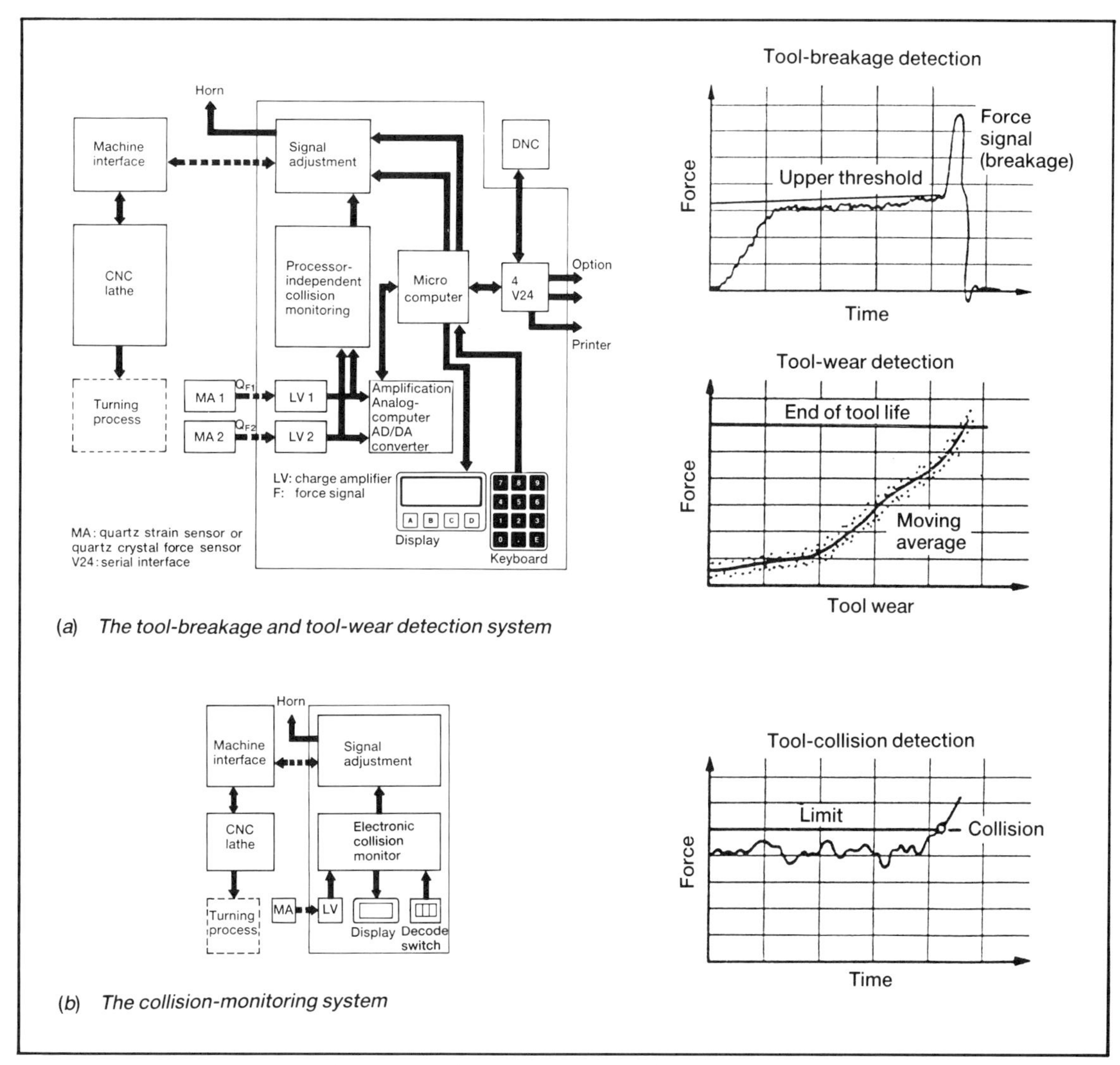

(a) *The tool-breakage and tool-wear detection system*

(b) *The collision-monitoring system*

Fig. 3.13 Tool-condition monitoring system (Courtesy of Krupp Widia)

tool-wear circuitry is similar to that for tool-breakage detection and they have recently been combined into a single unit. The wear monitor is again a digital system; it is based upon the rise in the force level that occurs over the lifetime of the cutting edge and operates by checking the rise against the force differential measured on a reference cutting edge operated until blunt. The wear threshold is obtained by adding the initial force value of a new cutting edge to the force differential of the reference tool. Once this limit is exceeded, the system indicates that the tool is worn and should be replaced. A typical graph of the tool-wear pattern produced during the tool's useful life is shown in Fig. 3.13a.

The statistics for tool collisions are quite frightening; it has been shown that 72% of all collisions are caused by operator error and that 26% can be traced back to faults in the CNC electrical system. The German survey referred to earlier found that a typical collision results in £7500 worth of damage to the machine tool, not including the consequential costs. This factor alone substantiates the necessity for investment in a collision-monitoring system!

The operating principle, as applied to a turning centre, is that the strain transducer measures the deformation in the turret during machining and then compares it with a preset threshold value. If the measurement exceeds the threshold, a command is given within three milliseconds to stop the feed motor, and no damage to the machine or workpiece occurs. A schematic diagram of typical circuitry and a graph of the expected forces during a collision are shown in Fig. 3.13b.

The photographs in Fig 3.14 show how easy it is to feed in new data to the monitors, and how compact and accessible they are. They can be conveniently positioned close to the machine control unit or at a remote cell-control station in a flexible manufacturing system.

Mention has been made of the piezoelectric sensor, but how does it operate and what design principle is it based upon? It is hoped that the following sections will illuminate the principle sufficiently for an understanding of why they are incorporated into monitoring systems.

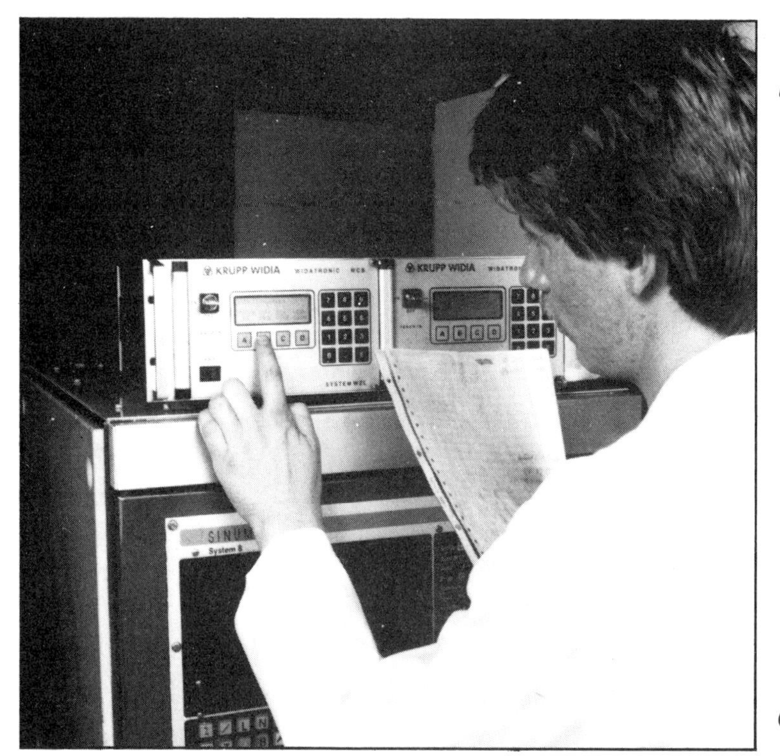

Fig. 3.14 Typical tool-condition monitoring equipment for unmanned machining (Courtesy of Krupp Widia)

(a) Programming the levels of wear and breakage for individual tools

(b) A turning centre equipped with wear and breakage tool-condition monitoring using the Widaflex modular tooling system

Piezoelectric sensors

Electrical and electronic devices are increasingly being adopted for the measurement of mechanical quantities; these measuring methods are characterised by their high sensitivity and resolution. The sensors require the minimum of power input, as they involve no acceleration or frictional forces. Also, the electrical signal, which is proportional to the measured value, can be amplified to such an extent that even minute values can be measured. Sensors of this type are known as 'active elements'.

One type of active sensor utilises the piezoelectric effect, by which electrical charges are formed on the surfaces of certain crystals when they are subjected to mechanical stress. Crystals that exhibit this effect include quartz, tourmaline and barium titanate. The reverse process, in which the crystal is deformed by application of an electric charge, is known as the reciprocal piezoelectric effect. The direct effect is utilised in the measurement of pressures and forces, whereas the reciprocal effect is the one used to power quartz clocks and watches. As the electrical charges are produced with virtually no delay, piezoelectric sensors can be used in high-speed operations. Quartz exhibits only a very small dependence on temperature and a high modulus of elasticity, so it is used extensively for measuring applications.

The piezoelectric effect

Fig. 3.15a shows an idealised hexagonal crystal of quartz. The arrangement of atoms in the hexagonal unit cell is shown in Fig. 3.15b. Positively-charged silicon ions alternate with negatively-charged oxygen ions at the corners of these hexagons. Silicon, being a tetravalent element, will bond with two bivalent oxygens, and each corner of the hexagon can be thought of as having a quadruple charge.

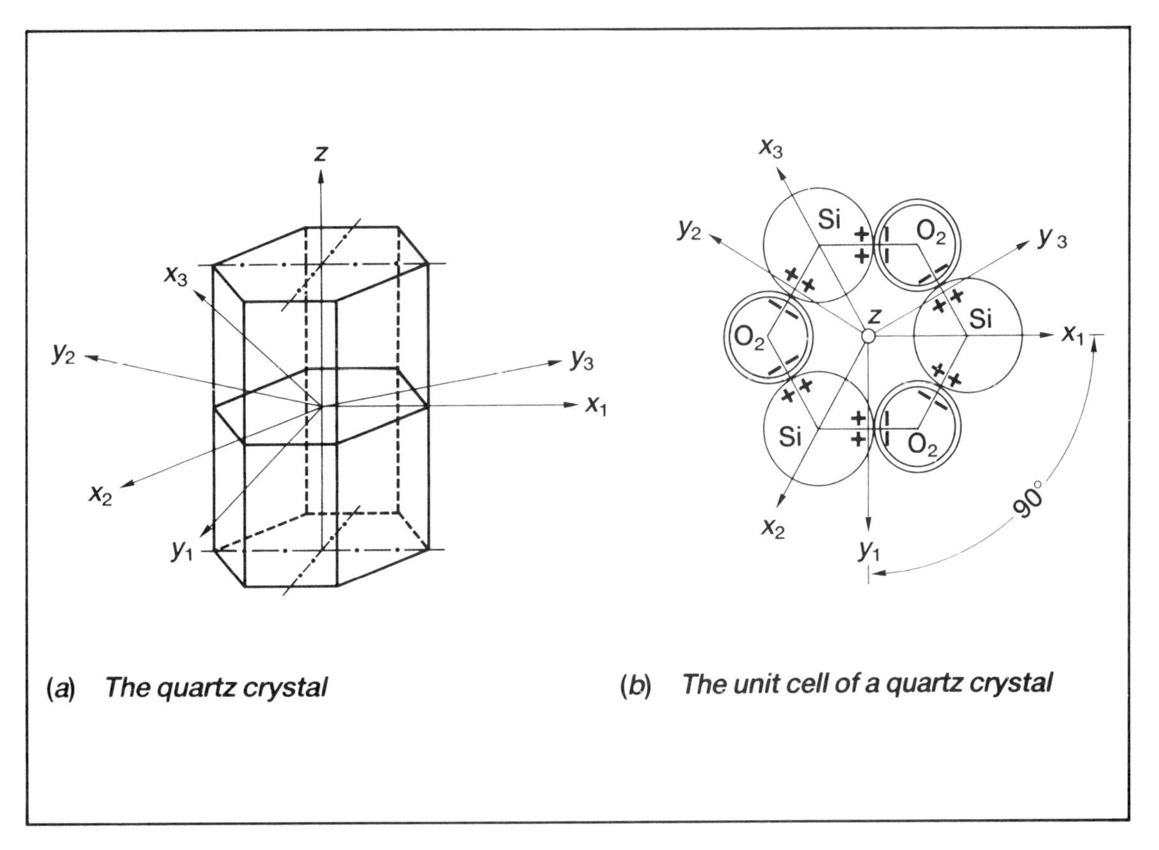

Fig. 3.15 The structure of quartz (Courtesy of Krupp Widia)

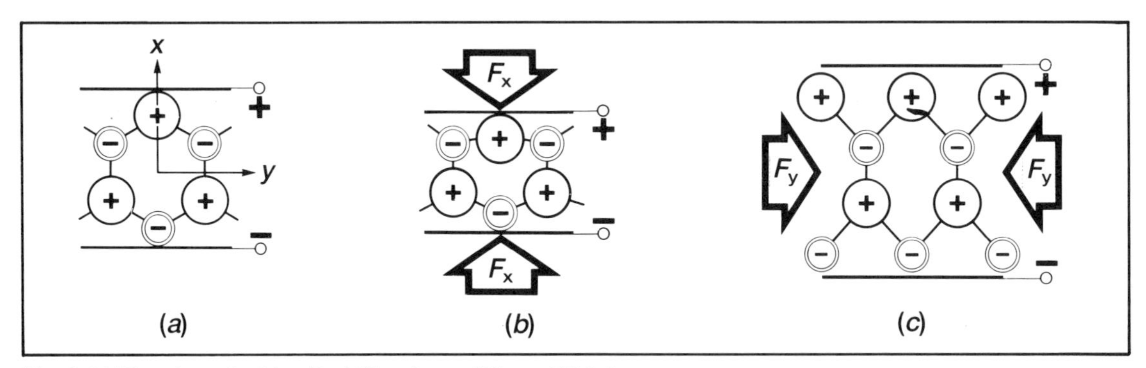

Fig. 3.16 The piezoelectric effect (Courtesy of Krupp Widia)

Fig. 3.16a illustrates how several such unit cells would be arranged in a quartz platelet cut or polished parallel to the base of a hexagonal crystal prism. The diagram also shows the x and y-axes, and the thicker straight lines represent the reflecting surfaces of the quartz platelet, which are generally silver-coated. If no mechanical stress is imposed on the crystal, its end faces are electrically neutral. If, however, a force is applied in the x-direction as illustrated in Fig. 3.16b, there is a net movement of the negatively-charged oxygen atoms (shown as the smaller circles) towards the top surface, and of the positively-charged silicon atoms towards the bottom surface. Therefore, an electric potential develops, and the crystal end faces become charged to an extent that is proportional to the applied force F. This is known as the longitudinal piezoelectric effect. A change in the charge also occurs if a force is applied in the y-direction as shown in Fig. 3.16c; this is known as the transverse piezoelectric effect. In this type of mechanical loading, the positively-charged silicon atoms are moved closer to the upper face of the crystal, whilst the negatively-charged oxygen atoms are moved closer to the bottom face. The electrical charges produced are opposite to those produced in the longitudinal effect, but are still on the faces perpendicular to the x-axis.

Strain pickups

Quartz-crystal piezoelectric strain pickups can be used for the indirect measurement of forces in machine tools. The term 'indirect' is used because the stresses set up during machining produce strain, which is what is measured by the pickup. The strain-measuring device shown in Fig. 3.17a consists essentially of two elements and two surfaces.

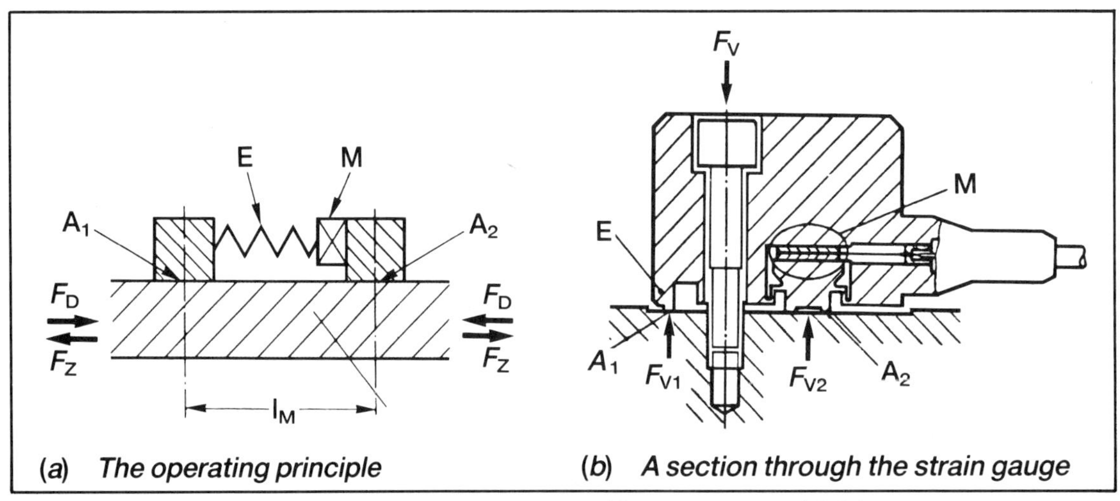

Fig. 3.17 The strain gauge's operating principle and design (Courtesy of Krupp Widia)

- E is an elastic element which converts strain into proportional forces.
- M is a force-measuring element, which measures the forces produced by the elastic element.
- A_1 and A_2 are two surfaces on the strain pickup which are placed at a specific distance apart on the specimen under load; they transmit its strain to the elastic element by friction.

The force-measuring element of the strain pickup, M, measures shear by utilising the piezoelectric effect. The elastic element is subjected to bending because there is hardly any measuring point on the machine which is subjected to a true tensile or compressive stress. Thus the strain at the measuring point is produced by a combination of tensile, compressive and bending stresses, with bending often being of the greatest magnitude.

Fig. 3.17b shows a cross-sectional diagram of the strain pickup. The electrode in the force-measuring element M carries a charge which is proportional to the strain between the two surfaces A_1 and A_2 of the specimen under load. The polarity of the charge is such that an extension of the specimen under load (i.e. a tensile stress) causes a negative charge to be produced on the centre electrode, giving rise to a positive output voltage at the succeeding charge amplifier. This signal can then be processed to allow the desired action on the condition to be taken.

The advantages of using piezoelectric sensors are:

- They have very high sensitivity compared with strain gauges.
- They have low hysteresis and small linearity errors.
- They are easy to install, and can be retrofitted to existing machines.
- The charge amplifier can easily be reset to zero.

The main disadvantage compared with strain gauges is that they are more expensive to manufacture.

The photograph in Fig. 3.18 shows an in-tool transducer mounted on the turret of a turning centre; it shows how compact and neat the pickup and support bracket are when in use.

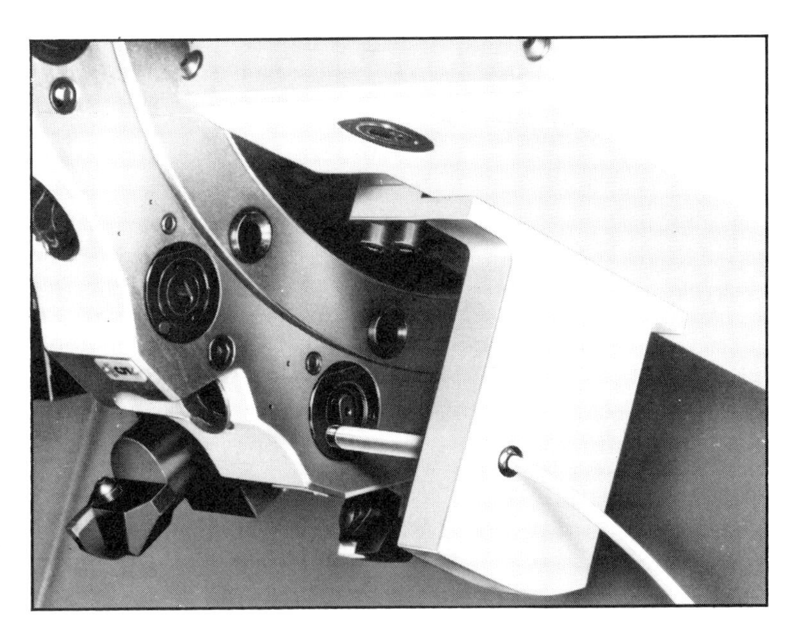

Fig. 3.18 A piezoelectric strain transducer on a turning centre turret showing its compact pickup and support bracket (Courtesy of Krupp Widia)

The other method of tool-condition monitoring which is becoming popular is the one using the acoustic emission principle previously mentioned; a discussion of this system will be given in the next section.

3.3.3 Tool-condition monitoring using the acoustic emission principle

The acoustic emission principle was described in section 3.2.4, which explained that the cutting of a workpiece material causes a release of energy from the deformed or destroyed solid, which is propagated in the form of an acoustic pulse. In normal conditions of milling or interrupted turning, where intermittent cutting occurs, the acoustic wave emitted is dependent on the tools, the workpiece's materials and shape, and on the cutting conditions. An acoustic emission system must be designed in such a manner that it can ignore the effects of these variables and only accept the acoustic waves emitted from tool breakages. By identifying the acoustic emission from tool breakage during machining, say, the monitor will recognise a tool breaking (or a close variant, e.g. chip crack, damage, rejected material), then stop the machine almost instantly and activate an alarm display. As the system detects tool breakage during machining, it prevents the occurrence of the secondary accidents that might follow from a broken tool. The fact that systems based on acoustic emission utilise sound propagation characteristics means that the lower limit of tool diameter for breakage to be detected may vary, depending on the type of tool-holder used.

The acoustic emission signal is picked up by a piezoelectric strain transducer fitted at some convenient position on the turning-centre turret housing (Fig. 3.19a). By using a pulsor as illustrated in Fig. 3.6b, an artificial signal can be generated which simulates the vibrations associated with a tool breakage. The pulsor enables the engineer to position the acoustic emission sensor accurately in the optimum position. It also allows the amplifier gain to be easily set, giving the monitoring system a 'degree of tuning' for the expected tool-cutting conditions.

Setting the sensor sensitivity

As schematically shown in Fig. 3.6b, the pulsor is attached to the workpiece, which is connected to the sensor via the vice, fixture, indexer, rotary table, etc. It is made to emit the simulated signal for tool breakage. As this simulated signal travels from the workpiece to the sensor through the vice and table, etc, the acoustic emission signal will be damped, owing to impedance. The acoustic emission sensor is mounted at a convenient position and the amplifier gain adjusted to detect the damped signal; this adjustment gives the required acoustic emission sensitivity. This procedure means that tool breakage may be reliably detected, since the signal will be at the same level as the simulated breakage signal transmitted from the pulsor.

Monitoring of tools

Once the acoustic emission sensor has been 'tuned' to the cutting application by using the pulsor, and the amplifier gain (acoustic emission sensitivity) has been set, the pulsor is removed from the workpiece and machining can commence (Fig. 3.6c). The tool monitor will monitor the tool's status continuously during its part of the CNC program through the acoustic emission sensor. If the tool fails, the alarm is instantly activated, and if any tool abnormality is detected then cutting stops immediately.

A typical machining-centre application of the acoustic emission principle is shown in the diagram in Fig. 3.19b. In this case, the acoustic emission waves travel through the cutter, spindle and casing, rather than through the workpiece, etc, as previously described.

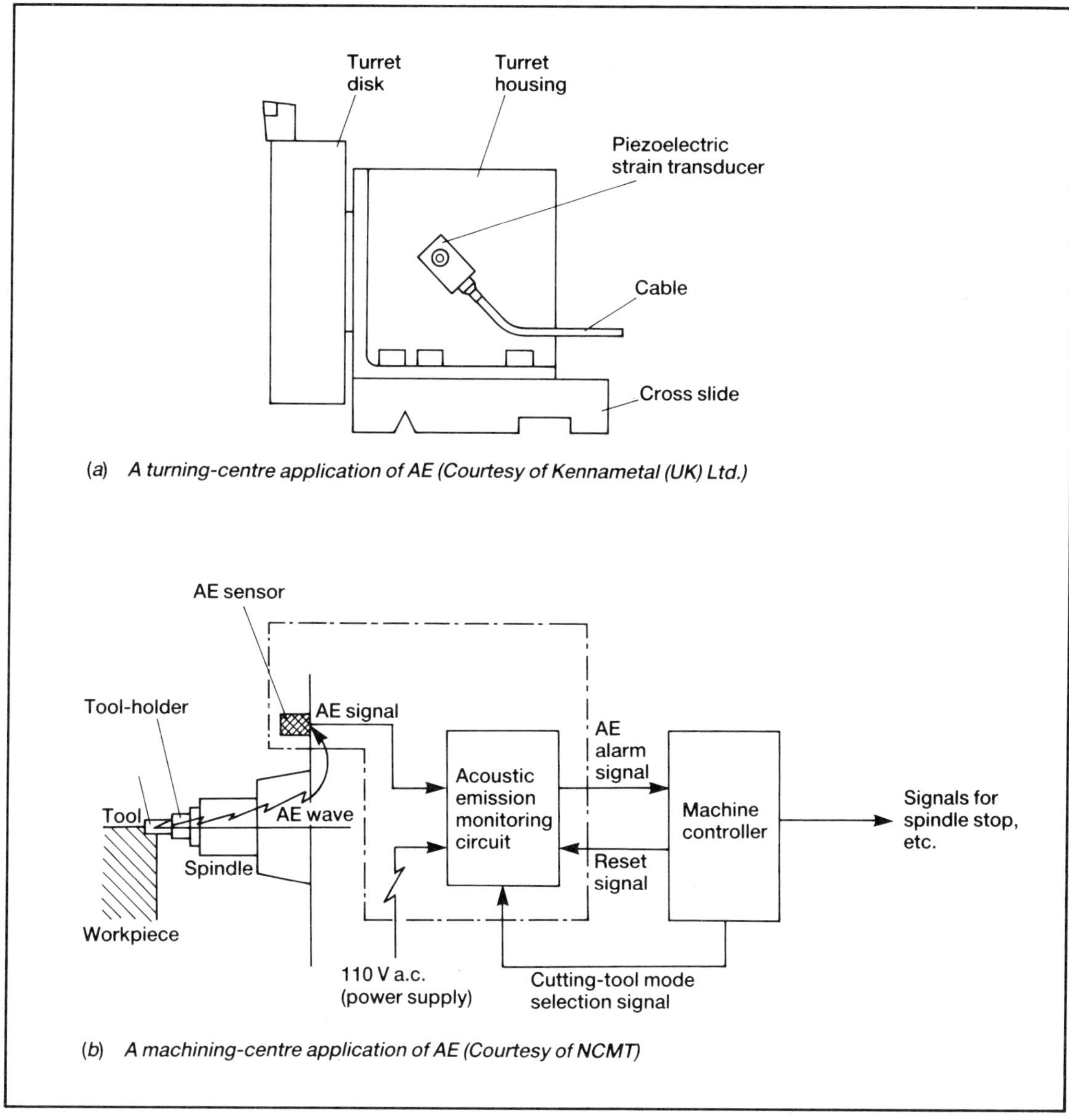

(*a*) *A turning-centre application of AE (Courtesy of Kennametal (UK) Ltd.)*

(*b*) *A machining-centre application of AE (Courtesy of NCMT)*

Fig. 3.19 Typical applications of the acoustic emission principle for tool-condition monitoring

An acoustic emission tool monitor and its associated pulsor and sensor can be seen in Fig. 3.20. It is worth emphasising that the acoustic emission sensor's compact design means that it can easily be positioned on the machine tool with only simple connections; this gives it a high versatility of fitting on a range of machine tools. The pulsor and sensor levels are adjusted by a simply push-button control that increases or decreases the level of sensitivity of the monitoring instrument. With this relatively simple instrumentation and robust, inexpensive, 'workshop-hardened' piece of equipment, a company can incorporate tool-condition monitoring with little inconvenience, allowing an unmanned machining activity to be initiated with confidence.

3.4 Adaptive control on turning and machining centres

The term 'adaptive control' has been coined to describe the in-process adjustment of a machine tool's operating parameters, for example the spindle speed and feedrate, based upon actual process characteristics. Therefore the primary objective of any adaptive

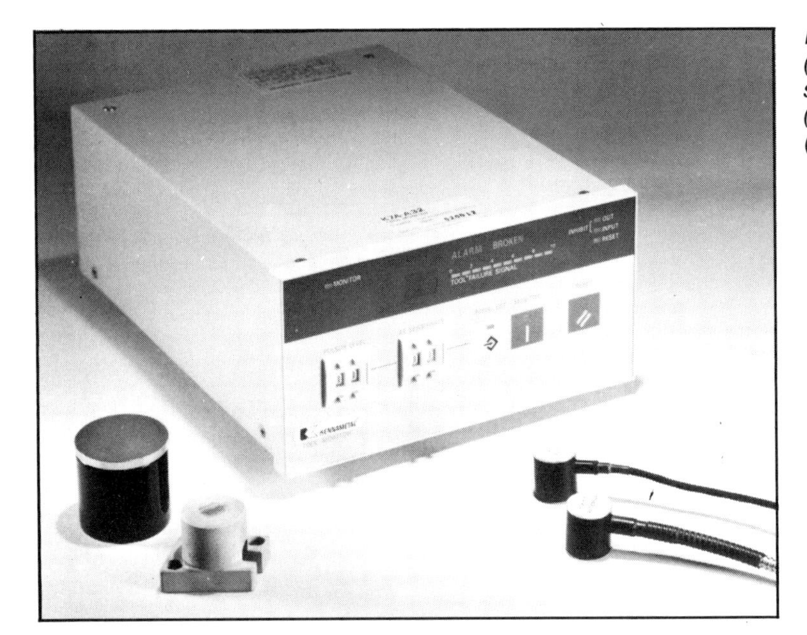

Fig. 3.20 Acoustic tool monitoring (ATM) for detecting breakage of small (12.5mm) rotating tools (Courtesy of Kennametal (UK) Ltd.)

control application is to have a control system which is capable of providing an automatic means of continuously adjusting speed and feedrate on the machine, whilst producing a workpiece at the lowest possible cost.

If a part program is written for a component on a machine tool without the benefit of adaptive control, the programmer will inevitably select the speed and feedrates with a 'fair safety margin', in order to allow for machining under any difficult conditions which may occur. This is all very laudable, but the conservative speeds and feeds that are chosen will be reflected in longer cycle times per part and, of course, a higher economic cost of each item and a longer subsequent pay-back on the investment, when compared with using an adaptive control system. The cost of installing an adaptive control system on a machine tool is approximately 5% of the machine' s purchase price, and the benefits that arise from adopting such a system mean that this can easily be financially justified.

Many other benefits can also be gained from incorporating an adaptive control system on a machine tool when it is new, or by retrofitting. These benefits, and the design and operating principles of the systems, are described in the following sections.

3.4.1 The benefits to be gained from using adaptive control

The principal benefits of using an adaptive control system are:

- The main spindle motor is protected from overload.
- Damage to the workpiece or cutter is prevented.
- Optimal stock-removal rates are possible, under steady-state machining conditions.
- By maintaining a constant cutting power, cutting force and feed force, it is possible to optimise the tool life.
- When unpredictable air gaps occur whilst cutting, the fastest possible travel can be used.
- The operator's experience, or the program's efficiency, may differ for different cutting operations, but with adaptive control this 'technical gap' can be eliminated.
- There is no overshooting of the permitted cutting power during re-entry into the material whilst machining a part under irregular conditions.

3.4.2 The design of an adaptive control system

When a programmer writes a CNC program, it is not possible to choose optimum values for speed, feedrate etc. because of variables such as tool wear, non-homogeneous materials and variations in dimensions of rough stock. It is normal for a programmer to choose median values for these factors in any calculations, but even here problems may arise: typical problems are caused by variations in the tolerance on a casting, say, or by the air gaps that must be traversed in such castings – doubts remain as to whether these gaps should be taken at rapid-traverse or normal feeds. Whatever route chosen in writing a program for a part, the programmer faces the quandary of whether to utilise high rates and sacrifice some tool-cutting predictability but increase workpiece throughput, or to use 'normal' rates and produce a dependable cutting action but with lower production cycle times. The adaptive control systems remove the need for this 'guessing game'.

Adaptive control systems fall into two main groups, adaptive control optimisation and adaptive control constraint; these are schematically represented in Fig. 3.21a.

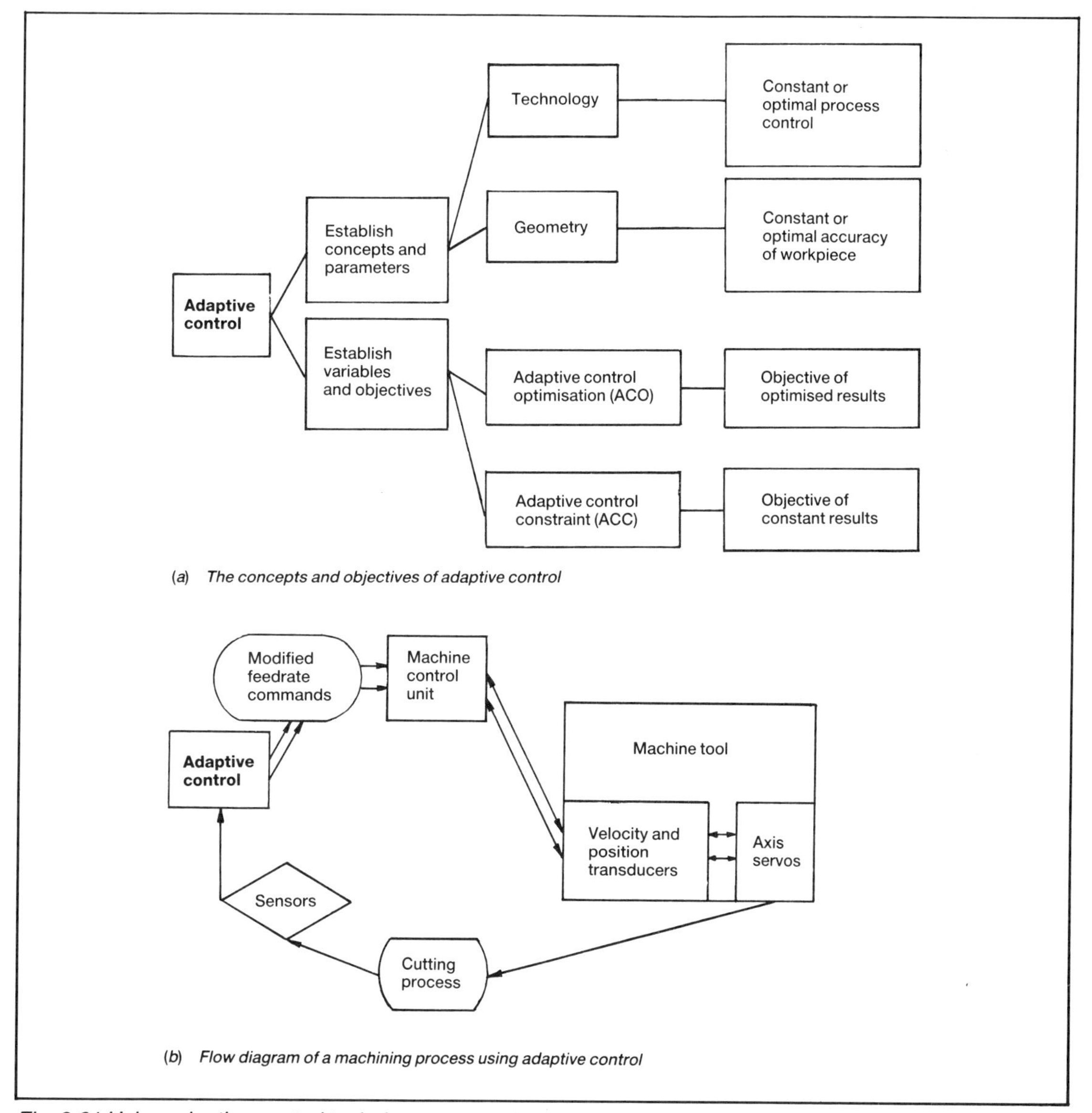

(a) *The concepts and objectives of adaptive control*

(b) *Flow diagram of a machining process using adaptive control*

Fig. 3.21 Using adaptive control techniques on machine tools

The earlier system to be developed was adaptive control optimisation, which required some form of in-process measurement to assess such factors as tool wear-rates, spindle deflection, and vibration. However, it was found to be rather difficult to measure such factors as tool wear under production environments. This is not to say that the early optimisation systems did not fulfil the 'ideals' of adaptive control, but they were prone to problems of troublesome sensors, and programming was made more complicated. The adaptive control constraint systems, on the other hand, were simpler to use and of lower cost. These systems sense the torque (the rotational cutting force) only and adjust the feedrate. The early constraint systems were known as 'feed-only' systems, but as they were developed and further refined, they became better-known as 'torque-controlled' systems. These sophisticated systems, with their unique sensory circuits and computation methods, measured the net cutting torque, then compare it with the torque limits established for the cutting tool being used. The appropriate control actions – a feedrate reduction – is then taken automatically, whilst keeping within the maximum torque and power limits of the motor. If a condition arises where the feedrate falls below a preset limit, a new tool – a 'sister tool' – is called up to complete the machining operation. This makes up a feedback loop in which continuous monitoring by the sensors, and updating of the machine control unit using adaptive control, produce optimal cutting conditions for the combination of tool and workpiece. This characteristic feedback loop is illustrated in the flow diagram in Fig. 3.21b. To explain the principle of adaptive control more clearly, a pictorial respesentation of an example of adaptive control constraint – torque controlled machining – is given in Fig. 3.22, along with a summary of some of the advantages of its use – in this instance, for a machining centre. The diagram illustrates the machining of a rough casting, exaggerating the effects of the unpredictable stock-removal rate. It shows that when the depth of cut is large, the control system senses the torque level increase, and the feedrate over-ride is activated. This over-riding of the programmed feedrate decreases the feed for this large depth of cut and

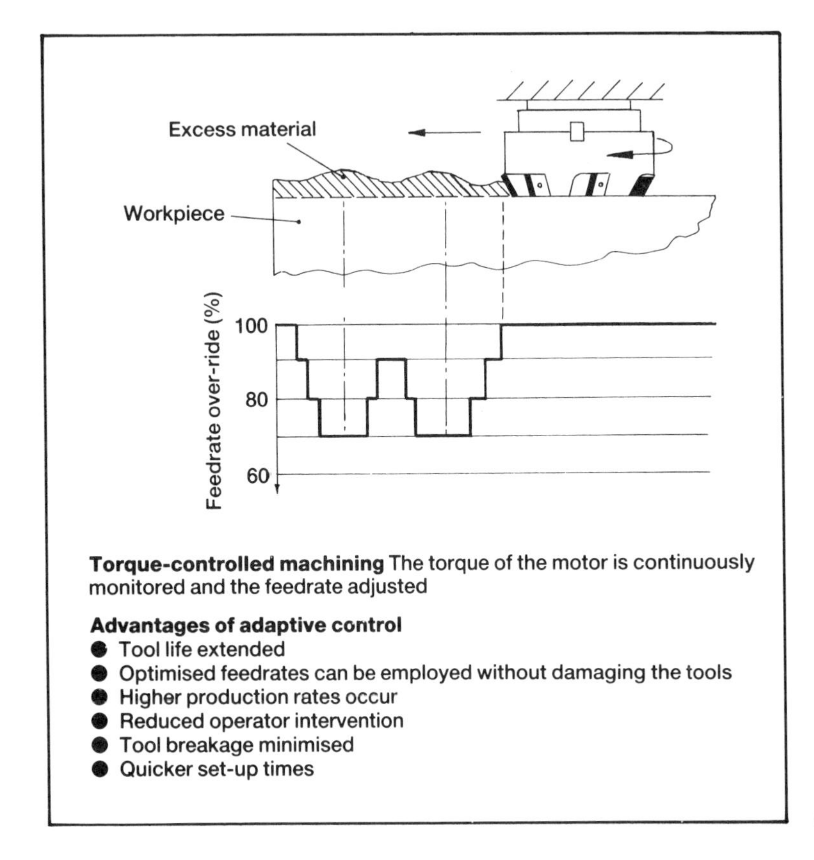

Torque-controlled machining The torque of the motor is continuously monitored and the feedrate adjusted

Advantages of adaptive control

- Tool life extended
- Optimised feedrates can be employed without damaging the tools
- Higher production rates occur
- Reduced operator intervention
- Tool breakage minimised
- Quicker set-up times

Fig. 3.22 Adaptive control and its benefits to the machining process

then increases it again as the depth of cut lessens. Thus adaptive control produces optimal cutting-tool efficiency by decreasing the chip loading on each cutting edge, and there is the added bonus of gaining a longer tool life and therefore better machine-tool utilisation.

It is possible to use the adaptive control system as a surface-sensing probe using any tool – for purposes of workpiece interrogation and fixture alignment, for example. However, the sensitivity and accuracy of this method are somewhat less than those of a probing system (such as the touch-trigger probe) specifically designed for the surface sensing of workpieces.

It should now be apparent to the reader that, for little extra financial expenditure, the benefits offered by adaptive control allow for a fast pay-back on the investment, and optimise the machining and tool characteristics. The final section in this chapter will describe some of the ways of implementing condition-monitoring systems on both 'stand-alone' machine tools and flexible manufacturing systems.

3.5 Condition-monitoring systems for stand-alone machines and flexible manufacturing systems

Most tool-condition monitoring systems available are designed with the 'building-block' approach to automated machining requirements. A company need not invest heavily in a large-scale system in one go, but can gain experience in a step-by-step manner, whilst looking at the short and long-term objectives of implementing an unmanned machining capability. Immediate benefits will accrue from introducing some form of condition monitoring to a 'stand-alone' machine tool, and it will be possible to tailor the system specifically to the company's operating requirements.

An example of this 'building-block' philosophy can be seen in Fig. 3.23, where a tool-offset sensor and a tool-condition sensor are incorporated neatly into an unmanned cell for a 'lights-out' machining capability. The tool-offset sensor allows deviations from one tool to another to be accommodated, overcoming the problems that might otherwise result in a costly stream of rejected parts. The tool-offset sensor (shown in the upper, righthand corner of Fig. 3.23) visually locates a new cutting tool and compares it with a known location. Once this has been achieved, the sensor sends a signal to the machine

Fig. 3.23 A turning centre fitted with KV tooling, GMF robot for part loading, tool-condition sensor and tool-offset sensor (Courtesy of Kennametal (UK) Ltd.)

control unit to make any offset corrections necessary. The inspection position is programmed into the CNC program, and is underneath the sensor's sealed camera. Once a tool is in the field of view of the camera, the sensor analyses where the cutting edge is, then compares the position with where the program says it should be, and transmits the resulting difference of these two positions to the machine tool's CNC tool-offset table for updating. The accuracy and repeatability of measurements are to within ±0.01mm.

The only limitation with this tool-setting system is that tools with a round insert geometry cannot be inspected; but verification of threading and grooving tools, and all other tool types, can be made. Up to four camera inputs can be accommodated, controlling four display panels and CNC controllers, using just one tool-offset sensor. These tool-offset display panels provide full progamming ability and visual verification of the camera operation: with a tool in position to be sensed by the camera, an area of the tool nose 2½ mm square is visible. The *X* and *Z*-offsets are compared against the programmed positions and transmitted to the machine control unit.

The tool-condition sensor is immediately below the tool-offset sensor in the console, as shown in Fig. 3.23. Its purpose is to monitor tool wear and breakage, using transducers mounted in the tool block to measure the cutting forces. These sensors may be used to monitor all types of cutting tools used on turning centres, and they never require adjustment from one operation to another.

The sensor operates as follows. Based on a gradual increase of cutting forces with machining time, the tool-condition sensor will calculate and display the actual amount of tool wear. If there is a sudden increase in the forces, indicating that the tool is broken, this is sensed. The sensor immediately signals to the machine control unit to retract the tool from the cutting zone, to initiate a tool change if a 'sister tool' is present. If, however, the tool does not break but progressively wears until the cutting pressure increases to a predetermined threshold point, the sensor signals automatically for a tool change.

A variety of displays can be selected, and these provide an immediate report of the tools' condition for single and dual-turret operations. The display shows, in real time, the actual cutting forces, the relative wear level, elapsed time in cut, and the tool's status (whether it is a new, worn or broken tool). Signals that the tool is worn or broken are transmitted to the machine control unit by an optically-isolated interface and may be used by the controller to initiate the appropriate recovery sequence, such as motion hold, safe retract, change tool, check offsets, change part or continue. Up to 60 unique tools can be monitored with this system. It also allows for a wide range of turret tool-holding capacities.

In any unmanned stand-alone machine or flexible manufacturing system, it is paramount that the overall metal-cutting efficiency and productivity is measured and analysed. A piece of equipment such as a data logger productivity monitor (Fig. 3.24), which can be simply connected to the machine or machines, fulfils these requirements admirably. It has the facility to generate reports detailing up-time, complete parts, average cycle-time, maintenance time, set-up time and many other day-to-day machine-related activities and gives a full analysis of the data for the machine or machines being monitored. The datelogger can monitor up to 99 separate programmed machine-related operations. The programming is self-prompting for ease of use. Remote printers can be connected to the datalogger to produce 'hard-copy' reports, and the information-collection ability can be expanded by connecting it to a company's mainframe computer. Another feature of such a monitoring system is that it can be programmed to initiate automatic tool changing at preset times if need be, or to alert operators or maintenance personnel to problems needing their attention.

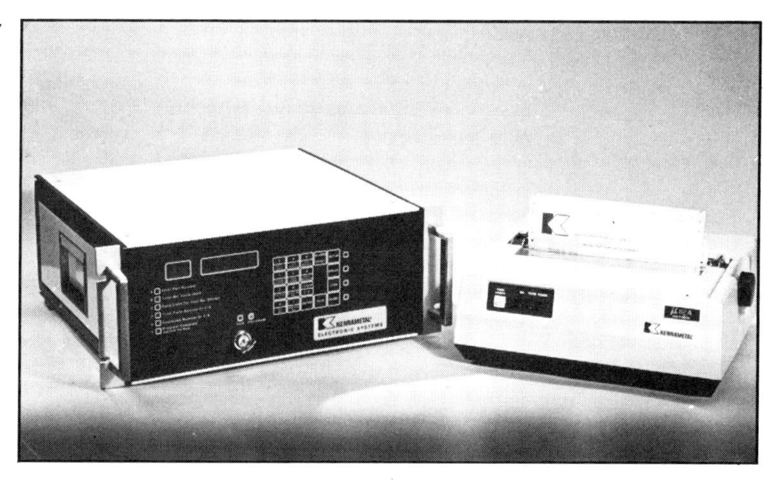

Fig. 3.24 A datalogger productivity monitor with printer (Courtesy of Kennametal (UK) Ltd.)

Finally, there is the complete monitoring system coupled to unmanned machine tools or flexible manufacturing systems. This allows a company to machine in an untended 'lights-out' condition successfully in the confidence of knowing that if a problem arises a monitoring system is available to take the appropriate action to recover the situation. This approach is depicted in the turning-centre monitoring configuration shown in Fig. 3.25, which could show either a highly-sophisicated stand-alone machine, or just part of a large cell, or flexible manufacturing system.

The next chapter will consider the problems of monitoring and identifying cutting tools for these highly automated machine tools.

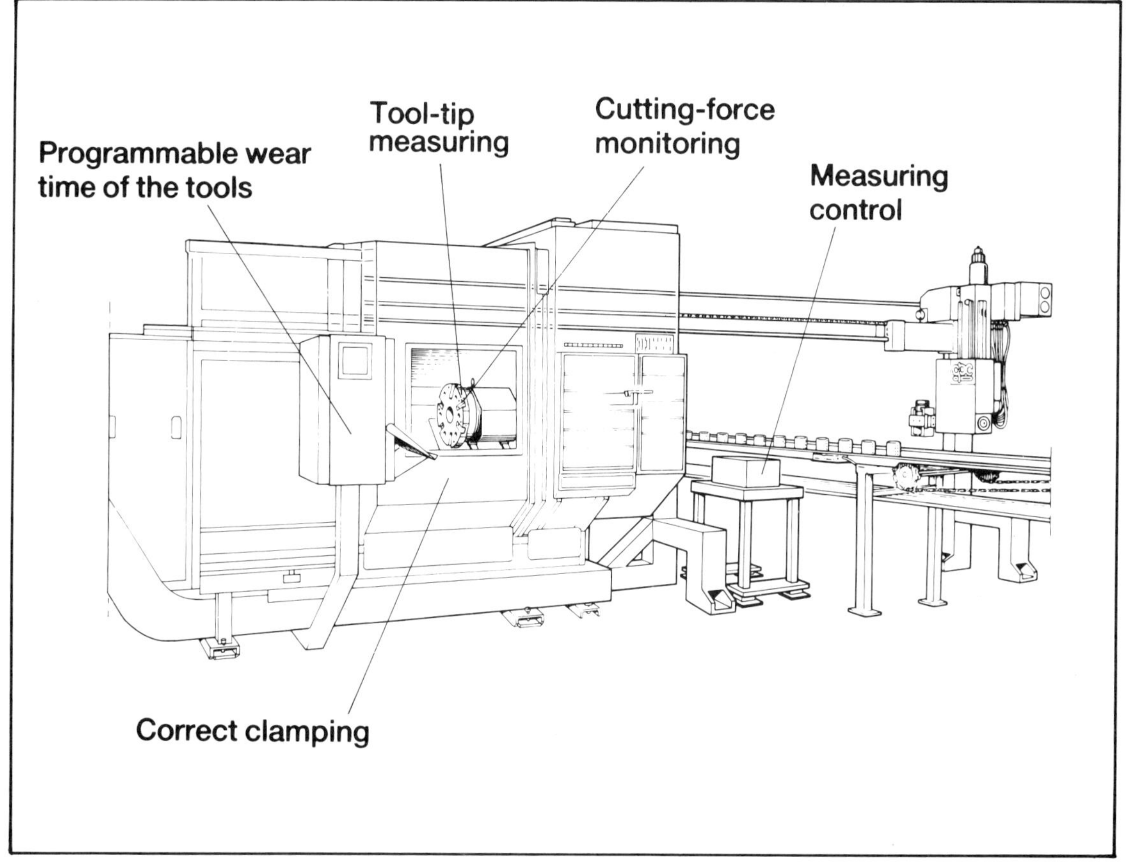

Fig. 3.25 Some of the monitoring systems available for use on highly-sophisticated turning centres (Courtesy of SMG (UK) Ltd.)

Chapter 4
Tool management

As a result of the introduction of CNC machines there has been a trend towards smaller batches, and this has meant that tool management has become increasingly important in machining operations. Previously, in the traditional production workshop environment, the supervision of tool management was generally left to the machine operator. Beside each machine would be situated a limited kit of tools which were maintained and replenished with spares and consumables through the operator's liaison with the tool storeman. A skilled operator's main responsibility was to select these tools and devise cutting techniques and data to manufacture the required components. This enabled the process planner or part programmer to treat the machine and the operator – plus the tool kit – as a single, self-maintaining system with a well-established performance. This allowed work to be allocated to specific machines whilst leaving the detailed definition of the process to the operator. This system is illustrated in Fig. 4.1a.

The increase in the diversity of work on CNC machines, which has come about as a result of the flexibility of manufacture and the reduction in economic batch size, has changed the pattern of working. In order to cope with this diversity of work, the individual machines have each acquired a very large complement of tools. A situation soon develops in which neither the operator nor the part programmer is sufficiently in control to accept responsibility for the range of tooling dedicated to any particular machine tool. So, as a result of the introduction of CNC machines, the organisation of the tooling normally changes to one in which:

- The process is defined in detail in a separate area, remote from the shop floor.
- The machining program and tool list are drawn up. Later they are sent down to the machine for each job via the tool-kitting area, with the process data and tooling fully defined – there may be some doubts about the quality of the definition.
- The machine operator is asked to run the program with the minimum of alteration.
- As the batch sizes become smaller, the operator comes under increasing pressure to run the given program without alteration. This tends to lead to 'conservative cutting', with the result that conditions for optimum metal cutting are not used.

The whole operation becomes critically dependent on the ability of the tool-kitting area to supply and support the part programmer's specified tooling requirements. This

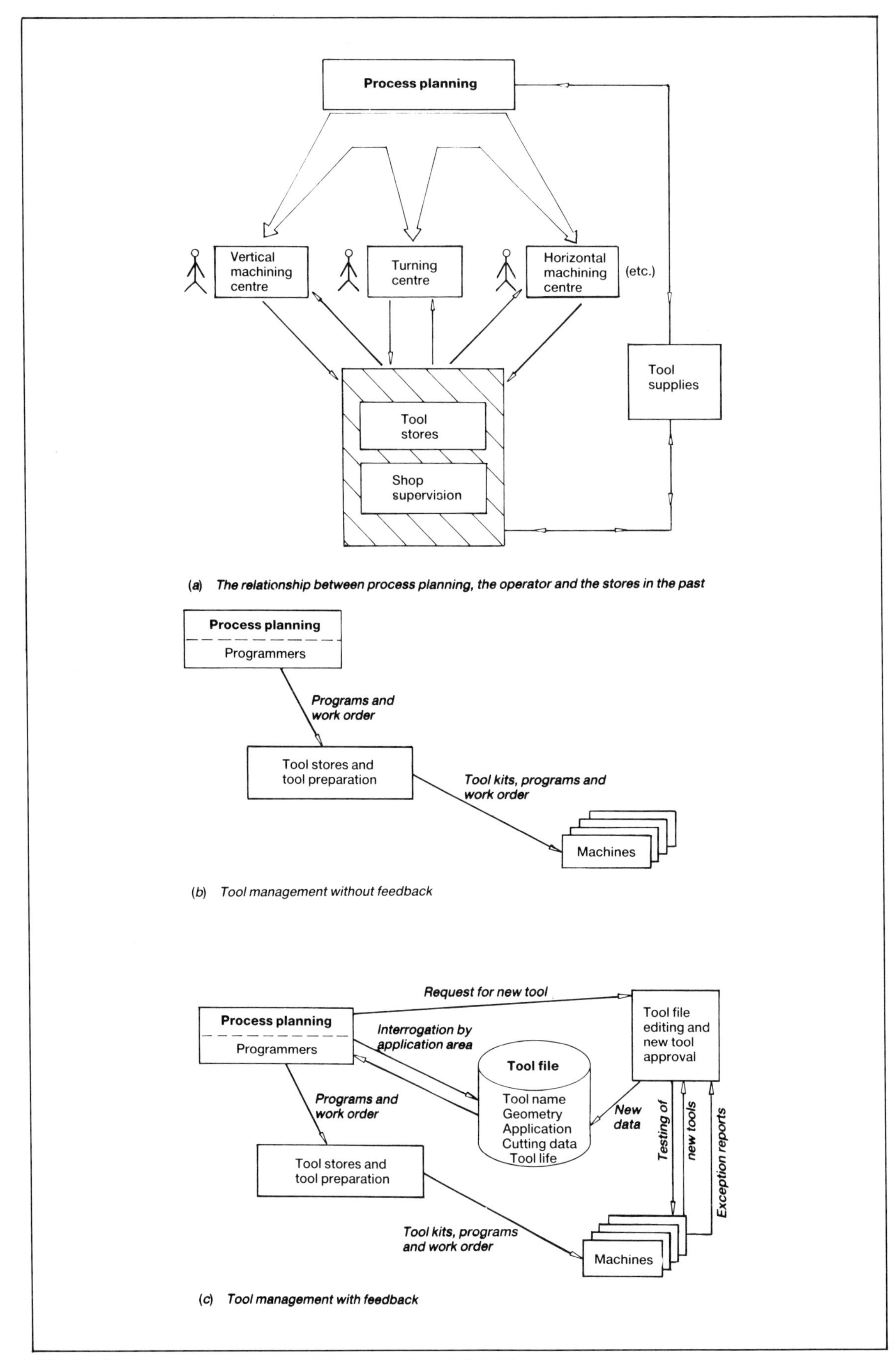

Fig. 4.1 Tool-management systems (Courtesy of Sandvik (UK) Ltd.)

unsatisfactory and ineffective tool-management system is shown schematically in Fig. 4.1b. The main problem with this type of system is that there is no feeding back of the experience gained from machining the parts, and this is obviously undesirable. The result is that the part programmer is oblivious of any problems encountered during machining of the part, causing further lack of awareness in the tool-kitting area and critical loss of tool-management support.

To overcome the problems associated with the lack of information received by the part programmer and the tool-kitting area, a loop may be introduced into the tool-management system to allow feedback from the operators. This feedback can be for either the whole shop or for each section of machines, in which case it is normal to centralise the system around a tool file which is managed by a file editor, as shown in Fig. 4.1c. A tool file may be either manual or computer-based, but it is normally organised so that it is accessible to the process engineer, the part programmer, the tool storeman, the purchasing department, the machine operator, the file editor and management, as necessary. The tool file must contain all the information relevant to the needs of all these individuals for every tool. Each file entry is likely to contain the following information:

- Tool identification code, usually with a cross-reference to the supplier's code.
- Application area – tools with applications in different areas may have many entries in the file.
- Cutting data for the materials that are expected to be encountered during machining.
- Tool geometry and dimensions.
- Expected tool life.
- Spare parts.
- Usage.

The value of such a tool file will, of course, depend upon the data from which it is compiled. Usually, such a file is built up progressively, so that it includes entries for only those tools that have been proven to be most productive in their area of application on the machines. Any cutting data and estimates of expected tool life should be taken from information established in actual use on the machines.

Leaving aside the discussion of tool files until later in the chapter, it is worth emphasising that a new competitive situation develops as more CNC machines are installed. The criterion of relative performance compared with the company's previous practice becomes less important when one looks for an absolute standard or the theoretical optimum cutting performance that may be attained. In previous chapters we have seen that CNC machines seem to be greatly underutilised: after allowing for unworked time, maintenance, setting-up, loading, etc., a machine tool may typically spend only 6% of its time actually cutting metal (Fig. 4.2)! Manufacturing is a business that relies on the 'value-added' concept; that is, raw materials are bought in at one price and are processed through a manufacturing system before being sold at a higher price. If the raw material and processing costs are less than the selling price, the manufacturer will operate at a profit, resulting in a potential for company growth. A company's potential for growth and its profitability are directly related to how efficiently the materials, manpower and equipment assigned to a process perform. The profitability of the total manufacturing system will therefore depend on the costs related to tooling.. Tool management and documentation are a way of measuring and controlling tool costs, and are therefore a way of understanding and quantifying the manufacturing performance.

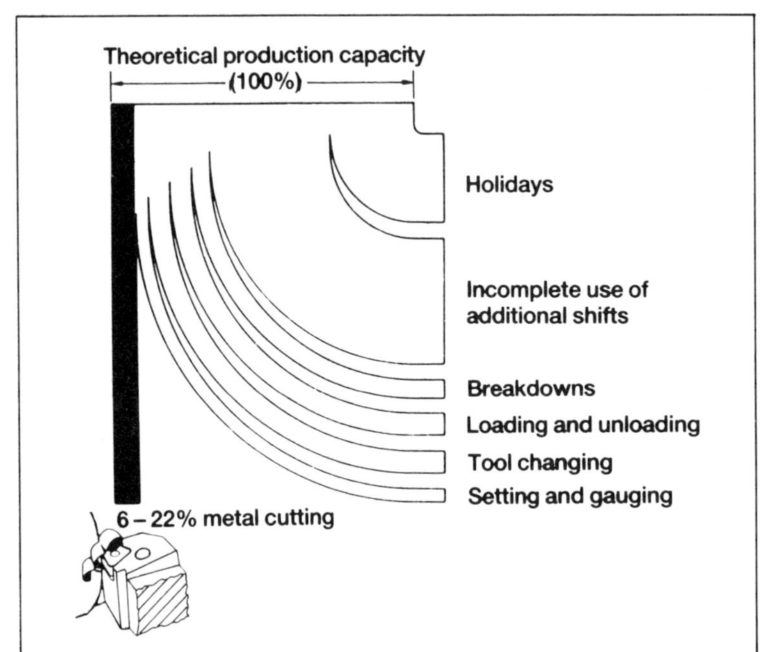

Fig. 4.2 The relationship between the actual metal cutting per year and the theoretical production capacity (Courtesy of Sandvik (UK) Ltd.)

Tool management and documentation is more than just assigning numbers to tools and then filling them into a book; it involves analysis of the overall manufacturing system, from the customer's needs to the manufacturer's shop-floor requirements. A good tool-management system should continuously review and revise its data in response to changing manufacturing technology and customer needs. A system such as this is more than simply an esoteric exercise in the gathering of data: it can lead to a reduction in tool inventories by 35% or, by improving selection of speed and feedrates, it can improve metal-cutting performance (in terms of costs) by between 25% and 35%. In fact, the data generated may lead to a reappraisal of the whole manufacturing system, resulting in a 50% improvement in utilisation of the machine shop's capacity.

As was said above, in some shops machine operators are left to choose which tooling is to be used, and in others programmers decide the tooling requirements. In others, though, a much worse situation may occur in which it is left to the purchasing personnel, who may not even be engineers, to make the final decisions on the tooling acquisitions. And even if they are engineers, they are probably too far removed from the 'cutting face' to be able to select the appropriate tooling! This highlights a basic problem often found when implementing a system for tool management and documentation, which is that, traditionally, no individual manager or department has had full accountability and responsibility for the selection of tools and their performance. Even when the responsibility is given to one person, they may be in the wrong place, or not have enough technical information to act effectively within the organisation. Yet it can be done. This is the theme of the next section, which shows the interaction that occurs throughout the manufacturing process as a whole, as illustrated in Fig. 4.3a.

4.1 Creating a tool-management and documentation database

4.1.1 Production requirements

In order to establish the 'true' production requirements of a company, cooperation between the sales and production departments is essential. The first requirement is an understanding of the manufacturing load, with work being classified as either 'job shop' (i.e. one or two-off special workpieces), 'small batch' (up to perhaps 50 workpieces), or 'volume production' (more than 50 workpieces, say). These classifications of batch size will obviously vary depending upon a company's production requirements. At any level above the 'job shop' size, similar products can be grouped into families according to their dimensions, tolerances, workpiece materials, etc. This method of allocating workpieces into groups is often termed the 'group technology' approach to machining.

It is vitally important that the sales and marketing personnel are aware of the company's patterns of manufacture and their capabilities if the company is to be able to respond properly to customer needs. The sales force will be able to relate a customer's requirements to the standard range, and the manufacturer will be able to 'fine tune' even small production runs for maximum efficiency. Understanding the manufacturing process for standardised items in the company's range allows the same optimum conditions of manufacture to be utilised when 'modified standards', or even 'specials', are required. This flexibility and ability to cater for unique customer needs may open up new markets.

4.1.2 Perishable and capital equipment

Many tooling companies produce standard forms to allow manufacturers to compile data on both their perishable and capital equipment needs. It is necessary to gather the data together because the performances of both categories are interdependent.

In gathering data on the perishable tooling, the company must analyse the entire toolflow system, including the use of inventory cards, the maximum and minimum tooling levels, the quantities of new and used tooling, and the tool-storage requirements. As a preliminary data-gathering exercise, all the items in stock should be listed, together with the number now in stock, the number used in the last 12 months of manufacture, and the last price paid for the item.

The next step is to review the stock list for tool obsolescence by checking to see which items have not been used during the last 12 months and can be replaced by an ANSI standard, for example. Any tools falling into this category can be considered as obsolete: as mentioned in Chapter 1, it has been shown that in the USA – and, no doubt, in most other technologically-advanced countries – a typical industrial inventory contains on average, about 50% obsolete tooling. This obsolescence can be regarded as money 'tied up' and doing nothing for the company's profitability.

The remaining items of tooling on the stock list, which are not obsolete, should then be reviewed. The carbide insert grades of tooling, for example, should be grouped according to their grade and size; once grouped in this way, it is possible to order large quantities at a time, enabling the company to exert some 'leverage' and to obtain substantial cost advantages as a result. If tooling information of this nature is compiled in a control book and regularly updated and reviewed, future tooling decisions will be speeded up.

The compiling of information on the capital equipment in the factory usually begins with a preliminary identification of the machine, using a 'brass tag' numbering system to give its location within the plant. A list is then compiled, giving every machine's power capacity, number of spindles, present tooling and operating condition, together with the current and past operations performed by the machine. Organising information about the capital equipment available produces a number of benefits, including a knowledge of the basic machine characteristics thus ensuring that the best machine can be selected for a particular job, and that an operation is performed using the optimum parameters. The knowledge of such simple facts as where a certain machine is located within the factory may lead to a better utilisation of floor space and plant capacity, and allow an accurate determination of the potential for either increased automation, or the combination of machines or operations to help solve the problem of poor work-in-progress. Even the maintenance department will benefit from this knowledge, since it will allow a more efficient maintenance schedule to be written and the preventive maintenance procedures to be improved.

The knowledge gained from a study of the perishable and capital equipment allows for better process planning (the plan of action for the manufacturing of a certain part) and production planning (the best use of a factory's resources for a particular workload).

4.1.3 The workforce

When a company considers the role of the workforce in tool management and documentation, there are three aspects to be discussed: education, involvement and review. Firstly, the workforce should be educated as to what must be done in the tool-management programme, and why. Secondly, they must be involved in the programme, not only in its application to the normal production duties, but also providing an input about tool performance and new ideas for its application. Lastly, a continuous review is required so that the programme does not stagnate; this allows new ideas and technologies to be incorporated as and when they become available. This type of 'fluid interaction' and continuing improvement of the system is of paramount importance to the sustained growth and profitability of a company (Fig. 4.3b).

Active involvement of the workforce in the tool-management and documentation programme gives rise to other benefits, including reductions in the amount of material used and in handling time, together with fewer errors and an increase in the accuracy of manufacture. These benefits can be indirectly related to a return of the workforce's 'pride in workmanship', brought about by the increased confidence of having the tooling applications under control. An analysis of the types of labour involved in the manufacturing process can aid decisions on any future additions of automation and flexible manufacturing systems (FMS).

4.1.4 The 'history' of a manufacturing system

An overall 'history' of a process can be developed by combining information gathered in analysing production requirements, perishable equipment and manpower. In a new factory, the 'history' will be short, but even here it will still contain the basic information discussed above, and the organisation of this knowledge will allow the manufacturer to plan to meet future manufacturing goals.

The analysis of the data contained in the manufacturing history requires a consideration of the projected markets for the products and of the priorities of the customers in those markets. As examples of typical priorities in different markets, customers in the aerospace and defence markets value high reliability and quality above mere cost considerations, whereas costs are an increasingly important factor in the automotive field, where extremely high production rates and the competitive marketplace mean that an economic costing is imperative.

If there has been an analysis of the overall manufacturing system, this should highlight areas where reductions can be made, for example in the tooling complexity, and allow the true cost of operations to be accurately determined. In such areas, significant savings

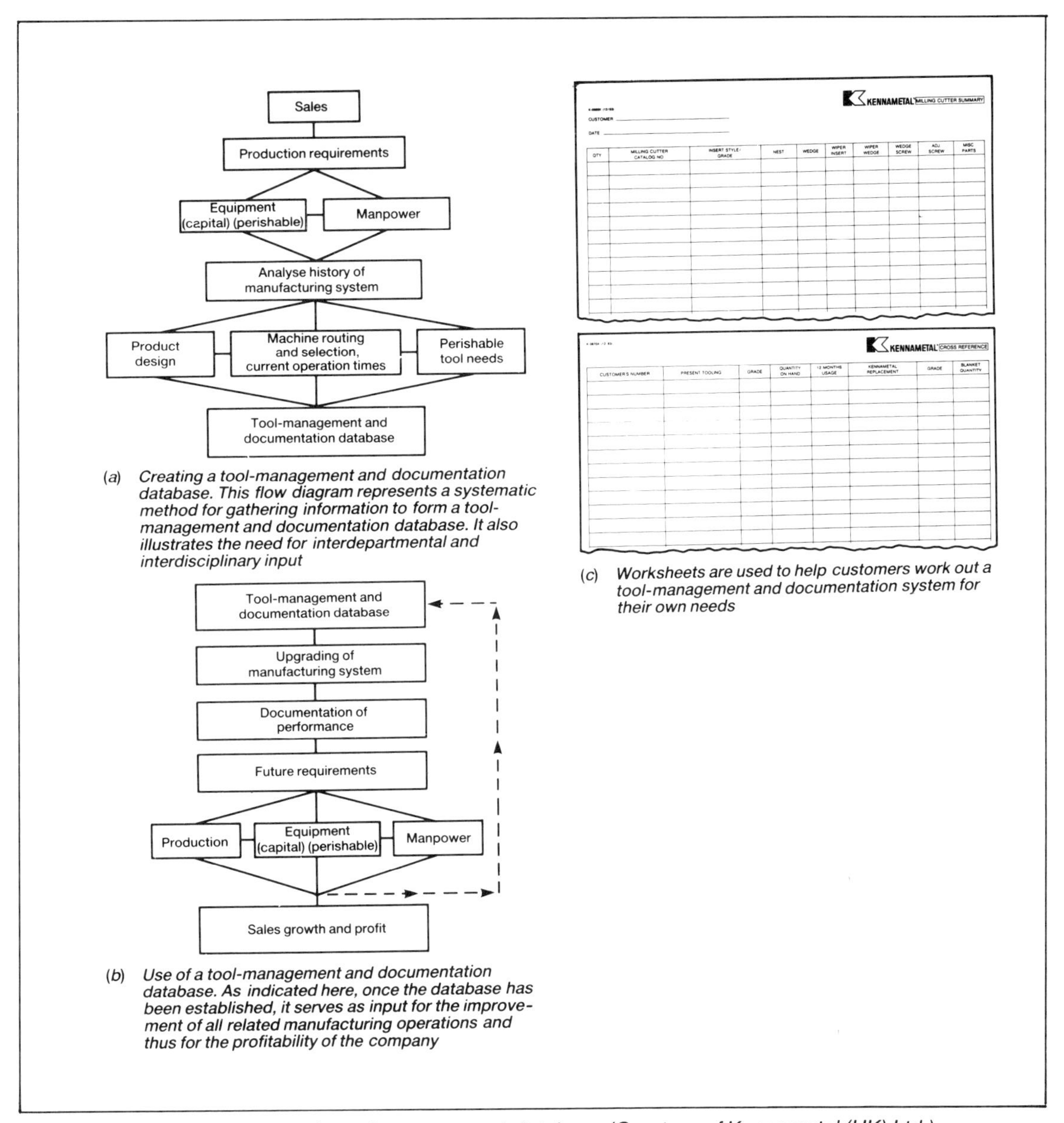

(a) *Creating a tool-management and documentation database. This flow diagram represents a systematic method for gathering information to form a tool-management and documentation database. It also illustrates the need for interdepartmental and interdisciplinary input*

(b) *Use of a tool-management and documentation database. As indicated here, once the database has been established, it serves as input for the improvement of all related manufacturing operations and thus for the profitability of the company*

(c) *Worksheets are used to help customers work out a tool-management and documentation system for their own needs*

Fig. 4.3 The creation and use of a tool-management database (Courtesy of Kennametal (UK) Ltd.)

may be made by introducing the 'just-in-time' philosophy, which eliminates the need for excessive stock to be kept to cover for manufacturing problems. Another advantage of using this approach is that it can help to identify the money-making products so that they can be fitted into the overall plan for manufacturing and the marketplace.

Once the 'history' has been analysed for inclusion in the management and documentation database, the next step is to analyse the actual manufacturing operations themselves. The needs of product design, machine routing and selection, and current operation times are interdependent, and they must be viewed as being inseparably connected. For example, if a company decides that a change is required in a product's design, then this usually means that changes must also be made in the tooling needed and the machine used, and there will also be changes in the time required for actual machining. Similarly, rerouting a part to a different machine may result in a need for changes in product design and the tools required to manufacture the part. The key point to remember in deciding on combinations of these factors, is that the goal is to produce the highest-quality product yet at the lowest possible cost. With this goal in mind, components should be designed for the highest production capability yet still retain the aesthetic and functional qualities that their designer intended. Ways of achieving this will be considered below.

4.1.5 Designing components for production

The product's function is, of course, paramount in any initial design consideration. However, the design can be altered to make the product more efficient to manufacture, or to create a family of parts that may be produced on a particular machine or series of machines. It is also possible to change tolerances in the component's design, or even, under certain circumstances, to use a different material, in order to eliminate problems in manufacture, so long as such changes do not affect the product's function. Sometimes changes in design can allow raw material of near-net shape to be used, which may mean that some operations may be entirely eliminated as a result.

4.1.6 Further aspects of tool management

The more time workpieces spend on the shop floor in a 'work-in-progress' situation, the greater are the manufacturing costs. A study of the routing of parts around the shop floor and of the process of selecting machines and the time spent actually performing operations, produces useful data on the costs related to labour, scrap, reworking and, of course, work-in-progress, which are the major indirect costs of manufacturing.

In an analysis of the machining itself, a number of factors must be considered, some of which have been stated previously; they include the machine's condition and spindle-power rating, the tool geometries and cutting grades that are required, and the part's material and design. Some operational data related to the specific component's requirements is also necessary, such as feeds and speeds, depths of cut, part fixturing or chucking, the number of parts in the run, and the operator's familiarity with the part and with the machine and its operation.

So far we have mainly been concerned with ways of achieving productivity gains through increased metal-removal rates, but it should be understood that there are other methods of achieving similar economies of production. Referring to Fig. 4.2 again, the industry averages show that metal cutting occurs for only approximately 6% of the time. If a reduction in the tool set-up time, tool procurement time, waiting time and all other sources of down-time can be made, then a significant contribution to increased productivity will result.

Once the review of tooling needs has been completed and the consolidation of all the perishable tools has been achieved, so that obsolete tooling has been eliminated and inventories reduced, perishable tooling requirements will be dictated by the specific application. It is only after a full review of a product has been done, including its design stage, that process analysis (in conjunction with cross-referencing of tooling grades and use of geometries which provide a broad range of cutting capability), can be used to reduce the number of perishable tools for the entire manufacturing process. If frequent reviews occur, then assurance is provided that the machine tools are operating at production rates which reflect the continuing progress of 'state-of-the-art' tooling.

A database centred around a tool-management and documentation system provides an overview of the whole manufacturing system. Therefore new tooling, machine tools and manpower can be simply and quantitatively compared with that which already exists in the workshop. If a new insert grade is purchased and tested on one operation, for example, then the results obtained can be quickly compared with those of the previous set-up. Also, by consulting the database a company can find similar cutting applications for the same grade of insert, and comparisons of the expected life times of different inserts for a given application will prove whether a change is desirable. The same procedures for feasibility comparisons find ready application on manual machines, flexible manufacturing systems and lines.

Troubleshooting will be simplified by using a tool database, and the lessons learned on one operation may be applied to others. The collated and extensive database provides a vast amount of information which allows engineering changes to be implemented and evaluated easily. Since the data on manufacturing the company's products is simplified, organised and stored, it may be retrieved at a later date when similar operations are required. This produces savings in scrap and in the testing of component runs. Another aspect of the database's usefulness is in the resourcing of manpower training. The information contained in the database tool file means that the company can use it as a reference source for the existing staff and as a training aid for new workers. The other feature, which is often overlooked when the need for a tool-management system is rejected, is that, once established and documented on the database, the standard practices are not affected if key personnel leave the company. Therefore the skills can be retained once acquired, and product quality is not affected as new staff are given company training.

So far, the database could have been simply collated onto worksheets as shown in Fig. 4.3c. This becomes rather unwieldy after a time as the amount of tooling information stored grows almost exponentially. Therefore, there is a need to 'computerise' this extensive tool data. A computerised tool-management system not only retains a large memory bank of information that may be accessed quickly and retrieved instantly, it also calculates interrelationships between relevant data and reduces paperwork to a minimum. Paperwork has an inevitable tendency to slow-down the whole system as well as increasing the likelihood of errors, but the computerised database eliminates this problem. The functioning of such systems will be discussed in section 4.6.

4.2 Building up the tool file

The principal users of the tool file within a manufacturing organisation are the process engineer and part programmer, and perhaps the tool-presetting operator and storeman. It was mentioned in section 4.1.6 that new tools would only be added to the tool file after a proper investigation of the need for them, assuming that such tooling was not already listed. By accepting this limitation on the number of different tools, a company can be assured that the tools called up for use in the manufacturing process are available and backed up with spares, since the tool storeman also has access to the file.

An important feature of any tool file is the cutting data and machining times listed. These are known to be actually achievable and will be those expected to be used during component manufacture. More specifically, they are the ones used to calculate a quotation price for the product for a customer's appraisal.

The editor of the tool file has a key function to play in the acquisition of tooling data; when building up the tool file they have to:

- Scrutinise any reported deviations from the recorded cutting data and tool life.
- Investigate claims of higher productivity ratings for new tooling.

- 'Weed out' tooling made obsolete by the addition of new tools.
- Investigate new tooling that may be available for the latest products to be manufactured.

The systematic accumulation of knowledge in the tool file for each section of the manufacturing operation ensures that the cutting performance will continuously improve. This improvement may be considered to be analogous to improving the skill of an operator on a conventional machine tool, but with even more flexibility, as the system is able to cope with much more diverse and complex tooling situations. This completely eliminates the 'hiccups' that occur whenever a new part programmer is employed – or even a tool storeman or machine operator, for that matter!

As the structure of tool files is of necessity highly complex and interrelated in nature, it is not possible within the confines of this chapter to show the operation of the system in detail. However, an appreciation of a simple format can be obtained from the manual tool file illustrated in Fig. 4.4. There is a separate record for each tool, and even for different inserts for the same tool. In the example in Fig. 4.4, there are four fields of tooling information:

- The tools are built up from modular elements, which are the 'key' to effective tool management and control, as they allow the widest range of tooling for a wide range of machines to be offered from the minimum number of elements held in stock. Hence, a tool for a given application may be assembled from different modular elements to suit a range of machine tools. An example, showing the build-up of a milling cutter which can be fitted to three different machine tools, is shown in Fig. 4.4a.
- The materials-requirements planning system and the tool stores must support the tools in the tool file with such items as spares, consumables and back-up tooling. In order to achieve this aim, the tool file record includes details of the build-up of each tool and usually gives the stores location for each part, using a 'key' notation.
- Certain 'steering' comments, based on shop-floor experience, are included to enable the process or planning engineer to select the correct tool (in this case, the correct shape of cutting insert) for the application.
- The basic cutting data has been arranged ready for inclusion into the CNC program. The data in Fig 4.4d is organised according to the component to be machined and the cutting data, but the optimum organisation of data will vary from one company to another, depending on their needs. Of course, all the data listed reflects the company's actual experience; in particular, it includes the results of any optimisation exercises carried out previously in the machine shop.

The following paragraphs give some idea of the practical aspects of starting up a system of tool files.

Whenever such a system is initiated, the important point to bear in mind is to start small and keep the information to be included 'sound'. Having accepted this principle, a company may start to build up the tool file over a month, say, and include practical test data for maybe a hundred or so of the most popular tools. The information now held on file will flow through the tool-management system, and it will begin to highlight the requirements of the system users and will drive the file's further development. On the other hand, a company may recommend a more rapid and 'heavy-handed' approach to the acquisition of data and decide to start by putting all the data on, for example, 2000 tools, on file, using provisional data instead of well-proven information. In this case, the probable outcome will be that there are so many 'dud' answers that the system is discredited even before it gets into full swing!

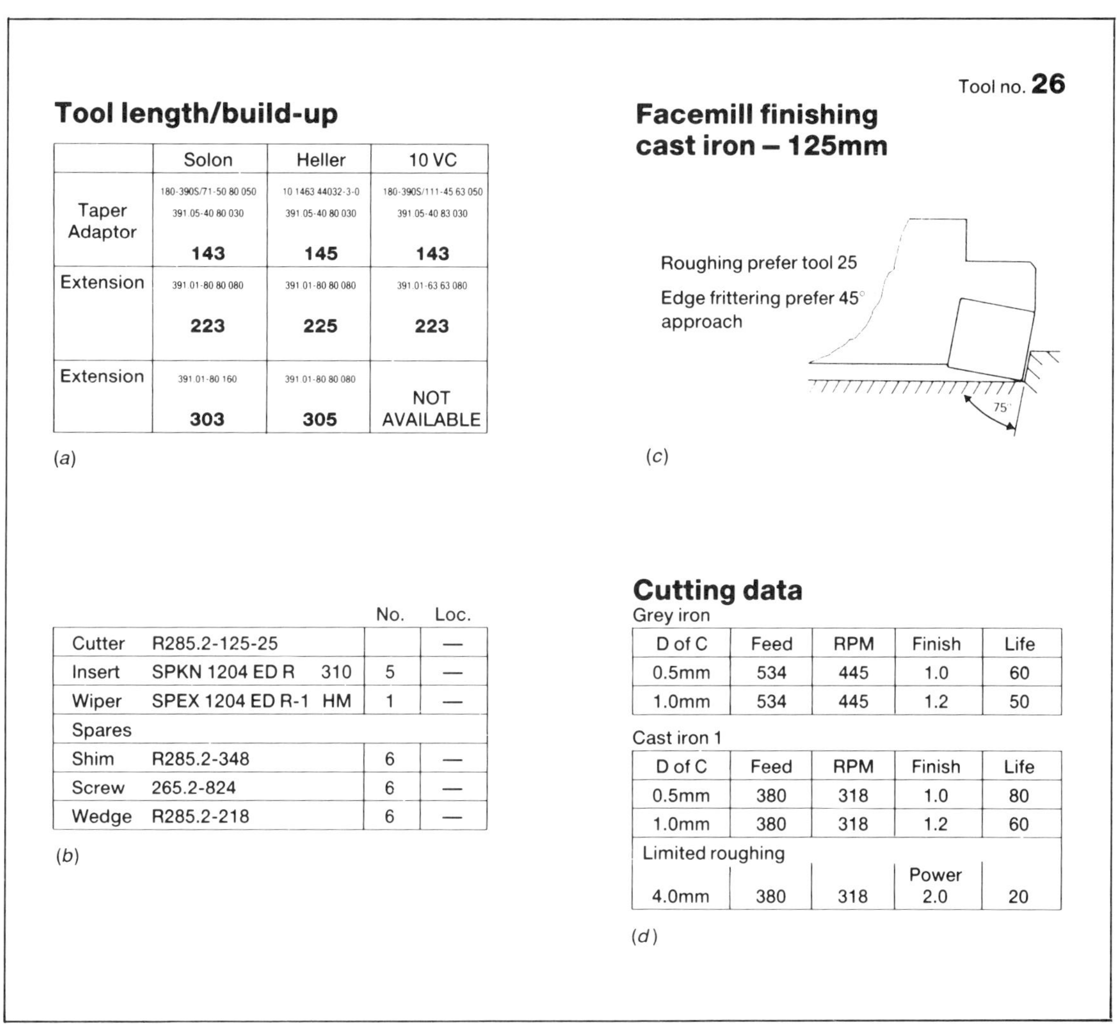

Tool no. **26**

Tool length/build-up

	Solon	Heller	10 VC
Taper Adaptor	180-390S/71-50 80 050 391.05-40 80 030 **143**	10 1463 44032-3-0 391.05-40 80 030 **145**	180-390S/111-45 63 050 391.05-40 83 030 **143**
Extension	391.01-80 80 080 **223**	391.01-80 80 080 **225**	391.01-63 63 080 **223**
Extension	391.01-80 160 **303**	391.01-80 80 080 **305**	NOT AVAILABLE

(*a*)

		No.	Loc.
Cutter	R285.2-125-25		—
Insert	SPKN 1204 ED R 310	5	—
Wiper	SPEX 1204 ED R-1 HM	1	—
Spares			
Shim	R285.2-348	6	—
Screw	265.2-824	6	—
Wedge	R285.2-218	6	—

(*b*)

Facemill finishing cast iron – 125mm

(*c*)

Cutting data

Grey iron

D of C	Feed	RPM	Finish	Life
0.5mm	534	445	1.0	60
1.0mm	534	445	1.2	50

Cast iron 1

D of C	Feed	RPM	Finish	Life
0.5mm	380	318	1.0	80
1.0mm	380	318	1.2	60
Limited roughing				
4.0mm	380	318	Power 2.0	20

(*d*)

Fig. 4.4 Typical tool file data for milling cutters (Courtesy of Sandvik (UK) Ltd.)

Obtaining meaningful test results and tool assessments does not necessarily demand extra effort from the company, merely the organisation of efforts already being made by the company's tooling engineers and those of the tool suppliers. Often, most of the information that the company's tooling engineers gather concerning tool testing and troubleshooting will 'evaporate' once the current batch is completed, simply because there is no framework in which this information may be recorded. So all the time and painstaking effort needed to gather 'sound' tooling information is disregarded and the information discarded, and the company must once again begin to 'reinvent the wheel' for the next product!

The tool-kitting area's service to the machine shop must be based upon the assurance that the completed kits are dependable, whilst providing the maximum security from a limited budget for tooling stock. As only finite tool stocks are normally available, the aim is to use the same modular tools over a range of machines; this is one of the major reasons why 'modular quick-change' tooling has become so popular recently. Another key factor in any tooling requisition is the quality of the tools used. A function of the tool-kitting area is to monitor the quality of incoming tooling from completed runs in the shop and to prepare tools going out in the form of new kits. As batch sizes get smaller, the flow of kits speeds up. The effectiveness of the tool-kitting personnel will be inversely related to the number of tooling items and standard they must control. This is a further argument in favour of a factory-wide standardisation of tooling inventories.

Summing up, two main points emerge:

- Linking every tool with its application technology in such a manner that it is the most productive tools that are chosen for new jobs and not the old ones, just because they have been previously used and are known to be supported by the tool stores, will result in the optimum cutting conditions.
- It is essential to formulate a rationalised standard of supportable tooling for the whole factory, and to refer to this standard when purchasing any new tooling or machines in order for the whole system to operate effectively.

So far, the information on tool-management systems has been principally concerned with justification for its adoption by a company and the philosophy behind its usage. But what are the practical realities of setting-up and running such a system whilst simultaneously keeping track of tools in use in the various machines throughout the factory? This day-to-day running of the critical area of tool kitting will be the subject of the next section.

4.3 Methods used in the presetting of tools

In its crudest form, tool presetting is achieved on machining centres by positioning the cutter in the spindle and slowly jogging it down until it touches a setting block. The axis position is then noted, and this value is entered into the tool table. The same procedure is adopted for all tooling of a semi-qualified nature. It should be apparent to the reader that this simple method of obtaining tool-offset values can hardly be called 'tool management', since it has several disadvantages: it is labour-intensive, it ties up cycle time, it is rather inaccurate, and it sets only one offset dimension. On turning centres, the method of determining the offsets is different, but similar limitations still apply.

4.3.1 Presetting tools off the machine tool

Various methods have been developed in order to eliminate the disadvantages of tool presetting on the machine tool. Most independent tool-presetting devices are of either the mechanical, the optical or the electrical type. The simplest versions use mechanical transducers of the dial-gauge type; a major disadvantage of this method of tool setting is that the values obtained by the transducers cannot be fed into a processor for either automatic tool-table updating or hard-copy reporting. A more advanced version of the mechanical presetting technique is shown in Fig. 4.5; this gives a digital read-out of the offset by sensing the position with respect to the linear scales on the instrument. Once this displacement of the axis has been zeroed, it may be used for hard-copy recording or tool labelling, or it may be fed to the tool-offset table on the CNC. Many of the more sophisticated tool presetters tend to be of the optical type, with linear scales for position monitoring. These produce a focused image, allowing the position of the cutting edge of the tool to be determined with a high degree of accuracy.

Some users consider that it is important whether the spindle of the tool-presetting device is horizontal or vertical if the preset offset values are to be reliable. They argue that if a tool with a large unsupported overhang is placed horizontally in the tool-presetting device, the cantilever effect will cause discrepancies between the tool's preset offset value and the actual value that occurs when it is loaded into a vertical spindle. This criticism is unsubstantiated and is rather contentious; by its very nature the tooling

Fig. 4.5 A digital read-out tool-presetting device (Courtesy of Kennametal (UK) Ltd.)

must have considerable rigidity, which tends to nullify this argument! Of more importance to the operator is the method used to load heavy cutters with their adaptors into the tool presetter, as many will argue that vertical cutter loading requires less physical effort. This may seem a rather dubious argument, but having loaded and preset many heavy cutters in both spindle arrangements in the past, the author feels that it has some merit.

A typical example of the range of sophisticated tool-presetting devices currently available is shown in Fig. 4.6. In this machine a large optical magnification is possible, allowing accurate focusing in the view screen, and hence the determination of very accurate presetting values. The information and its identifying code (which may be preset using the keyboard) are recorded onto either a hard-copy printer or paper tape as in this example. The punched paper tape may then be taken to the machine tool and fed into the tape reader which will automatically update the tool offsets in the tool table of the CNC – for diameter and length on a machining centre, say, or the X and Z-positions plus the tool-nose radius in the case of a turning centre.

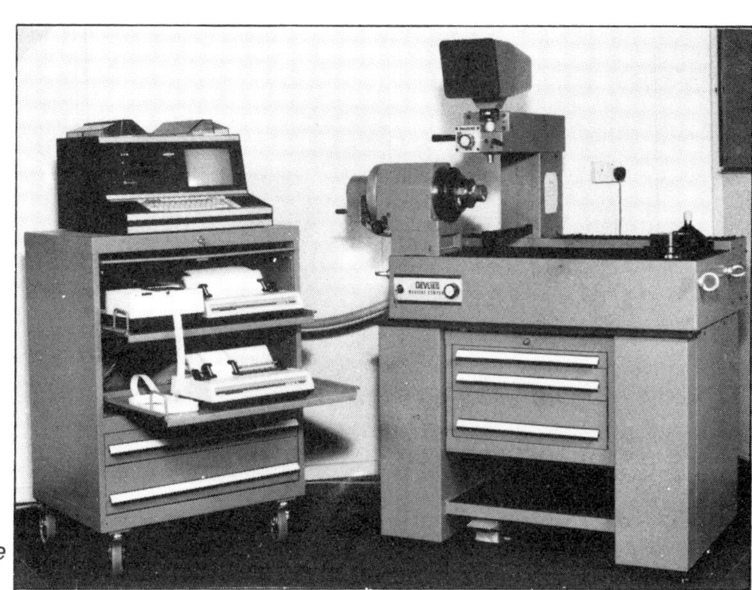

Fig. 4.6 Optical tool-presetting equipment with facilities for tool-data generation and collection and for down-loading to a CNC machine (Courtesy of Devlieg)

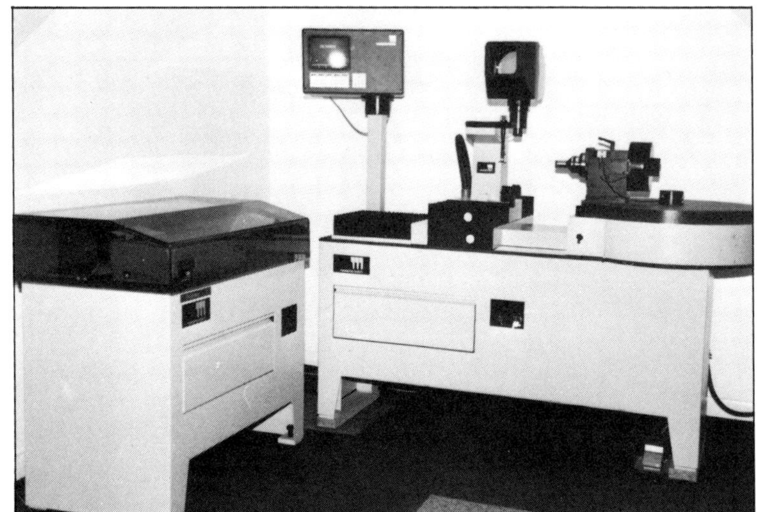

Fig. 4.7 Sophisticated tool-presetting equipment (optical tool setting) can have automatic tool loading and tool-offset updating to CNC on the machine (Courtesy of Messma-Kelch-Robot and Polstore Ltd.)

The photograph in Fig. 4.7 shows a tool presetter of considerable adaptability compared with many currently available. Again, it is of the optical type, with linear scales and electromagnetic clutches for releasing the two axes (as, incidently, had the tool presetter shown in Fig. 4.6). With this equipment, the tool management can be conducted at a very high level of 'machine awareness', as labelling can be easily employed for tool identification and a direct numerical control (DNC) link established between the presetter and the various machine tools for tool-table updating. It is even possible to arrange for the tools and holders to be automatically loaded into the presetter and then transported to the machine tools in an FMS environment, with the minimum of human involvement. In this way, the many machine tools in an FMS can be serviced with thousands of tools during the life of the system from one highly automated presetting station. This number is not hypothetical but fact: three machining centres in an FMS with additional tool magazines could easily contain upwards of 1000 tools, for example! The major advantage of an automatic delivery system to an FMS is that the tool-carrying capacity is not limited to those tools that can be accommodated in each magazine (a finite reserve) but has an inexhaustable supply that it can draw from for any machining requirement.

The advantages gained by presetting tooling off-machine encourage a fast pay-back on the investment (which might have been considerable, in the case of the more sophisticated presetting equipment). This type of equipment can only be financially justified with large tooling inventories as it represents an indirect cost and must be paid for in terms of increased machine-tool utilisation or greater flexibility in the machining operations that can be undertaken. In order to have some degree of tool-presetting capability on stand-alone machine tools – which are by far the most popular systems – on-machine presetting might be the answer, but it should be stressed that this is, at best, only 'barely' a tool-management system! On-machine tool management will be the subject of the next section, but it must be remembered that it does not achieve anything like the levels of flexibility, tool reporting and identification that can be attained with the off-line tool presetters. They are, however, considerably cheaper to purchase for new machines or for retrofitting to existing ones.

4.3.2 Presetting tools on the machine tool

On-machine tool-presetting devices fulfil a number of requirements, such as in-cycle checking and resetting of tool length and the detection of broken tools, at minimal extra cost. Typical examples of the on-machine approach on vertical and horizontal machining centres are shown in Figs. 4.8 and 4.9 respectively. Similar on-machine tool-presetting devices on turning centres were illustrated in Figs. 3.2b and 3.3b on pages 83 and 84. In both of the applications illustrated in Figs. 4.8 and 4.9, the tool sensors are used in conjunction with the machine-table movement. Tool lengths are measured and set automatically, and there is no need for operator intervention after the tools have been loaded into the tool magazine. The *X* and *Y*-dimensions can be programmed so that an offset is introduced when required, for example on milling cutters or boring tools.

Tool lengths can be checked at any time by simply programming the tool to touch the tool sensor before machining commences or before machining particularly close-toleranced features on the part, for example.

Fig. 4.8 Automatic tool-length offset updating and tool-breakage detection monitoring on a machining centre (Courtesy of Cincinnati Milacron)

Fig. 4.9 On-machine automatic tool-offset updating and tool-breakage detection on a horizontal machining centre (Courtesy of Cincinnati Milacron)

When the part program is written, a tolerance can be included in the tool-checking cycle which may be used to show up any significant variation from the previously established tool length. If the length is found to differ from its expected value by more than this tolerance, a warning message will be shown on the display screen of the CNC. This message stops the machine, and the operator is alerted to check the tool. On an FMS, however, both tool and workpiece can be replaced, to await later examination.

Typical sensor software usually utilises two special function commands to initiate the specially-developed cycles for tool-length setting and checking for tool breakage:

- *Tool-length setting cycle,* using the command 'G68', for example. This command moves the *X* and *Y*-axes to the sensor position, including the tool offset if programmed, at the machine's rapid rate. The *Z*-axis then advances towards the sensor at the fast feedrate until it touches the sensor; it then retracts, then re-approaches at the slower rate until contact is re-established. The *X* and *Y*-offsets are automatically stored in the memory along with the measured tool length, the tool retracts, and cutting can commence.
- *Tool-length checking cycle,* using the command 'G69', for example. In this case, the command moves the *X* and *Y*-axes at the machine's rapid rate to the sensor position, including the offset if applicable. The *Z*-axis advances at the rapid rate to within 5mm of the sensor, then slows to its fast feedrate until it touches the sensor, then it retracts, and then re-approaches at the same setting rate until contact is made once more. If the tool length is within the programmed tolerance, the new tool length is stored. If,

however, the tool is out of tolerance a message will be displayed on the screen so that appropriate action can be taken by the controller. Using a tool-rotation feature of programmable spindle orientation will enable the sensor to make a complete check on multi-toothed cutting tools.

This is an important feature with inserted-tooth milling cutters, as the tolerance on each insert will depend on its seating; the highest value is the offset recorded.(This is true for both on and off-machine tool presetting).

4.4 The tool store and presetting facility – a typical system

An area of the workshop needs to be set aside for the purposes of storing tools and associated equipment and for tool presetting. The advantages of having the tooling stored and preset in this manner are numerous.

Consumable tools can be tracked within the tool store, but when they go out into the machine shop, they are to all intents and purposes 'scrapped' as far as the stores are concerned. Returnable tools destined for eventual re-use can be 'tracked' around the machine shop by judicious labelling and the feedback of information to the stores. The most frequent case is that the tools are assembled into a composite form and issued as kits, which can be either consumable or returnable items. The groupings of tools in the kits are temporarily given one kit number to identify them for a particular job (i.e. as production kits) or for the eventual return to the tool store. Other advantages of such a tool-labelling and identification system are that it can be particularly useful for finding the nearest equivalent tools and that it is a means of further assistance in tool-rationalisation programmes.

Where tools are issued to operators individually, and not as assemblies or kits, the stores normally keep records to show the operator to whom they were issued, the machine on which they will be used, the number of tools issued, and the date of issue.

The advantage of having an area of the workshop dedicated to tool management is that tool kits can be made up ahead of the master schedule, so that they are ready for collection just before the manufacturing commences. A result of this feature is that the lead times are reduced, which is of prime importance to the company.

It is often the case that *all* tooling is assembled in the tool stores. This means that limit gauges, etc., as well as jigs and fixtures, are also the responsibility of the stores personnel. When this is the case, total packages are issued, containing cutting-tool kits and work-holding kits, together with the gauges necessary for metrological assessment. A tool store with responsibility for all these tooling aspects is the 'focal point' of the machine shop for all matters to do with tooling, whether they be the breakdown of previous kits, the issue of new kits, the calibration of gauges, or even purchase requirements for the latest tools available. Therefore the specialist personnel in the tool stores have a considerable responsibility in servicing the shop. They amass a large working knowledge of production tooling requirements, and their opinions should be sought before any new tooling purchases are decided upon.

An example of a well-planned tool store within a machine shop, with its presetting equipment strategically positioned, is shown in Fig. 4.10. The photograph shows how all the tools are held neatly in cabinets and are functionally laid out, so that the minimum of effort and time is spent in locating parts and assembling the tool kits. It is important for a

Fig. 4.10 A universal tool store and presetting area of some sophistication (Courtesy of Polstore Ltd.)

company to set up a facility such as this and not just consider it as an indirect cost which affects the profits, because it has been shown that the proper supervision and maintenance of tooling increases a company's profitability and helps to organise and assist the flow of work-in-progress through the shop. A result of setting up an efficient tool store is that items such as holders and cutters are less prone to damage and can therefore function at their optimum performance for much longer. When companies consider the money tied up in their tooling, which is considerable, they will see that it is worth looking after correctly!

Once a tool kit has been assembled and allocated to a particular operator and machine tool, it should be delivered with efficiency and speed. Tool 'taxis' are a good method for efficient tool delivery (Fig. 4.11). These can carry a large range of tooling to the machine, together with any ancilliary equipment necessary for the job, and allow for delivery with total tool protection and the minimum of effort.

4.4.1 Machine-tool spindle cleaning

As mentioned above, a company has substantial amounts of capital tied up in its tooling. It is important not only to maintain the tooling accordingly, but also to ensure that even such items as the machine spindles remain in good condition. An automatic taper cleaner has been developed for this purpose (Fig. 4.12a.). It can be held in the tool magazine of a machining centre or loaded manually into a CNC milling machine's spindle. Small particles of foreign matter are very adhesive and can adversely affect the spindle's accuracy (Fig. 4.12b); they can only be removed by solid means. Two basic types of automatic taper cleaners are available to remove this unwanted debris from the spindle; they differ in their drive systems, which use either an inertia mass system or a compressed air system.

The inertia mass system is the simpler technique, and it may be adapted to all types of milling machine spindles because the energy required for cleaning does not need to be supplied separately. The compressed air system, however, requires compressed air to be supplied through the machine's spindle hole at a pressure of at least 6 bar to enable the compressed-air motor to fulfil the cleaning requirements.

Fig. 4.11 A tool 'taxi' for tool-handling efficiency on machining centres (Courtesy of Kennametal (UK) Ltd.)

Both types of taper cleaners can be loaded into the tool magazine, and the cleaning process can be entered as part of the CNC program. The cleaning process may be programmed to occur before or after a specified operation, using programming instructions such as the following: *

t tool or location number
M06 tool change
G04 dwell time (approximately five seconds)
S rotational speed (approximately 1000rev/min)
M03 direction of rotation
M19 spindle orientation

Fig. 4.12a shows how an exchangeable taper sleeve (which is made of chamois leather) is used with an automatic taper cleaner of the inertia mass type. The cleaner rotates in the spindle owing to the presence of the inertial mass and the drive system, and this allows the taper to be efficiently and speedily cleaned. Cleaning produces a positive and accurate location for the tool-holder adaptors in the machine spindle, ready for cutting operations to restart once the automatic taper cleaner has been deposited back in the tool magazine.

*The instructions will obviously vary according to the CNC control system used, and the information needed could differ slightly for different machine tools.

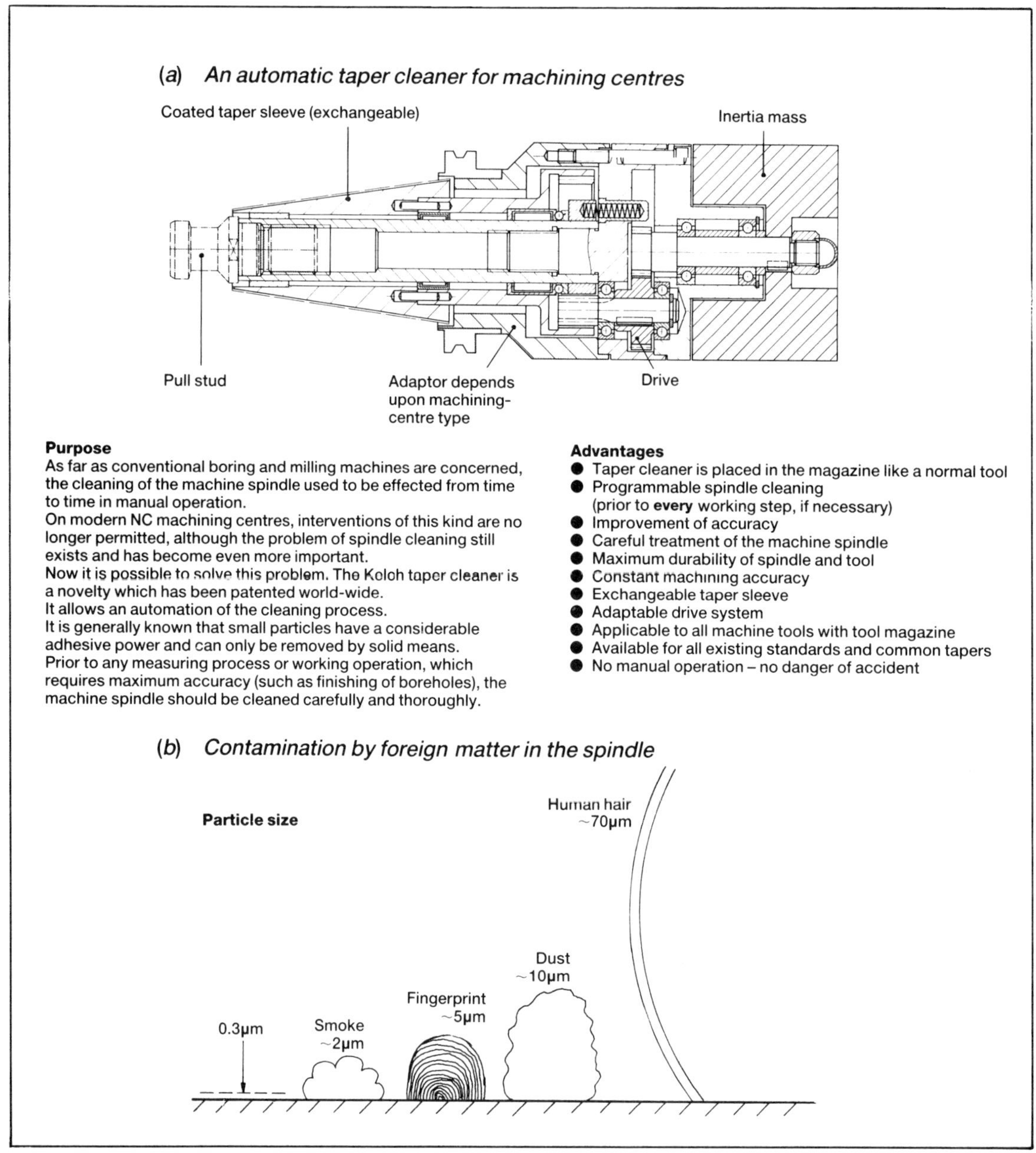

Fig. 4.12 Automatic spindle cleaning (Courtesy of Polstore Ltd.)

4.5 Tool-recognition systems – using 'intelligent' tooling

The growth in the range of tooling available for turning and machining centres in recent years has meant that there is a great diversity of cutting units and their adaptor configurations. Compounding the problem of the tool ranges available for metal-cutting operations is the large number of tools being used in the machine shop at any one time. Trying to keep track of the tooling movements is bad enough even with a sophisticated tool-management system with stand-alone machines, but if a complex FMS system is also in use, the demand for instantaneous tool delivery compounds the problem of identification, and it becomes immense.

In order to tackle the problems caused by the increasing level of automation and flexibility of approach to cutting-tool requirements, whilst satisfying the demands for tool security and for a tool-management system that is in control, tool-recognition systems have been developed in recent years. These so-called 'intelligent' tools carry discrete information in a coded form within the cutting units; this information might typically include the tool-offset, insert geometry, cutting insert material and an identification code to ensure that the correct tool is then automatically selected and inserted in the holder or spindle.

These 'intelligent' microchip memories, located on the tool are in the form of small integrated circuits, typically measuring only 4mm in length and 12mm in diameter. The miniaturised hybrid circuits on these small semiconductor chips can contain up to 512 bits of information, which is enough for even the most demanding and complex cutting applications.

These microchip capsules can be classified into several types and configurations; they may be either 'read-only' or 'read/write', and both of these types may be found in contact and non-contact varieties. Of late, most companies have favoured the read/write type, and of these, the non-contact versions are the most popular. The non-contact programmable tool systems sense and read the identification data across the air gap by electrical induction or a similar transmission method. The time taken for data reading and transmission is approximately 300msec, and the chip memory is of a non-volatile type. The information may be written onto the chip by a special computer, which may also be linked to other peripherals, such as a stand-alone computer or a local area network system, through an RS232/422 port or similar. This effectively increases the memory of the programmable chip so that it becomes almost inexhaustible, with billions of possible tool-code combinations. The chips have a high memory stability in the oil-laden and adverse environmental conditions which are present in the cutting area.

These non-contact programmable identification systems are very tolerant in terms of the sensing distance and any misalignment to the read head. For example, the transmission of data is secure up to a distance of 8mm with a permissible misalignment of up to ±5mm, together with a relative motion of 600mm/sec.

Fig. 4.13 shows an example of the contact type of programmable cutting units. These have read/write EEPROM data carriers, which contain large quantities of tool and machining data that can be accessed either manually or automatically. The operator-supervised manual insertion of data into the machine controller's memory via the memory chip in the cutting unit can be performed at the tool-presetting stage before the tool is loaded into the machine; this method eliminates tool-changing errors. Alternatively, in the fully-automated systems, the read cycle can be performed automatically whilst the tool is in the storage unit and before it is mechanically loaded into the machine tool. The coupling of the microchip and sensor head is very secure, and any foreign matter present during cutter-unit changeovers is not usually a problem. Thus, the security offered by both the contact and non-contact systems eliminates the need for further tool gauging in the machine tool and, for that matter, for measuring cuts to be taken and tool offsets modified as a result of these trials.

The read/write facility of the data-carrying implants in each cutting unit enables it to store, receive and output tooling information both during the presetting of the tool and after installation in the tool-storage unit of the machine. Data can be automatically read and written, in a two-way dialogue between the CNC and the tool, without the need for supervision by another computer. Any item of information retained by the cutting unit's implant can be read or overwritten at any stage during tool preparation, or in the machine. In this way the system provides an accurate, secure and efficient means of tool control and management.

Fig. 4.13 The read/write coding system for Block tool cutting units (Courtesy of Sandvik (UK) Ltd.)

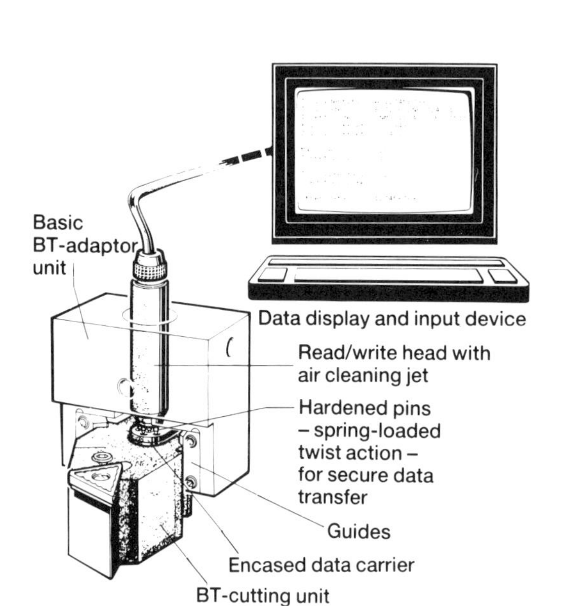

(*a*) *The read/write function of coded Block tools*

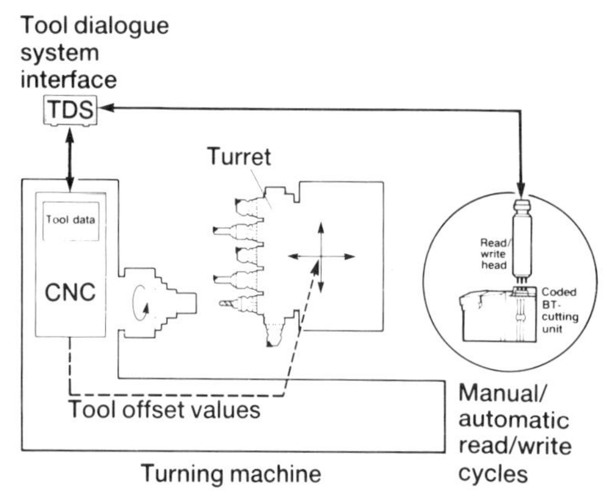

(*b*) *Tool data can be automatically down-loaded into the machine CNC before changing tools*

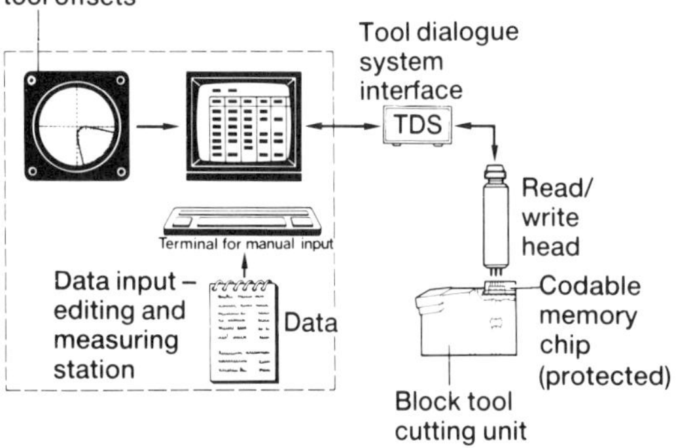

(*c*) *Writing data into the Block tool memory chip during tool preparation*

The coding system for the contact read/write programmable cutting unit shown in Fig. 4.13 comprises a read/write head and interface unit connected to the editing and measuring station, or to the machine's CNC unit. Fig. 4.13a shows how the data can be input and displayed on a compatible microcomputer and then down-loaded via the read/write head to the encapsulated data-carrying chip. Once the cutting unit is situated in the tool magazine (Fig. 4.13b), the tool data is read from each coded cutting unit by the automatic read/write head; alternatively, it may be input by a hand-held unit. All the data is processed by the CNC or programmable logic controller (PLC) which, when suitably programmed, ensures that each tool is inserted into its correct magazine position, and that the correct tool offsets are used when it is loaded, etc. Any of the data-carrying capsules in the cutting units can be updated in the machine by the CNC for general tool-management purposes.

Data can be transmitted to the non-contact read/write programmable cutting units in a similar manner (Fig. 4.14a). The down-loading of the information to the programmable chip in the tool is shown schematically in Fig. 4.14b; this information held in the chip can later be transmitted to the machine's CNC. A typical sequence of events in programming these read/write chips is as follows:

- After suitable tool assembly and presetting, the information is input through the keyboard of a data manager (which is an intelligent microprocessor-controlled system capable of communicating with all the machine tools and peripheral equipment).
- The converter processes the data, checking and evaluating the signals received from the data manager, and coding it into the form required for input into the programmable chip.
- The processed data is passed across the air gap to the tool's programmable chip, via the aerial in the read/write head.
- When the tool is loaded, the information held in the non-volatile memory of its programmable chip is ready by another read/write head, for example on the automatic gripper (Fig. 4.14c).
- The converter decodes the information from the gripper's read/write head and transmits it through the data manager to the machine's control unit.
- The cutter is placed into the spindle and can now be used with the confidence that the information initially programmed into the data manager is now being used to machine the workpiece.*
- Once the machining has been completed, the machine control unit returns the updated values to the data manager, including information on the tool life and tool condition, for example.
- The data manager transmits the new information to the programmable chip in the tool via the converter and read/write head, updating its information to the present status.
- The tool is stored in the magazine until required once more.

These programmable chips can be discretely positioned in the holders (as shown in Fig. 4.15) or in the cutting-unit heads in modular quick-change systems. They are fully encapsulated with high protection from swarf and coolant, and there are no external features for chips to snag on whilst cutting. The photograph (Fig. 4.15) shows how they are mounted on the cutting unit for turning or machining-centre applications. The chip implants can be seen in Fig. 4.16, where it can be appreciated how small they are. The read heads are also very compact, which means they can be situated as required at any

*This form of sophisticated tool-management is ideal for any 'lights-out' (i.e. unattended) machining on an FMS, ensuring that the correct cutter is being used to machine the right feature on the workpiece.

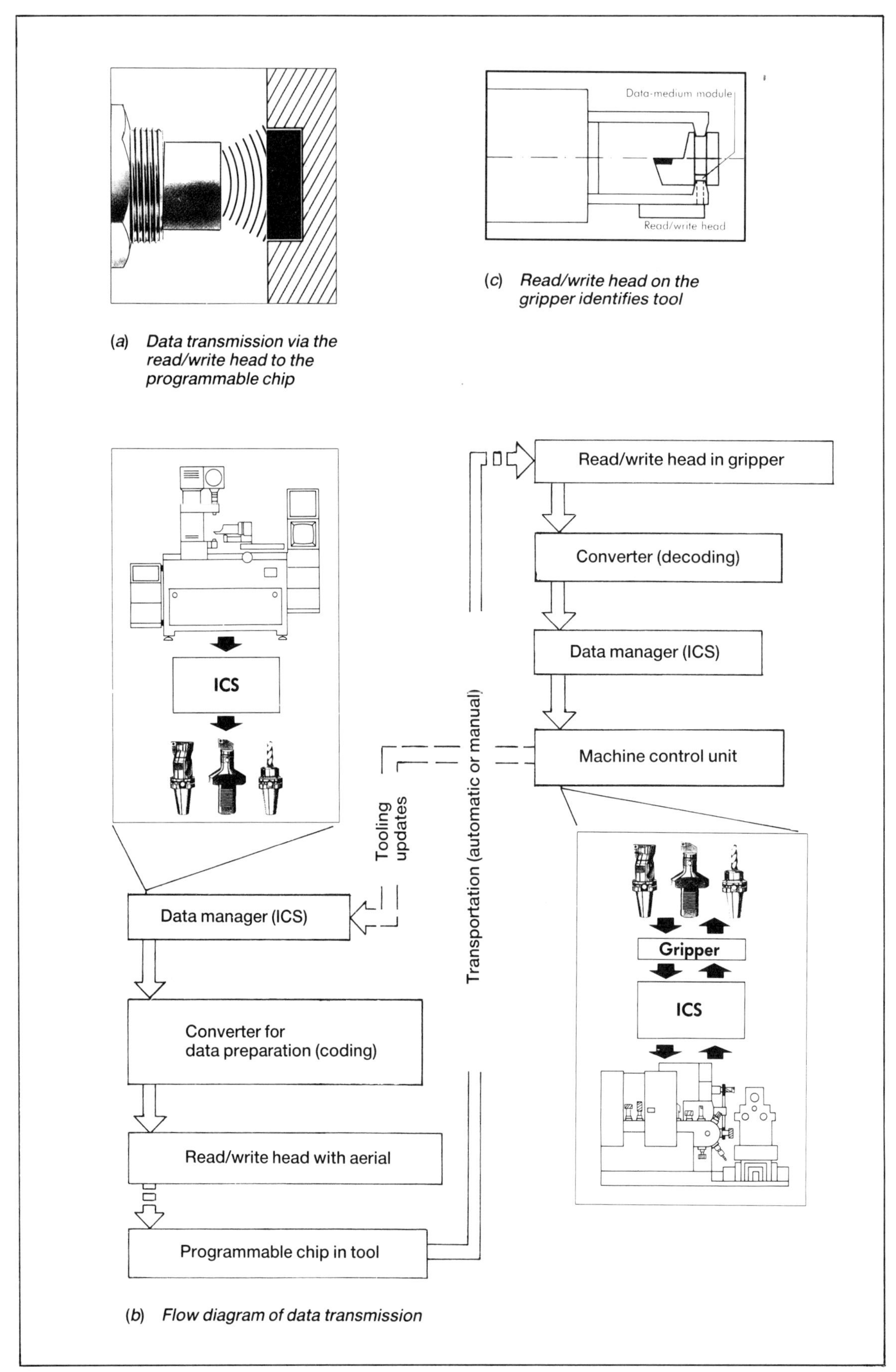

(*a*) *Data transmission via the read/write head to the programmable chip*

(*b*) *Flow diagram of data transmission*

(*c*) *Read/write head on the gripper identifies tool*

Fig. 4.14 Tool identification using programmable chips (Courtesy of Krupp Widia)

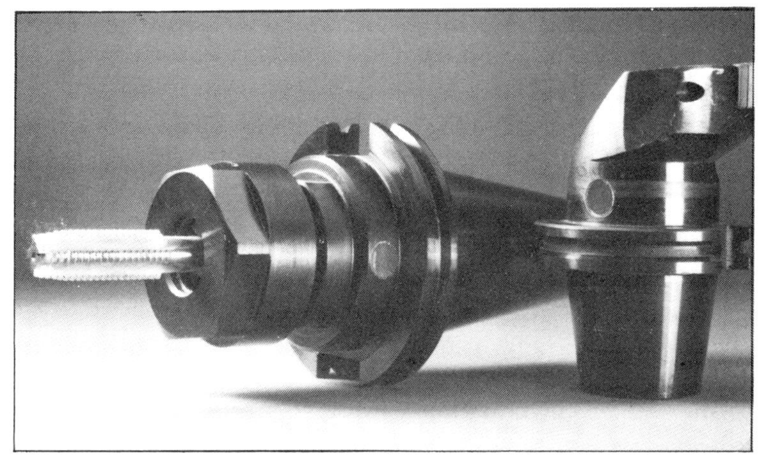

Fig. 4.15. Tooling with Microlog tool identification system using code-carrying capsules (Courtesy of Kennametal (UK) Ltd.)

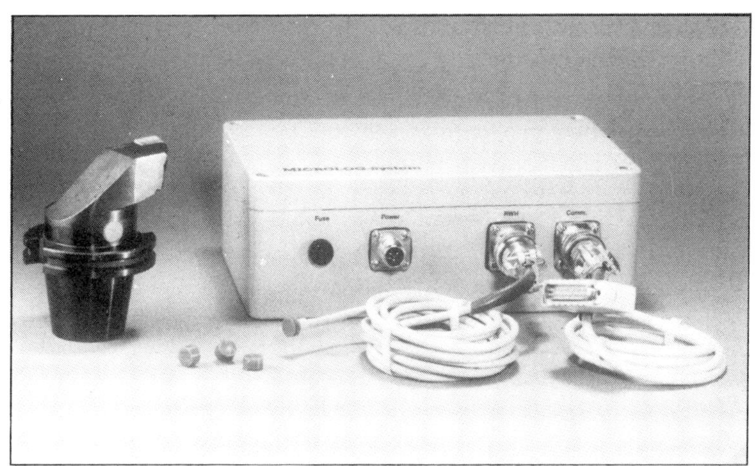

Fig. 4.16 The Microlog tool-identification system: the read head, the read unit, code-carrying capsules and a typical installation of a capsule in a tool (Courtesy of Kennametal (UK) Ltd.)

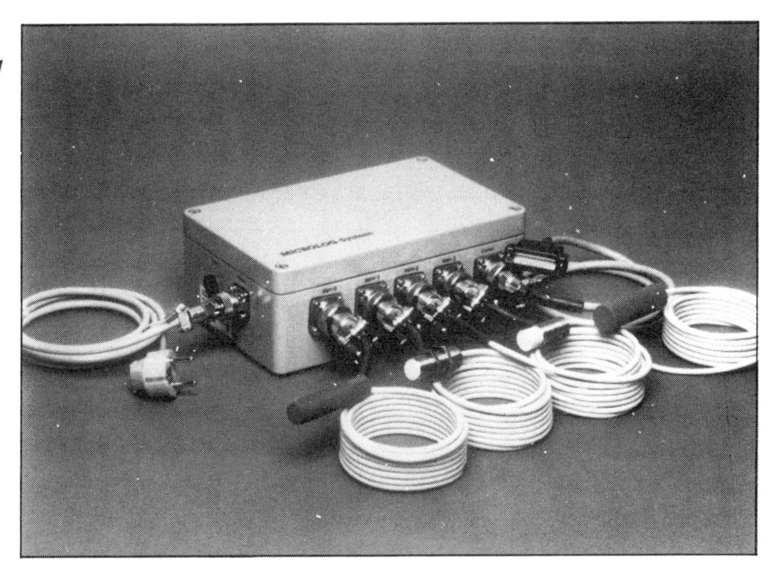

Fig. 4.17 The identification reader unit and read heads of the Microlog system (Courtesy of Kennametal (UK) Ltd.)

position on an automatic gripper's end-effector, etc. In fact up to four read heads may be accommodated by each system box (Fig. 4.17). This number may be increased by adding further boxes as necessary. Fig 4.17 also shows the RS232 connection for up or down-loading of data to peripheral devices, as mentioned above.

It should be clear to the reader that the security provided by using these programmable cutting units, either in modular quick-change form or with the conventional arrangement of cutter and adaptor, gives a whole new meaning to the term 'tool management', and that their use should be encouraged in any highly-automated manufacturing organisation. Their cost is minimal when compared with the benefits gained through their use, and even this cost should soon be recouped. Even when a company has stand-alone machines or conventional machine tools, there are occasions when it might be prudent to include such an 'intelligent' programmable tooling system in the company's tool-management scheme.

4.6 Computerised tool management

Any conventional paperwork tool-management system eventually reaches a limit in the amount of tool data that can be handled. These conventional systems are fine for companies that have a few stand-alone and conventional machine tools, but there is a real need for a more dynamic and interactive system, with a larger capacity for tool data, in companies using highly-automated machines or with a large tool capacity. A large industrial facility may often have literally thousands of tools to keep track of, and this is made worse by the need to control work-holding tooling and other equipment in order to keep production running smoothly. Once this level of tool control and accountability is reached, it is necessary for a company to invest in a tool-management system which is computer-based. The potential of such a system can be appreciated from the diagram in Fig. 4.18, which shows the interactions that might occur in an extreme case, a fully-automated computer-integrated manufacturing (CIM) facility. The benefit to the whole company should be obvious from the number of interested departments whose names appear on the diagram, from process planning through to machine-process monitoring: as one can see, it involves the whole manufacturing environment. At any instant, the status of the tooling throughout the whole production facility can be monitored, and hence controlled, by the personnel responsible for the tool-management system.

Tool management of this complexity and level of sophistication can equally well be applied in a conventional machine shop with little or no CNC machine tools, but with a large tooling range to keep track of and service. The vast amounts of paperwork generated by these shops simply to control the tooling on a hit-and-miss basis can be costly; and often the personnel involved are never quite sure whether they control the tooling system or it controls them!

Invariably, crises are reached, the problems are eventually sorted out, and the company 'staggers on' until the next crisis arises. Where conditions like this occur, a computerised tool-management system must be the answer to produce a smooth, controlled and directed tool flow throughout the company. Investment in such a system should bring immediate benefits to the company, in terms of increased production rates through efficient tool distribution and monitoring, reduced tooling inventories and more efficient utilisation of personnel. So, what are these computerised tool-management systems, and how do they operate? These questions will be answered in the next section, which discusses one of the popular systems currently available. It also briefly describes the general operating procedure and gives a schematic representation of another system.

4.6.1 The TOMAS computerised tool-management system

An example of a typical software-based tool-management system, known as TOMAS, is illustrated in Fig.4.19. The system has been designed to meet the needs of many different types of company, and has built-in password security at all levels. Every terminal has

Fig. 4.18 Tool management in computer-integrated manufacturing (CIM) (Courtesy of Sandvik (UK) Ltd.)

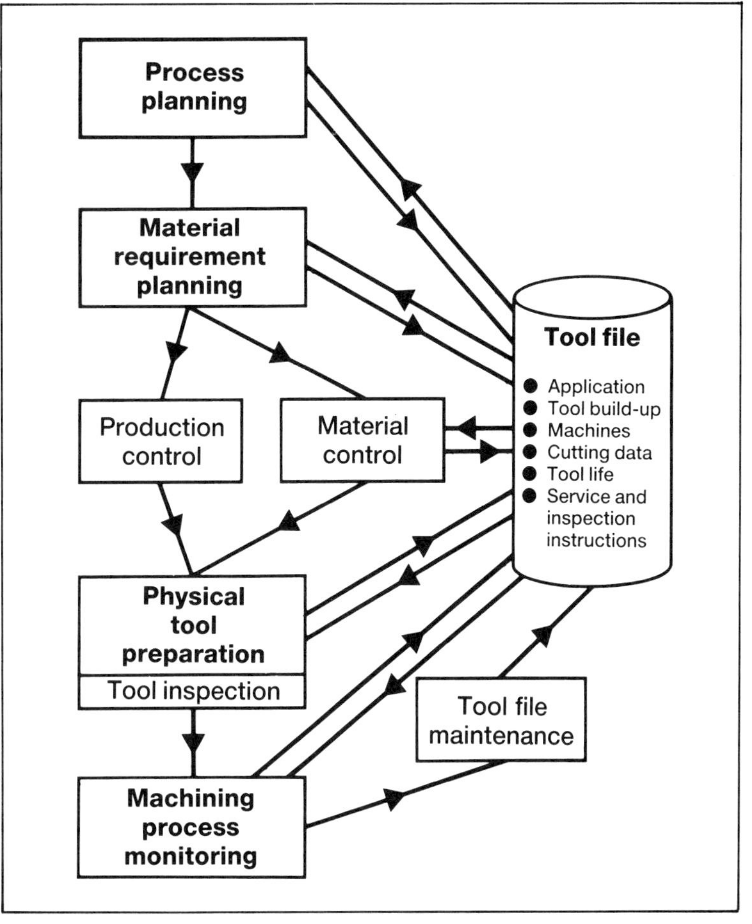

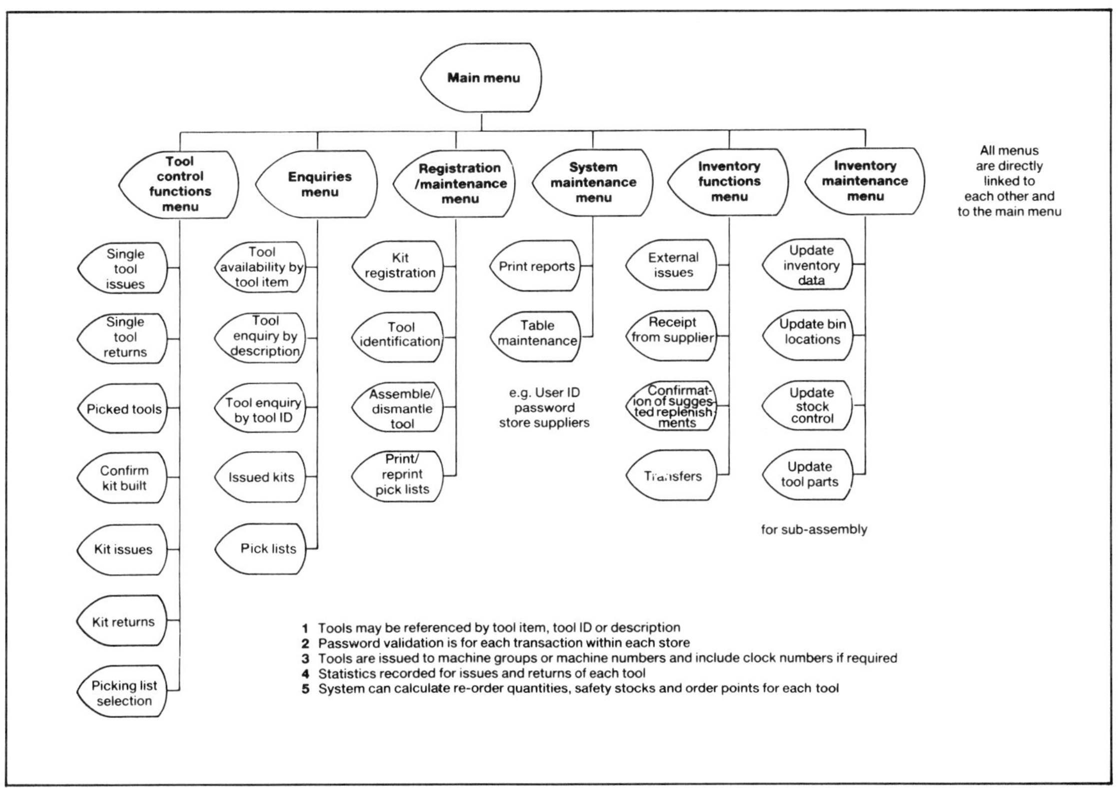

Fig. 4.19 A software-based tool-management system (TOMAS) – an overview chart (Courtesy of Sandvik (UK) Ltd.)

access to 'help' facilities and, to use a much abused term, the system is 'user-friendly'! This is an important characteristic, since most of the workforce who use the system will not be 'computer-literate' and will need guidance to gain the full potential from such a computer-based system. The hardware is conventional equipment, not a hybrid, so that the terminals are relatively inexpensive and sufficient of them can be purchased to provide the multi-user environment which is required in a factory.

The software works on the principle of using a series of menus that list the choices that can be made at any point:

- The main menu lists the major functions of the system and is used to gain access to the different sub-menus.
- The sub-menus list information relevant to the item chosen from the main menu – the tooling needs for example.
- Further sub-divisions are used for the actual tooling data.

There is full interaction between the menus so that changing the details in one of them has the effect of editing information related to this change in the other sub-menus throughout the tool-management system. This interaction is vital to the ability to make changes in real time, as needed in the operation of such a system.

Looking at the 'issues and returns' transactions for either a unique tool, an assembly or indeed a kit, the TOMAS system allows the tools to be issued either to a group of machines or against an employee's clock number. On its return to the tool store, the tooling is available for inspection, calibration, scrapping or simply booking-in for further use. Each tool has a unique coding, which may be used to gain access to its data; alternatively, a description or a tool item number may be used.

In many of the expensive and specialised machine tools now available it is becoming commonplace for the tooling to be replenished by self-changing magazines; this has the effect of potentially minimising downtime. To capitalise on these total tool-replenishment systems, the TOMAS software allows all the tools, both in the stores and on the machines, to be under the same stock-control disciplines, with automatic prompts for replenishments.

To produce the optimum level of tooling inventory, which avoids tool stock running out yet keeps stock levels small, the tooling availability must be predictable. This is the essence of any good tool-management system. To create the climate for constant and sustained tool delivery on the one hand, yet minimal tooling on the other, computerised tool management is essential. These systems remove the problem of over-use of tools, and the uncertainty of their return, by maintaining a comprehensive tool-identification system and stock-control appraisal.

Kits and assemblies may be identified and issued to specific jobs or machines by the software. The efficiency of the system is maintained through software generation of picking-tool lists which, in turn, should ensure that the correct tool is delivered to the machine. The basis of efficient tool management is: the better the tool environment is controlled, the higher the potential for preparation to be done correctly and in a non-critical timeframe.

An important feature of any tool-management system based on software is its ability to be interfaced to other systems already used within the company. It must obviously be interfaced to any process-planning software already in use, so that kits of tools are identified early on, and the process can be started at the earliest availability of the machine tools and materials. This allows the job to be scheduled into the overall workload in plenty of time.

Such a computerised tool-management system will provide the management with online access to up-to-date information covering all tools, both in the stores and on the shop floor. This provides a powerful mechanism for imposing management control through summaries and detailed reports giving current and historical data on the tooling investment, consumption, scrap levels and re-order quantities, for example.

Another sophisticated computerised tool-management system is Toolpro/Toolware, which is shown schematically in Fig. 4.20.

4.7 Tool location and inventory rules

One of the key elements in establishing the basic framework for a tool-management system is the tool location and inventory method established in the tool-preparation area. There are a few basic commonsense rules for setting up a tool-control system:

- Control the access to the stores.
- Organise the shelves and cabinets that are to store the tooling, giving them location names or tags, and arranging them in some logical sequence, either numerical or alphabetical (or both).
- Insist that all tooling issued or returned is recorded in the system.
- Involve the storekeeper from the outset of the tool-control project, as they know more about how the tooling is issued than anyone else in the company and can play a crucial part in providing critical information in the establishment of a successful tool database.
- All stores personnel must be trained. There should be at least two competent systems administrators for overall system control, and when running shifts there should always be someone on hand who understands the overall system.
- Guidelines must be established for returning tools to the stores for rework consideration, as scrap, or as new or used tooling.
- Purchasing procedures should be reviewed: now that the tooling is under control, piecemeal ordering should be the exception rather than the rule.
- Rules should be established for determining the different categories of tooling, possibly as follows:
  Perishable This refers to those tools that can be consumed by use, e.g. drills, endmills, taps, carbide inserts, etc.
  Durable This refers to those tools that are normally not consumed by use, e.g. tool-holders, collets, micrometers, fixtures, power tools, etc.
  Returnable This refers to those tools that are expected to be returned to the stores after use. For example, a perishable returnable tool could be an endmill that could be reground and re-used, and a durable returnable tool could be a set of micrometers or a tool-holder that requires storage in the stores after use.
  Non-returnable This refers to those tools that are not required to be returned after use, and to tools that are permanently assigned to a machine, department or employee. Examples of perishable non-returnable items could be a small-diameter drill or gloves. Durable non-returnable tools could be special tool-holders or fixtures assigned to a specific machine.
- Rules must be established for re-ordering the perishable and durable items, possibly as follows:

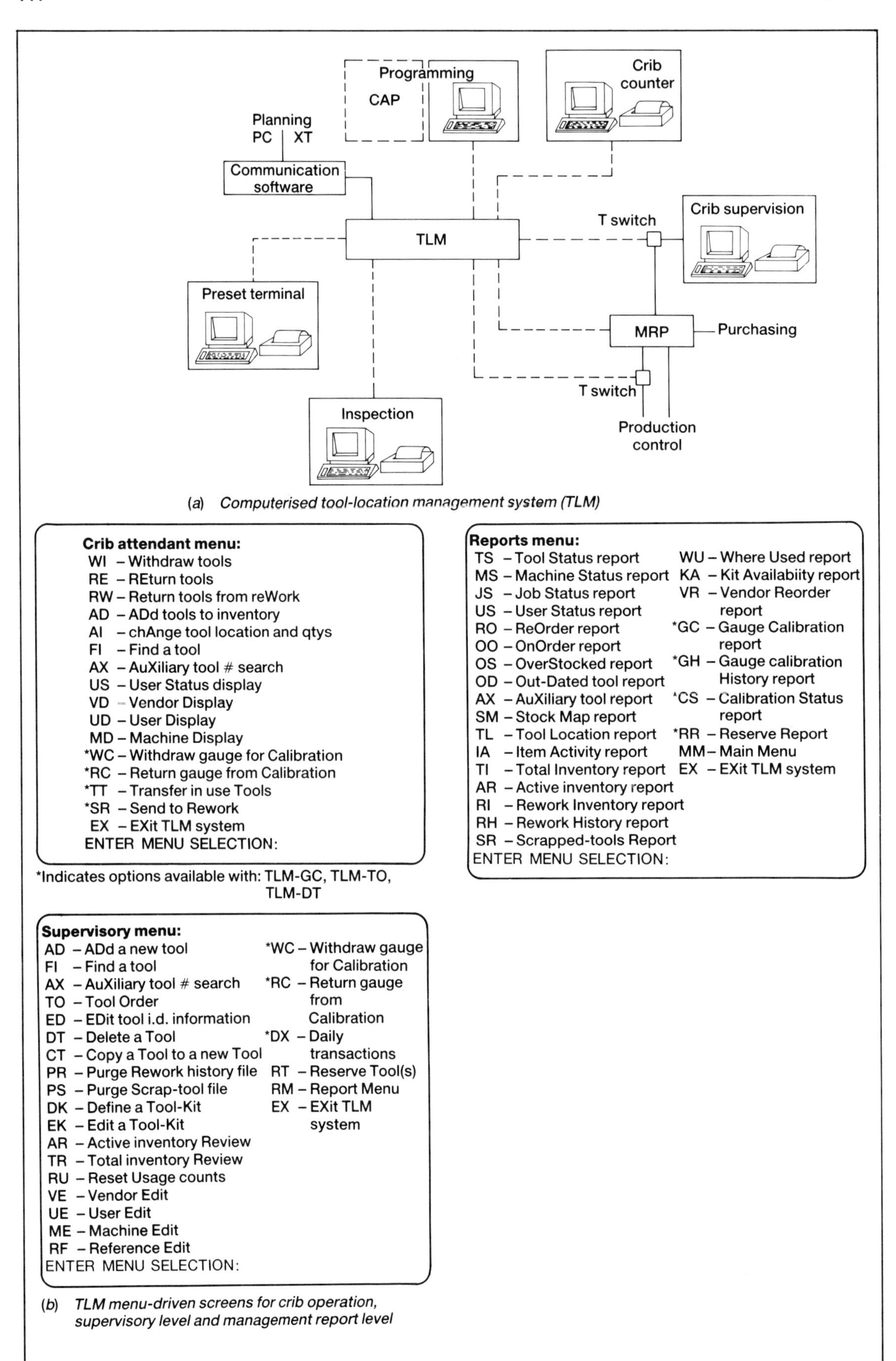

Fig. 4.20 Tool management with advanced software capabilities (Courtesy of Toolpro/Toolware)

For perishable tools When the total number of tools in the inventory is below, or equal to, the defined minimum quantity. In this case, used tools in the store are included in the inventory; tools in use are not usually included as part of the available inventory.
For durable tools When the total available (on-hand) number is below, or equal to, the defined minimum quantity. Both tools located in the stores and those on the shop floor are included in the total.

The comments above are suggestions for ways of establishing control in the stores and for obtaining data for a tool-management system. The final section in this chapter lists the benefits that a company can obtain by employing a tool-management system.

4.8 The overall benefits of a tool-management system

Through the correct implementation of a basic and competent tool-control system, the following benefits can be obtained:

- Manpower is conserved and training requirements minimised.
- The number of tools lost or misplaced is reduced.
- Timely and up-to-date information on tool usage is produced.
- Tool-inventory shortages are identified and prevented.
- The accuracy of the inventory is improved.
- Space requirements and overheads are reduced.
- Inventory levels and excess purchasing are reduced.
- Time spent on re-ordering, etc. and piecemeal purchasing are reduced.
- The record-keeping functions are consolidated.
- Tooling can be tracked and its availability within the shop monitored.
- Tools in rework can be tracked.
- A record of scrapped tools can be kept.
- The value of the total tool inventory and of tools in use can be assessed.
- Machine set-up, tool-return and withdrawal times are reduced.
- It is possible to pinpoint machining or over-use problems, by the employee, job, machine or department.
- The gauges and fixtures are identified and tracked.
- Obsolete tooling is identified and eliminated.
- Improper charge-outs, pilferage and losses can be minimised.
- It is possible to take advantage of tool kittings, presetting, storage techniques and other FMS concepts.
- It is possible to incorporate existing tool numbers and the current mode of operation into an automated system without making radical changes.

A tool-management system provides these benefits by allowing the operations to be easily reported, analysed and corrected, and allows for timely decisions concerning tools with the minimum of manpower and operational changes. So that the information required by a company can be obtained, the system should be organised to allow personnel responsible for the tools to record their activity. Two basic groups, or levels of responsibility, are desirable to provide the information about tooling which helps a company; the stores personnel and tool-supervisory personnel.

A company usually begins its tool-control system with a paperwork or simple computer system. This is developed further with more familiarity, and inspires confidence in the accuracy of the company's tool-control system reports. At this stage a decision is usually taken to explore other uses of the system, such as the control, tracking and recalibration of gauges and inspection tools, the ability to place tools on a 'reserve' or 'advance-need' status, the ability to track fixtures and dies, plus the replacement of complete tool assemblies and magazines, with interfacing to tool presetters and FMS control systems. At this point it is realised that the company needs to expand the tool-control system to allow for the simultaneous input of tool information by the store counters, supervisors and planners, and for the retrieval of reports. When this state of interaction is reached in a company, then it means that up-grading to a multi-user system is necessary in order to gain the full benefits of a highly efficient tool-management system as listed above.

The following chapter is included because cutting fluids play an integral part in the behaviours of cutting tools. The chapter discusses the types of cutting fluids available and likely applications of specific fluids. Further discussion of the role played by cutting fluids in determining the life of cutting tools and on methods of obtaining longer tool life and better work finishes, will be given in the final chapter, on machinability and surface integrity.

Chapter 5
Cutting and grinding fluids

The vast majority of machine tools involved in metal-removal operations use cutting fluids to assist in the cost-effective production of components to the required quality standards. Unfortunately, many machine-shop managers and production engineers view cutting fluids as something of a 'necessary evil', owing to problems associated with fluids in their machine shops. But it need not be this way. Correct selection, application and control of the cutting fluid will result in its becoming an integral part of a cost-efficient, well-organised and safe machine shop.

The purpose of this chapter is to examine cutting and grinding fluids – as grinding applications can also be done (with grinding attachments) on turning and machining centres. The use of fluids in practical engineering applications will be considered, to provide a better understanding that may turn a 'necessary evil' into a 'valued ally'.

5.1 Why use a cutting fluid?

Before getting involved in any sort of detail, it is perhaps sensible to consider exactly *why* almost every metal-removal operation carried out uses some form of cutting fluid.

There are two prime functions for any cutting fluid: cooling and lubrication.

Cooling Cutting-tool wear-rates have been shown to be heavily temperature-dependent; higher temperatures mean higher wear-rates. Simple cooling of the tool can significantly extend tool life.

Cooling of the workpiece is also important. When producing precision components the possible thermal effects cannot be ignored. Excessive heat can result in out-of-tolerance, or even distorted, workpieces. On a more practical note, where components are to be handled (unloaded) by operators there may well be complaints if the components are too hot!

The introduction of more sophisticated swarf conveyors and removal systems has contributed to increased metal-removal rates by effectively removing swarf from the machine tool. One tends to consider only the visible volume of the swarf and to forget that around 80% of the energy used in metal-cutting becomes heat in the swarf. Effective cooling of the swarf and its rapid removal from the machine ensure that this heat is not transferred into the machine-tool castings, causing thermal distortion and affecting precision.

Lubrication Lubrication has a beneficial effect in the reduction of 'built-up edge', especially when cutting the more ductile materials. In addition, by lubricating the passage of the chip over the rake face of the cutting tool, both frictional heat generation and tool wear-rates are reduced.

An equally important effect of lubrication between the cutting tool and the workpiece is to improve the surface finish of the machined part.

Other effects In addition to the cutting fluid's two primary functions, of cooling and lubrication, there are also a number of secondary functions. In some operations, it is important that the cutting fluid flushes swarf away from the work area into the removal system or tray. Also, some materials produce very fine swarf or dust which can be unpleasant, or even hazardous, to the operator. Here, flooding the machining zone with a cutting fluid acts to damp down the dust, keeping both the air and the machine's surfaces cleaner.

5.2 The 'ideal' cutting fluid

So far, the reasons *why* metal-removal operations require a cutting fluid have been discussed. Having accepted that a fluid is required, one must ensure that the fluid achieves its purpose, and that it does not cause any other problems. These conditions imply that there are many characteristics that the 'ideal' cutting fluid should have.

Optimum cooling and lubrication Clearly, the ideal fluid would have the optimum cooling and lubricating properties, to ensure the best cutting performance as measured by production rate, tool life and surface finish. But there are many other factors to consider.

Acceptability to the operator Not least amongst these other factors is acceptability to the machine operator. Depending on the machine's design, the machining operation and, indeed, the amount of care taken, all operators will have some degree of contact with the cutting fluid. Operators will consider the lubricant's overall performance, but even when the fluid is perfect in every other respect complaints are likely if, for example, the smell is unpleasant. The following features are likely to be of particular interest to the operator:

- *Smell* Ideally, the fluid should have no perceivable odour, but certainly it should not be objectionable.
- *Colour and clarity* Most operators prefer products which are perceived to be 'clean and fresh' throughout their life, and some prefer dye-coloured translucent products for this reason.
- *Misting* High-speed cutting and, more particularly, grinding operations tend to generate a mist. Occasionally these mists may be associated with dry throats or stinging eyes, and may lead to complaints. Although misting is largely dependent on the machine, operation, ventilation, etc., different fluids have different misting characteristics, and the ideal fluid should be non-misting.

- *Irritation to the skin and and eyes* Cutting fluids are sometimes associated with skin and eye irritations – itching, rashes, swelling, stinging, etc. Once again, fluid formulations which are 'kind and gentle' are preferred.

Long 'sump life' All machining fluids have a finite life. At some point the machine's cutting-fluid system must be completely emptied, cleaned, flushed and refilled. There are many reasons why the fluid might be regarded as 'dead' (see section 5.6), but obviously its life is an important economic consideration in terms of fluid usage, labour costs, down-time, etc. However, it is unwise to aim for the longest possible life without considering the implications. As the fluid approaches the end of its useful life, it becomes more likely to cause problems (e.g. smell, corrosion, operator complaints, etc.). It is always wise to 'service' the fluid so that these problems, which may well cause unplanned down-time to change the fluid, are avoided. Having voiced these warnings, there is no doubt whatsoever that some cutting-fluid formulations are capable of achieving significantly longer 'sump life' than others, so enabling planned maintenance over scheduled shut-down periods.

Corrosion protection All cutting fluids are formulated to provide corrosion protection to the machine tool and the workpiece during, and for a short time after, the cutting operation. Ideally, fluids should maintain their corrosion-protective properties throughout their useful life to avoid the potentially expensive problems of rusting the machine tool or producing rusty products liable to be rejected later by the customer.

Low foaming properties Some modern machine tools incorporate fluid systems that agitate the cutting fluid to such an extent that a foam spills over onto the floor. The ideal fluid will withstand swarf-washing jets, high-pressure fluid delivery, centrifuges, etc., even when prepared with the softest water supply.

Machine-tool compatibility No engineer wants to see their new (or, come to that, their old) machine tools being attacked by a cutting fluid. The ideal fluid should therefore have no detrimental effects on paint finishes, seal materials, etc.

Workpiece compatibility A wide range of workpiece materials is available to the design engineer. Ideally, it should be possible to machine all of these materials without the need to change the grade of cutting fluid at every material change designated to be necessary by a research and development department. Non-ferrous metals and their alloys are particularly susceptible to staining. In machine shops where many different alloys are cut, there is a real need to have a 'versatile' cutting fluid capable of handling all, or most, of the materials machined.

Water-supply compatibility Water-soluble cutting fluids should ideally be capable of being diluted with any water supply. Geographical variations in water condition, especially in water hardness, can be quite considerable. The ideal cutting fluid would therefore not cause the typical problems of foaming in soft waters or forming insoluble soaps in hard waters.

Freedom from tacky or gummy deposits As water-soluble fluids dry on a machine or component surface, the water content evaporates to leave a residue which is basically the product concentrate. This residue should ideally be light and wet. Gummy or tacky deposits collect swarf and debris, necessitating increased machine and component cleaning.

'Tramp oil' tolerance 'Tramp oil' is lubricating, or hydraulic, oil which leaks from the machine tool and contaminates the cutting fluid. Most modern machines are equipped with total-loss 'slideway' lubricating systems which can contaminate the cutting fluid with up to a litre of oil a day. The ideal cutting fluid would be capable of tolerating this contamination without any detrimental effects on its performance. Some cutting fluids are formulated to emulsify the tramp oil, whilst others have formulations that reject it, allowing the oil to float to the surface for removal by physical 'skimming'.

Cost-effectiveness Last, but certainly not least, the cutting fluid must, of course, be cost-effective. But what is a cost-effective cutting fluid? There was a time when the cost-effectiveness was simply judged in terms of the price per litre of the product concentrate. Fortunately, there are very few engineering companies who still take this attitude; most recognise that there are many factors which contribute to cost-efficiency. These include the dilution ratio, sump life, material versatility, tool life, component quality, safety aspects, and many others.

Unfortunately, having identified the features of the 'ideal' cutting fluid, one now has to face reality – there is no such product, nor is there ever likely to be! For it is impossible to combine all of these features at optimum levels in the same product.

However, all cutting fluids are *not* equal, and even apparently similar products may well perform in quite different ways! It is for the machine-shop managers (in conjunction with other interested departments – purchasing, health and safety, unions, etc.) to select a reputable supplier who is prepared to carry out the necessary survey work to recommend the best fluid (or fluids) for a particular manufacturing environment.

5.3 Types of cutting fluid

With well over 20 manufacturers of metal-cutting lubricants operating in the UK market alone, and with each of these having a large range of products, the selection of cutting fluids is certainly not easy for the end-user. The aim of this section is to differentiate between the types of product available, giving an indication of the features and application of each, so helping users of cutting fluids to understand their needs better.

5.3.1 Hand-applied lubricants

As the name suggests, these products are applied to the tool or workpiece by hand, for example by brush or 'squeeze bottle', or from an aerosol (Fig. 5.1). Usually these products have excellent cutting properties and provide the optimum in tool life and surface finish on the most difficult operations and materials.

Generally each manufacturer has a small range of products in various consistencies, for example a compound of grease or paste consistency for application by brush, a liquid for application by squeeze bottle or drip, and an aerosol spray for use where the convenience of spray application is required.

These types of products tend to be used either when the machine tool is not fitted with a cutting-fluid system, or (more usually) for selective application to arduous cutting operations (e.g. reaming, tapping and drilling), where the use of a water-soluble product would result in unacceptable tool life or surface finish.

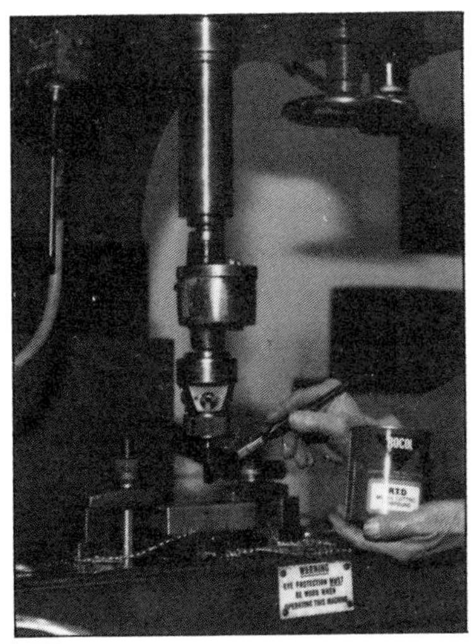

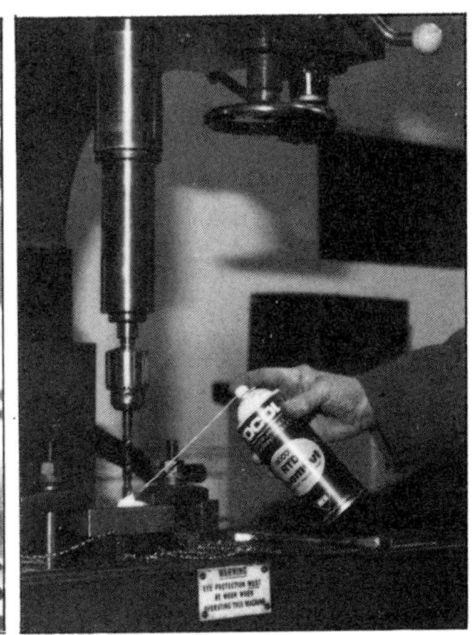

Fig. 5.1 Hand-applied cutting lubricants – for use in arduous cutting operations or where the machine has no 'flood' cutting-fluid system (Courtesy of Rocol Ltd.)

Recognising that interrupting the cycle of CNC machines is highly undesirable, some machine-tool manufacturers are willing to offer an automatic system which operates independently of the 'flood' cutting-fluid system, to inject a measured shot of fluid each time a severe operation (e.g. tapping) is carried out. The operation of this automatic system is included into the CNC program to eliminate any operator involvement.

5.3.2 Flood-applied lubricants

These products are applied in a 'flood' over the work area (Fig. 5.2) and are recirculated via a system of sump, pump, filters, pipework, nozzles, etc.

There are two distinct categories of products which are supplied by flood: 'neat' cutting fluids, where the product is used as supplied and is not diluted with water, and 'water-soluble' cutting fluids, where the product is supplied as a concentrate for dilution with water by the customer prior to use. This latter group can be further sub-divided into three groups: 'mineral-soluble' (emulsion), 'semi-synthetic' (micro-emulsion), and 'synthetic' (mineral-oil free). Thus there are four broad areas into which flood-applied products can be classified. Although the number of possible formulations within each division is almost infinite, it is worth taking a closer look at them, together with a new type of cutting fluid known as 'long-life' products.

Neat cutting fluids

These products are supplied ready for use and require no mixing or dilution. They contain a large proportion of refined mineral oil and, depending on their intended application, may contain lubricity or extreme-pressure (EP) additives (or both) to improve their cutting-performance characteristics.

Neat cutting fluids provide excellent lubricity but are relatively poor coolants: water-soluble cutting fluids provide three to four times their cooling ability.

The use of neat cutting fluids has been declining for some years, whilst the water-soluble cutting fluids have enjoyed a corresponding growth. Neat cutting fluids are now generally restricted to the older machine tools which were designed in such a way that the cutting fluid also provides lubrication to the moving parts of the machine tool (e.g. rotary-transfer automatics and multi-spindle automatic lathes). The demands and

Fig. 5.2 Flood-applied water-soluble cutting fluid: Rocol Ultracut 260 machining cast iron on KTM FM100 (Courtesy of Rocol Ltd.)

requirements of modern machine tools make the use of a neat oil highly undesirable, for the following reasons:

- Their relatively poor cooling ability makes them unsuitable for high-speed cutting operations, in which there is also a tendency for them to fume or smoke.
- They present a fire risk, making them unsuitable for untended production.
- They present a health and safety risk to the operators who have contact with them.
- Their high cost, compared with that of the water-soluble types, makes for less cost-competitive machining.

Since neat oils are rarely used in modern machine tools these days, they will not be discussed any further, and the rest of the chapter will concentrate on the water soluble fluid types.

Mineral-soluble oils, or emulsions

These form the familiar milky-white colour when mixed with water. In fact, 'soluble oil' is a misleading name, since oil is not water-soluble and therefore cannot form a solution! Rather, it forms an emulsion, or suspension, of oil in water. The oil particles formed by the concentration of emulsifier in mineral-soluble oils are of such a size that they reflect practically all incident light, and this gives the emulsion an opaque, milky-white appearance.

Soluble oils typically contain 40%–80% mineral oil along with an emulsifying agent. Water is obviously corrosive, and the water phase of a soluble oil emulsion is no different in this respect. It is, therefore, necessary to incorporate corrosion inhibitors so that the emulsion does not cause corrosion, either to the machine tools or to the components. Since bacteria and fungi can live in water-based fluids, almost all formulations also include a biocide to inhibit the growth of bacteria which could ultimately cause fluid failure. The oil phase of a soluble oil provides a degree of lubrication but no anti-weld properties. For high-performance cutting fluids, it is common, therefore, to include additives which improve the lubricity and provide an extreme-pressure activity.

Semi-synthetic fluids, or micro-emulsions

During the 1980s the use of this class of cutting fluid has become much more widespread because of certain advantages over the mineral-soluble oils.

By increasing the ratio of emulsifier to oil in the formulation, either by reducing the oil content or by increasing the level of emulsifiers, the product takes on different characteristics from those of the mineral-soluble oils. Because of the increased ratio of emulsifier to oil, the oil particles formed are significantly smaller than with mineral-soluble oils. These 'micro-emulsions', therefore, appear translucent or even transparent, owing to the fact that the oil particles are smaller than the wavelength of light (less than 0.5μm). This translucency is an obvious advantage where workpiece visibility is important to the machine operator.

In addition, the high level of emulsifiers in the product leaves some 'spare' capacity, which enables the micro-emulsion to emulsify any oil leakage from the machine, so keeping the machine cleaner and delaying the formation of a layer of tramp oil on the fluid surface (which may encourage the growth of bacteria). It should be noted that semi-synthetics will only emulsify tramp oil until the 'spare' emulsifiers are used up; after this time the oil will float on the cutting-fluid surface. Some formulations will emulsify only small amounts of oil, whilst others will emulsify much larger concentrations.

Semi-synthetic fluids are also more suitable than mineral-soluble oils for most grinding operations because of their lower oil content: high oil-content products tend to load the grinding wheels, making frequent re-dressing necessary.

The definition of semi-synthetics can cause some difficulty, but generally the oil content is much lower than with the mineral-soluble oils, say between 10% and 40%. Perhaps the best and easiest definition is this: 'A semi-synthetic forms a translucent emulsion and contains mineral oil.'

Additives for corrosion inhibition, bacterial control, lubricity and EP are used in the same manner as for mineral-soluble oils. Also, there is often the addition of a dye: translucent micro-emulsions can look a little 'watery', so to make the product look more acceptable, a coloured dye, such as green, blue or pink, may be added.

Although translucent micro-emulsions are formed initially, semi-synthetics do tend to go cloudy in use. They contain excess emulsifiers to ensure that a fine micro-emulsion of oil particles in water is formed. These 'spare' emulsifiers enable the micro-emulsion to absorb tramp oil. As the spare emulsifiers are used up on suspending the tramp oil, both the amount of oil in the emulsion and the oil particle size increase. This increase causes more incident light to be reflected, and results in the 'clouding' effect. 'Cloudiness' is not necessarily an indication that there is anything wrong with the fluid: it is merely an indication of the amount of oil absorbed by the fluid.

Synthetics

Synthetic cutting fluids contain no mineral oil. These fluids are a solution of chemicals, which usually include corrosion inhibitors, biocides and dyes, in water. In addition, they may contain such additions as synthetic lubricity additives and wetting agents.

Synthetic fluids form transparent solutions and, as a result, provide good visibility of the cutting operation.

In use, synthetic fluids require special attention in their application; because they contain no mineral oil, they tend not to leave a corrosion-protective oily film on machine surfaces. Because of this, it is essential to lubricate machines carefully. In addition to the lack of protection, there may be some effect on certain paint finishes and even a degradation of machine seals, as a result of the synthetic fluid entering the machine's lubrication system. As a result of these limitations, the use of synthetic fluids tends to be restricted to grinding applications. In grinding, the synthetic product usually offers real benefits in terms of wheel life (lack of loading), rapid fines settlement and low foaming properties when compared with other fluid types. In addition, the manufacturers of grinding machines are aware of the possible problems of synthetic fluids, and take special care in the machine's design.

'Long-life' products

The 1980s have seen the emergence and growth of a new breed of water-soluble products which, although they could be classified in the preceding groups (i.e. as mineral-soluble, semi-synthetic or synthetic), are probably best considered separately.

Variously described as 'long-life', 'new-generation' 'biostable' and 'biostatic', these products utilise the element boron to control bacterial growth without the need for the more traditional biocides used in 'conventional' fluids. It is included in the formulation as a boramide, which acts both to control bacteria and to inhibit corrosion. The outstanding feature of these products is their resistance to bacterial degradation, and they can offer impressive improvements in sump life over conventional fluids.

These fluids can be formulated with high, medium or no mineral oil content, and with lubricity and extreme-pressure additives for difficult cutting operations. The 'long-life' fluids are well suited to modern machine tools and centralised systems where regular fluid changes and unplanned down-time cannot be tolerated.

The distinguishing properties and formulations of the types of cutting fluid described in these sections are summarised in Fig. 5.3.

5.4 Application methods

With the exception of the 'hand-applied' metal-cutting lubricants covered in section 5.3.1, which are applied by brush, drip, squeeze bottle or aerosol, almost all cutting fluids are supplied to the machining zone by flood application (Fig. 5.2). To achieve a flood of fluid at the machining zone requires a recirculatory system consisting of sump, pump, filters, pipes, nozzles, etc. The system must be designed such that the fluid in the sump is allowed to cool down and be cleaned before being returned to the cutting zone. The flow of coolant should be at low pressure to reduce any possibility of foaming problems, but should be of high volume to absorb as much heat as possible.

A minority of machines are fitted with spray-mist systems, either as a supplement to the flood system, or occasionally as a substitute for it. Spray-mist systems are usually of the 'total-loss' type, supplied from a tank; they produce a fine spray of fluid droplets in the air which can be directed at the tool/workpiece interface through a flexible supply pipe.

Type	*Typical properties*	*Formulation*
Mineral-soluble oils	Not a solution	40%–80% oil
	Emulsion (suspension) of mineral oil in water	Emulsifier (usually non-phenolic)
	Milky-white appearance	Additives – corrosion inhibitors, biocide, lubricity, EP
Semi-synthetic oils	Micro-emulsion of mineral oil in water	10%–40% oil Emulsifier
	Translucent appearance	Additives–corrosion inhibitors, biocides, lubricity, EP, dye
Synthetic cutting and grinding fluids	Solution or micro-emulsion Transparent appearance	Oil-free Corrosion inhibitors Wetting agents Additives – biocides, lubricity, dye
'Long-life' cutting and grinding fluids	High, medium or no oil content	Mineral and vegetable oils Boramide Emulsifier Additives – lubricity, EP, dye

Fig. 5.3 Properties and formulations of water-soluble fluids

Spray-mist application is particularly useful on some milling and routing operations where, owing to the size of the component, it would be difficult to contain a flood of fluid within the machine tool.

Some sources claim that a superior 'refrigerant' effect is obtained by the spray application, which improves the cutting performance by lowering the tool temperature. However, this method of application is losing popularity because of the health and safety hazards associated with such sprays circulating through the atmosphere in a workshop.

5.5 Selection

The selection of the appropriate fluid, or fluids, for particular applications is best done in conjunction with qualified and experienced sales or technical personnel from the fluid supplier. There are many factors to be considered, and the purpose of this section is to help the fluid user to liaise effectively with the fluid supplier. These factors include the workpiece material, the cutting operation, the machine tool, and the condition of the water supply.

5.5.1 The workpiece material

Most machine shops handle a wide range of materials, including both ferrous and non-ferrous metals, and non-metallics. There are two aspects of workpiece material which require consideration: the machinability and the compatibility of the fluid and the workpiece.

Machinability The more 'difficult-to-cut' materials will require the selection of an appropriate fluid to ensure that tool life and surface-finish requirements are met. Such 'difficult' materials will usually require the selection of a product with extreme-pressure additives or lubricity additives (or both).

It is also important to consider the severity of the machining operation, since a fluid with no 'performance' additives may be acceptable for single-point cutting of difficult stainless steel, for example, yet may not be acceptable for multi-point cutting of ductile aluminium alloys.

Compatibility of fluid and workpiece Almost all cutting fluids are compatible (i.e. non-staining) with ferrous materials and provide some degree of residual corrosion protection to the workpiece after machining. But non-ferrous alloys of copper and aluminium may be susceptible to staining or corrosion in certain contact conditions with some cutting fluids. Where such alloys are being machined, the following points should be considered:

- What does the fluid manufacturer recommend?
- Is the fluid manufacturer prepared to conduct pre-sale compatibility tests?
- What are the possible effects of contact with different material types (workpiece materials, fixtures, machine elements, etc.)?
- How quickly can the components be cleaned?
- Are components stacked, trapping fluid between faces?
- What are the implications of any staining?

5.5.2 The cutting operation

Clearly, there is a wide variation in the severity of the many different cutting operations. Even within any one operation, the severity will depend on variations in cutting speed, feedrate and depth of cut. For the purposes of simplicity, though, a useful 'rule of thumb' is to divide operations into two groups: 'multi-point', e.g. milling, tapping, reaming and sawing, and 'single-point', e.g. turning. In general, the multi-point operations are considered to be the most severe, and therefore require the use of a product with good lubricity and extreme-pressure properties.

5.5.3 The machine tool

It has already been noted that some machine types are not ideally suited to the use of water-soluble cutting fluids as they rely to some extent on the cutting fluid for lubrication of their parts. Where a machine has historically been run on neat oils, it is wise to consider its manufacturer's recommendations before introducing a water-soluble fluid. Some machine manufacturers, most notably German companies (e.g. Traub and Maho), make recommendations on the fluid types to be used in their machines. The main criteria for approval are that the fluid should be of a relatively high mineral-oil content and that it should reject (as opposed to absorb or emulsify), the slideway oil contamination. When in doubt, both the machine manufacturer and the cutting-fluid supplier should be contacted for advice if problems are to be avoided – and the warranties unaffected!

5.5.4 The condition of the water supply

Water makes up between 90% and 99% of the water-soluble emulsion in the machine, depending on the dilution. It should be readily accepted, then, that the condition of that water is likely to affect the performance of the products.

Mains tap water, of course, is virtually bacteria-free, but the water from storage tanks and their like may well be infected and result in premature failure of the emulsion due to bacterial degradation.

Hard water, containing between 200ppm and 400ppm of total hardness salts, may cause two problems:

- Some cutting fluids contain ingredients which can react with the hard-water salts to form an insoluble soap scum. In some machines this can cause sufficient build-up to block filters, pipes or pumps.
- Corrosion-protective properties may be reduced, necessitating the use of a stronger emulsion to avoid corrosion.

For these reasons it may be wise to avoid bore-hole water (which is frequently very hard). Similarly, if the mains water supply is in excess of around 250ppm hardness, the fluid supplier should be consulted for their recommendation.

Soft water, with less than about 100ppm hardness, may result in excessive foaming problems in certain combinations of machine tool and cutting fluid.

5.6 Problems

A variety of problems can occur during use of water-soluble cutting fluids. Often these problems are due to a lack of maintenance of the fluid, most notably of its 'dilution ratio'. If any problem arises, it is of course advisable to seek assistance from the supplier of the particular fluid. However, the following sections include comments on the most commonly encountered causes of problems and a 'troubleshooting' guide, and provide general guidance.

5.6.1. Bacteria and fungus problems

Most cutting-fluid failures are due either to bacterial degradation causing the fluid to smell unpleasantly rancid or to become corrosive, or to solid fungal growths blocking the fluid system. Bacteria and fungus can develop in any water-based cutting fluid, and there are a number of factors which determine the time interval between fluid changes necessitated by bacterial or fungal failure.

At fluid changes, the residual infected fluid and contaminated swarf and debris left in the machine will put the fresh fluid under immediate attack. Ineffective cleaning is likely to result in rapid infection, since strains of bacteria will have developed which can best survive in the particular cutting-fluid and system conditions (aeration, oil contamination, etc.). So unless systems are properly cleaned prior to filling with fresh fluid, the coolant cannot achieve its optimum life potential.

Proprietary system cleaners are available which provide an easy and effective method of cleaning. They are simply added to the old fluid several hours before the clean-out and, in most cases, the machine can be operated as normal. At the end of the treatment period, all areas of the system in contact with the fluid will have been treated by the cleaning action and biocidal properties of the system cleaner, ensuring the best possible start for the fresh fluid.

Biocides used in 'conventional' cutting fluids are depleted as they kill bacteria. Proprietary biocides are available which can be added to the fluid during use to 'top up' the biocide protection and so lengthen the useful life of the product. Where the problems recur despite good housekeeping, however, it may be necessary to consider a change of fluid type to overcome them

5.6.2 Problems of dilution control

To provide its optimum performance, any cutting fluid must be maintained within the recommended dilution range. If the fluid is too strong, it will be more likely to cause operator irritation, leave thick, tacky or gummy deposits, foam excessively, and affect machine paint, seals, etc. If it is too weak, it will be more likely to corrode the machine or the workpiece, suffer rapid bacterial degradation, become corrosive or cause fungal blockages, and provide insufficient lubricating properties, leading to poor tool life, surface finish, etc.

5.6.3 Problems of product selection

If the wrong product has been selected for a given application, the following are the most likely problems: poor tool life or surface finish, water incompatibility causing soap blockages or foaming, machine-tool problems with paint, seals, etc., and staining of non-ferrous workpiece materials.

Fig. 5.4 provides a summary of the causes of the most common problems concerning cutting fluids, with suggestions as to how they may be remedied.

5.7 Care and control

All cutting fluids benefit from proper care and control, both before and during use; only when properly maintained will they provide optimum life and performance.

5.7.1 Storage

Cutting-fluid concentrates are generally supplied to the customer in 200-litre (45-gallon) steel barrels. Larger customers may prefer bulk-tanker deliveries, and smaller users may purchase 25-litre (5-gallon) plastic or steel drums.

Concentrates should be stored in an environment which does not allow freezing, since this might adversely affect the stability of the concentrate. Similarly, if barrels are stored outdoors they should not be stored 'bung-end up': any water that might collect on the barrel's top will, owing to temperature changes, be sucked into the barrel – even through the unopened top! These small amounts of water may destabilise the concentrate. Destabilisation caused by freezing or water contamination can result in a complete separation of the concentrate into distinct layers. The product is completely unusable in this form, and it is unlikely to remix adequately.

These notes on freezing and water contamination are for general advice. Where in doubt, the manufacturer of the product would be consulted for specific guidance.

5.7.2 Preparation of the emulsion

Correct preparation of the emulsion is essential if the cutting fluid is to provide its optimum performance. By preference, the water supply should not be excessively hard (less than 300ppm total hardness), and it should be of drinking quality, that is, not infected with bacteria.

Problem	*Possible cause*	*Remedy*
Rancid smell	Bacterial degradation	Clean system with a proprietary system cleaner and refill at correct dilution
Persistent or frequent rancid smell	Contamination by resistant bacteria	Ensure removal of swarf conveyors, etc. during cleaning, and consider addition of biocide to fluid during use or changing to a more resistant fluid
Corrosion of machine tool	Overdilution	Adjust dilution
	Bacterial degradation	Clean system with a proprietary system cleaner, and refill at correct dilution
	Fluid in use leaves insufficient protective film	Increase machine maintenance or lubrication, or consider changing to a different product type
Corrosion of workpieces	Overdilution	Adjust dilution
	Bacterial degradation	Clean system with a proprietary system cleaner, and refill at correct dilution
	Excessive storage period	Apply proprietary corrosion protective prior to storage
Blocked system	Fungal growth	Physical removal of all deposits, clean with proprietary system cleaner, and refill at correct dilution
	Hard water soap	Check water hardness and fluid compatibility, and consider softened water or alternative fluid
Persistently blocked system	Persistent fungal growth	Consider changing fluid type
Foaming	Underdilution	Adjust dilution
	Soft water or high-pressure system	Consider addition of anti-foam additives, hardening of water or alternative fluid type
Paint softening or stripping from the machine tool	Underdilution	Adjust dilution
	Aggressive fluid	Consider alternative fluid type
	Poor-quality paint	Repaint with good-quality paint
Operator irritation	Underdilution	Adjust dilution, and **seek medical advice**
	Excessive contact, or poor hygiene	Introduce protective clothing, barrier cream and good washing facilities, and **seek medical advice**
	Excessive additions of biocide	Control dosage rates carefully
	Sensitive individual	Consider move to 'dry' work
Poor tool life or surface finish	Overdilution	Adjust dilution
	Unsuitable product	Consider use of different product type

Fig. 5.4 Troubleshooting guide

Hand mixing Hand mixing can be used to prepare small volumes of emulsion. The method is as follows:

- Choose a suitable mixing vessel (not the machine sump, and not a galvanised container).
- Fill with the measured amount of water.
- Slowly add the measured amount of concentrate whilst stirring continuously, from top to bottom, until a consistent emulsion is formed.
- Add the mixture to the machine sump.

Automatic mixing For larger volumes of emulsion it is preferable to use a purpose-designed automatic fluid mixer, such as the one shown in Fig. 5.5. Some cutting-fluid manufacturers supply these units at nominal cost to users of their fluids.

The best types of mixer incorporate the following features:

- They can connect directly to a standard water tap.
- They can screw directly into 200-litre barrels.
- There is a screw adjustment for the dilution ratio.
- They incorporate a non-return valve to ensure that the emulsion cannot leak back into the barrel's contents.

Most fluid mixers work on the venturi principle; because of this, there will be a degree of fluctuation of the dilution should there be a change in the water pressure. For this reason it is preferable to connect the mixing units to a water supply which is not subject to pressure variations due to other equipment drawing water intermittently.

Local water authority regulations should be observed in the installation of these units. These regulations may well require the use of a header tank with an air gap, in order to prevent any possibility of contamination of the mains.

Fig. 5.5 The Rocol automatic fluid mixer (Courtesy of Rocol Ltd.)

5.7.3 Machine cleaning

It is important that cutting-fluid systems are correctly cleaned before fresh cutting fluid is introduced if the best performance is to be obtained. Such cleaning should include the following stages:

- Physical removal of all deposits, swarf, etc.
- Treat the system with an appropriate system cleaner (Fig. 5.6). This should be added to either the old cutting fluid or to clean water for the specified time period as recommended by the manufacturer.
- Remove guards, swarf conveyors, etc. as necessary to allow effective cleaning of inaccessible areas. (Fig. 5.7 shows a machine that can be used to empty the sump for this purpose.)
- Flush with clean water.
- Refill with fresh emulsion at the correct dilution.

5.7.4 Maintenance of the fluid during use

Once in the freshly-cleaned machine, the fluid should be checked on a regular basis. The following procedures will help to prolong its life:

- Check the dilution regularly using a refractometer to ensure that the dilution remains within the recommended range (Figs. 5.8 and 5.9). Where the fluid is underdiluted, never add water but rather add weak emulsion; where the fluid is overdiluted, do not add concentrate but rather add strong emulsion.
- Do not wait for the fluid level to fall below the pump before topping up: topping up regularly with fresh fluid assists in controlling bacterial growth.

Fig. 5.6 Addition of a proprietary system cleaner to the old cutting fluid several hours prior to cleaning out (Courtesy of Rocol Ltd.)

Fig. 5.7 The Rocol Victor sump vacuum cleaner empties sumps quickly and cleans effectively. Simple attachments enable waste fluid to be filled directly into 200-litre barrels for disposal (Courtesy of Rocol Ltd.)

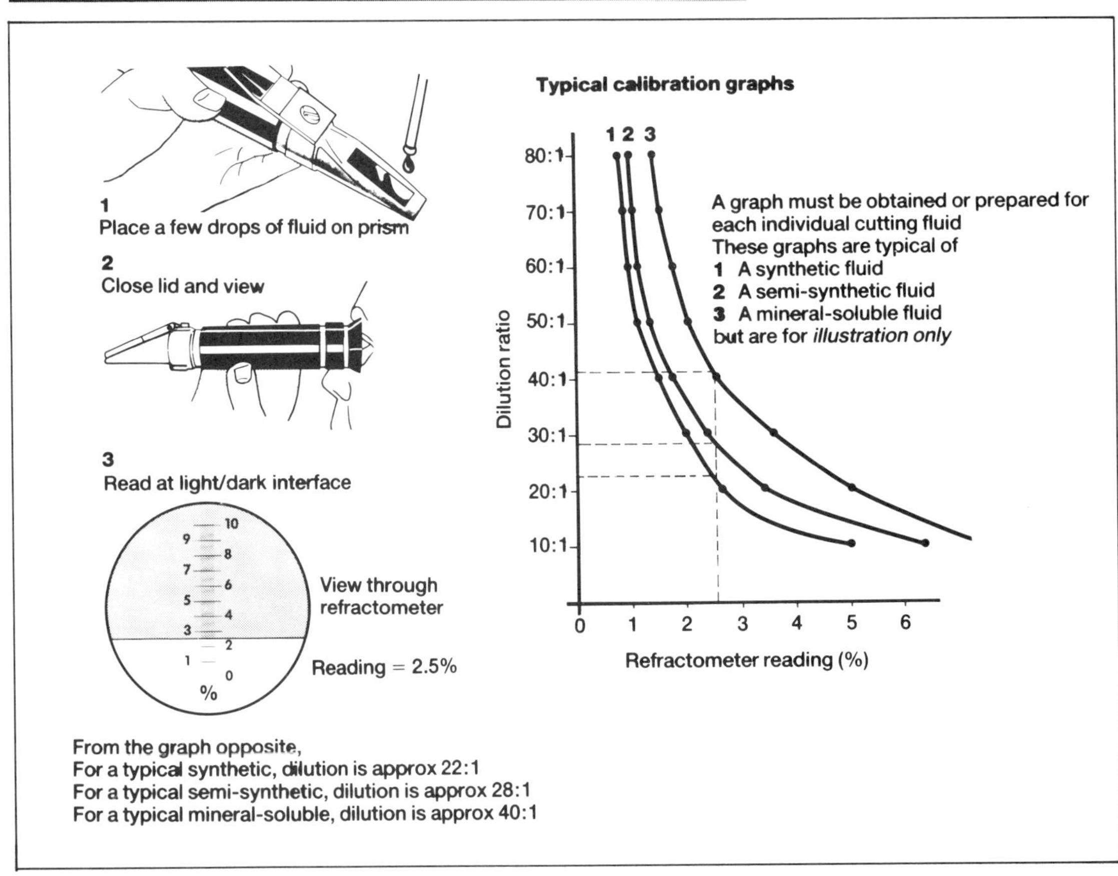

Fig. 5.8 How to check cutting-fluid dilution using a refractometer (Courtesy of Rocol Ltd.)

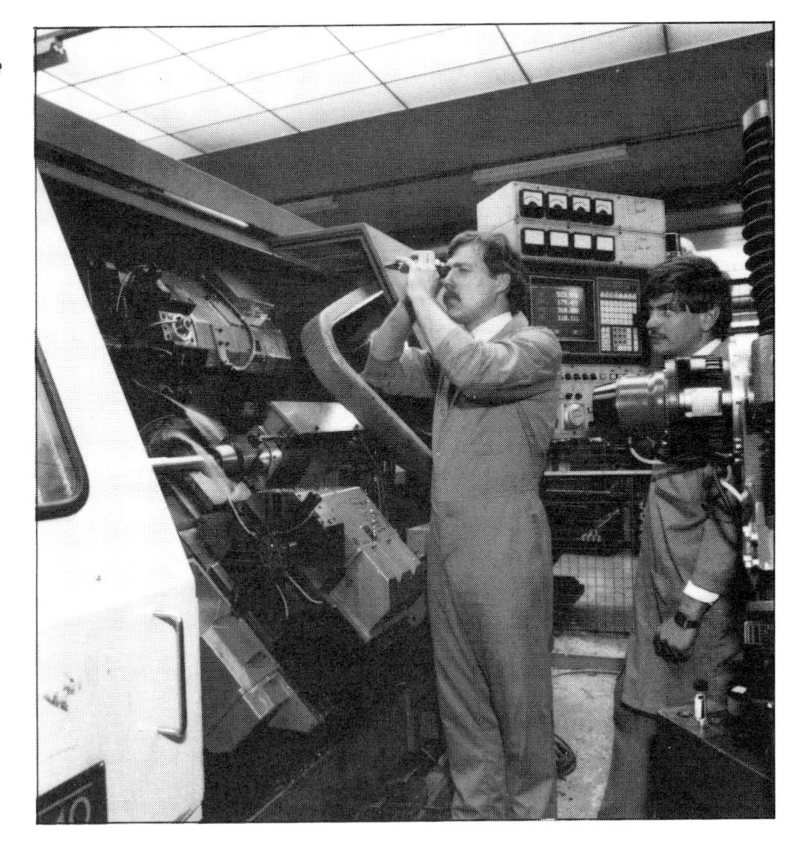

Fig. 5.9 On-site dilution control: using a refractometer to measure dilution (Courtesy of Rocol Ltd.)

- Do not allow large amounts of swarf to build up on machines, as this may encourage bacterial growth and corrosion.
- Do not allow the cutting-fluid sump to be used as a dustbin for cigarette ends, paper cups, rubbish, etc.
- Avoid excessive leakage of oil from the machine, especially during maintenance.

More sophisticated condition monitoring is, of course, possible and may be appropriate where the consequences of fluid failure are serious (e.g. in large centralised systems). Some cutting-fluid manufacturers offer this detailed monitoring as a service. More detailed checks that can be done on-site include dilution control, monitoring of the fluid's pH with colour indicator sticks or electronic meters, and monitoring of bacterial infection using specially-designed test slides. Checks that need to be done in a laboratory include monitoring of the corrosion-protective properties and the levels of the important ingredients, e.g. biocides.

5.7.5 Disposal

No waste cutting fluid should be disposed into the public drainage system or rivers except with the express permission of the local water authority. Some synthetic cutting fluids may be acceptable for disposal in this way, if they are not excessively contaminated with oil or metal particles and if they can be further diluted before being discharged.

In general, mineral-oil containing cutting-fluids must be disposed of by separation of the oil-phase using 'acid-splitting' or 'ultrafiltration' techniques. These techniques are relatively expensive, and are generally only justified where large volumes of fluid are to be disposed of regularly. The majority of cutting-fluid users, therefore, entrust the disposal of their waste fluids to a reputable waste contractor.

5.8 Health and safety

No-one employed in manufacturing industry can be unaware of the increased importance of, and interest in, the subject of health and safety. In the UK there is now much legislation which places demands on the manufacturer, supplier, employer and employee alike. At the time of writing (1988), the draft legislation *Control of Substances Hazardous to Health* (COSHH) is before Parliament; it is likely to be implemented in 1989.

So what aspects of cutting fluids does the user need to consider? Probably the most important are product Health and Safety Information Sheets, the formulation, hazard labelling and dilution control of the product, and emulsion-condition monitoring.

5.8.1 Product Health and Safety Information Sheets

Before any new product is introduced, its Health and Safety Sheet should be obtained from the manufacturer or supplier and studied by all interested parties. Should there be any areas of uncertainty, these should be clarified with the supplier before the sheet is filed for future reference. It should be remembered that this sheet should be readily obtainable in the event of an accident or emergency.

5.8.2 Formulation of the product

A description of the main ingredients of the product should be included on the Health and Safety Sheet. This description is likely to be in generalised terms, since manufacturers need to guard their formulations from falling into the hands of their competitors! For this reason, most manufacturers will be extremely reluctant to divulge the full chemical formulation. Indeed, even if the formulation were provided, how many engineers would be able or willing to interpret its safety implications?

It will not usually be necessary for the user to know the full chemical formulation, but where it is required, a sensible compromise is for the cutting-fluid manufacturer to supply full details to the company doctor in medical confidence. In this way, the confidentiality of the formulation is protected, and the user gets a qualified assessment of the acceptability of the product, rather than an incomprehensible list of chemical names.

There are, however, some ingredients which deserve special attention and which the user should be aware of.

Cresylic or phenolic products

Almost all the early soluble oils contained 'cresylic' or 'phenolic' compounds which acted as emulsion couplers and biocides. Such cutting fluids can usually be readily identified by their strong disinfectant-like odour. In recent years, however, these products have tended to go out of favour, for two main reasons. Firstly, their strong smell does tend to cling to clothing and, whilst it is relatively acceptable in the workplace it is not so acceptable in public transport, public houses or the home! Secondly, phenolic compounds are associated with an irritant potential to operators.

These products, therefore, now account for only a small proportion of the market, although (because there are still some 'die-hards' who prefer not to change) certain manufacturers do still offer phenolic products as part of their range.

Products which contain no such compounds may refer to this fact on their data sheets; they may be described as 'phenol-free' or 'cresylic acid-free', for example.

Mineral oil

The majority of cutting fluids contain at least a small percentage of mineral oil in their concentrate form. Reputable manufacturers will use only mineral oils which are appropriate to the application. Such oils are highly refined to reduce the content of aromatic carbon significantly, as this is associated with adverse skin reactions. Products that use oils from acceptable refining techniques are likely to be described as 'severely hydro-treated' or 'solvent-refined'.

Sodium nitrite

This acts as an extremely effective and relatively inexpensive corrosion inhibitor in cutting fluids. Until the late 1970s and early 1980s, most water-soluble products contained some sodium nitrite.

However, research has shown that in products which contain both sodium nitrite and amines (e.g. diethanolamine and triethanolamine), there is the possibility of a reaction between these ingredients. This reaction produces a group of chemicals known as 'nitrosamines'. Some nitrosamines, including the types isolated from cutting fluids, have been shown to be carcinogenic in rats and hamsters (though not in mice). Whilst there is no specific evidence of the effect of nitrosamines on humans, it is clearly wise to consider them as potential human carcinogens. The main area of concern is with synthetic grinding fluids, the most simple of which contain large proportions of sodium nitrite, amines and water, along with other ingredients at much lower levels.

In response to concern in the marketplace, cutting-fluid manufacturers have pursued a policy of 'substitution', i.e. removal of either the nitrite or the amine from their formulations. Most commonly it is the sodium nitrite that is omitted, although it is thought by some authorities that nitrogen can be absorbed from the atmosphere to form small quantities of nitrosamines by reaction with amines even in nitrite-free formulations.

Because of this trend towards 'nitrite-free' formulations, the user can be forgiven for believing that it was the nitrite itself that was hazardous. Sodium nitrite is, in fact, toxic to humans if eaten in large amounts, but it is unlikely that this would occur in any cutting-fluid applications. Indeed, sodium nitrite is used at low levels as a preservative in canned vegetables and meats, when it is classified as E250.

5.8.3 Hazard labelling

In the UK, the Classification, Packaging and Labelling of Dangerous Substances Regulations Act (1984) applies to cutting fluids and related products. All product containers must carry the appropriate warning symbol and risk and safety phrases on a label, correctly sized to suit the particular pack size. Firstly, it must be pointed out that some cutting fluids are not required to carry hazard labelling, either because they contain no ingredients listed as hazardous, or because they contain them at a level below the limit requiring labelling for that particular ingredient. Secondly, it is the 'product supplied' (i.e. the concentrate) which requires labelling. In most cases, the level of any hazardous ingredients at working dilutions is well below that requiring labelling.

The classifications most usually applied to cutting fluids are:

- *Irritant* due to alkaline surfactants.
- *Harmful* due to sodium nitrite at levels between 1% and 5%.
- *Toxic* due to sodium nitrite at levels greater than 5%.

5.8.4 Dilution control

Cutting fluids are more likely to cause operator irritation (to the skin, eyes, cuts, etc.) when used at excessive concentrations. It is therefore important not only that the fluids are mixed at the correct dilution, but that they are maintained at the correct dilution during use in the machine tool (see section 5.7.4). Advice on the maximum allowable concentration and the recommended method for checking should be sought from the fluid supplier or manufacturer.

5.8.5 Emulsion-condition monitoring

During use, cutting fluids become contaminated with various materials, e.g. lubricating and hydraulic oils and fine swarf particles – and sometimes with general waste, including cigarette ends, sweet papers, orange peel, etc. Many of these contaminants are likely to increase the irritancy of the fluid, or to introduce and encourage bacterial infection. Clearly, then, it is wise to consider the degree of contamination, even when there is no apparent problem with the fluid, to ensure that poor fluid condition does not increase the potential for health problems.

5.8.6 Health and Safety Executive publications

A number of very useful health and safety publications are produced by the Health and Safety Executive and available through HMSO in the UK. These publications include wall posters, pocket cards and detailed reports. It is strongly recommended that all users of cutting fluids should obtain the relevant publications, display the posters and ensure that the machine operators are fully aware of the potential hazards and precautionary measures required.

This chapter includes little discussion of the effects of cutting fluids on the cutting tools, or of how they affect the workpiece's machinability and its resulting surface integrity. The final chapter, however, presents an enlarged review of these subjects.

Chapter 6
The machinability and surface integrity of engineering components

The term 'machinability' is an ambiguous term, which can have a variety of different meanings, depending on an engineer's requirements. Two commonly used definitions of machinability are: 'the totality of all the properties of a work material which affect the cutting process', and 'the relative ease of producing satisfactory products by chip-forming methods'.

Even these definitions still lack sufficient precision to be of much use, and there is no agreed measurable quantity for machinability testing. Why is this so? The chemical composition of the workpiece, its microstructure, heat treatment, purity, etc. are all variables that affect the machinability. The workpiece itself will determine such characteristics of machinability as tool wear, chip formation and surface finish, but they are also considerably dependent upon the machining application.

Fig. 6.1 summarises the parameters and characteristics that affect the results of the machining process; the wear, forces, chip-forming properties and surface integrity are dependent on the operation used (e.g. turning, milling, etc.) and on the input variables (cutting-tool material, auxiliary materials, cutting data, etc.). From the viewpoint of Fig. 6.1 it is only possible to improve, worsen or generally change the end results of the machining process, not the machinability itself. There is a certain logic with this viewpoint over either of the 'definitions' given above. If one increases the cutting speed, say, and thus produces a better chip flow, the increased speed leads to higher wear-rates on the tool. This corresponds to both better and poorer machinability at one and the same time, depending upon which of the two 'definitions' is used! Obviously, such assessments of good or bad machinability are subjective and depend upon which criterion is considered the more important. Hence we shall take the view that a material's machinability is determined by its production process and the heat treatment it receives. Thus, machinability can only be affected by microstructural transformations in the material caused by the machining process, if at all.

In fact, some cutting-tool companies produce machinability charts that are similar to the time-temperature-transformation (TTT) charts used in material heat treatment and from which machinability ratings can be obtained, classified by ease or difficulty of machining.

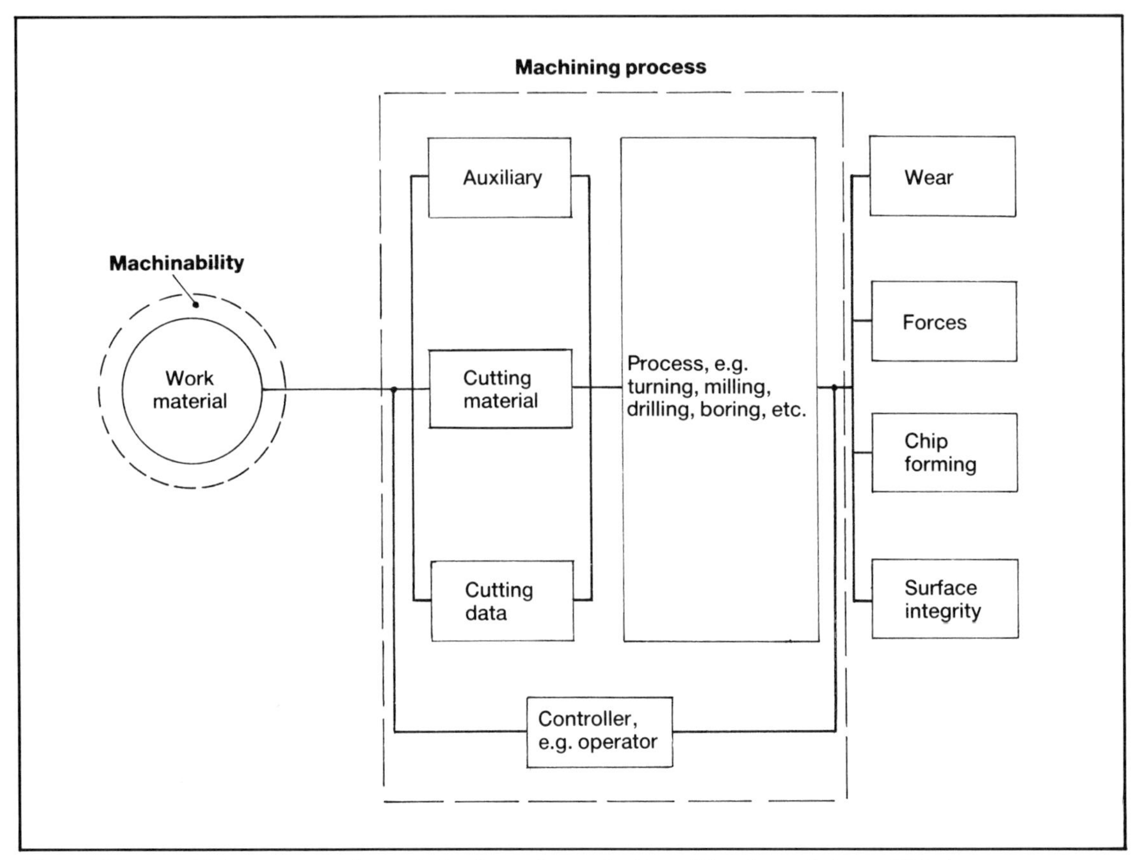

Fig. 6.1 The machining process shown in the form of a control loop (Courtesy of Krupp Widia)

If machinability testing is so important, what methods are used in determining machining data and how is the testing carried out? This will be theme of the following section.

6.1 Test methods used for assessing machinability

The assessment of a material's machinability can be done using either of two types of test: machining tests and non-machining tests. There is a further sub-division of the machining tests into ranking and absolute tests, as illustrated in Fig. 6.2; all the non-machining tests are ranking tests. The ranking tests are often termed 'short' tests, whereas the absolute ones are known as 'long' tests. By their very nature, the 'short' tests (either machining or non-machining) merely indicate the relative machinabilities of two or more different combinations of tool and workpiece. The 'long' tests, on the other hand, produce a more complete picture of the expected conditions for various combinations of tool and workpiece, but they are obviously more time-consuming and costly to develop and carry out. A very brief review of each type will be given below; further information is available from the publications listed in the Bibliography.

6.1.1 Machining tests

Ranking – fast assessment methods

The rapid facing test This turning test consists of facing off a workpiece, preferably with a large diameter, using a high-speed steel tool. The machinability is assessed by the distance the tool will travel radially outwards from the bar's centre before breaking. This

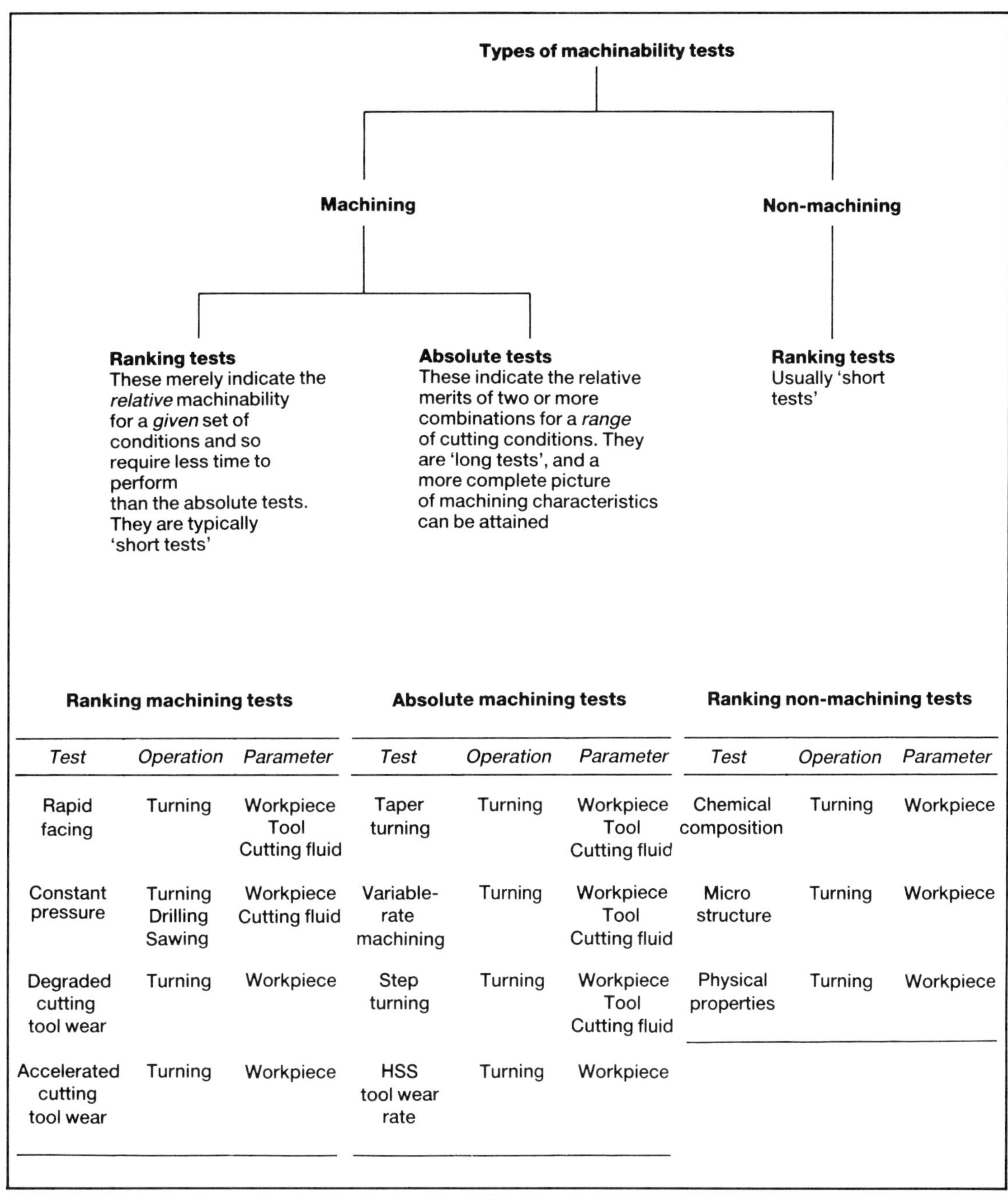

Ranking machining tests

Test	*Operation*	*Parameter*
Rapid facing	Turning	Workpiece Tool Cutting fluid
Constant pressure	Turning Drilling Sawing	Workpiece Cutting fluid
Degraded cutting tool wear	Turning	Workpiece
Accelerated cutting tool wear	Turning	Workpiece

Absolute machining tests

Test	*Operation*	*Parameter*
Taper turning	Turning	Workpiece Tool Cutting fluid
Variable-rate machining	Turning	Workpiece Tool Cutting fluid
Step turning	Turning	Workpiece Tool Cutting fluid
HSS tool wear rate	Turning	Workpiece

Ranking non-machining tests

Test	*Operation*	*Parameter*
Chemical composition	Turning	Workpiece
Micro structure	Turning	Workpiece
Physical properties	Turning	Workpiece

Fig. 6.2 Test methods used for assessing machinability (After B. Mills and A. H. Redford)

end-point is compared with the distance for tool failure using a reference material. (The fact that the end-point is determined by catastrophic failure of the tool is the main reason for using high-speed steel tools.) Although the test quickly measures one criterion that a machinability rating can be based upon, it suffers from a number of limitations. Firstly, the material's diameter may be smaller than that which one would ideally like to use for the test; secondly if the material's structure is not homogeneous, then the test only indicates the properties over the diameter range used for the test. This latter problem of material inhomogeneity can be lessened by boring out the material at the bar's centre.

The constant-pressure test This well-known test can be done using a variety of methods of machining assessment. In turning, for example, machinability is assessed using a predetermined tool geometry and a constant feed force. The technique has been used to some effect on the machining of free-cutting steels. The test is basically a measure of the friction between chip and tool, which is related to the specific cutting temperatures generated whilst machining and its effects on tool wear-rate.

Normally a lathe must be modified so that the feed force can be held constant, in order to obtain the relevant data. However, the same data can be gathered on a normal lathe by using a tool-force dynamometer to measure the feed force and plotting a graph of feed force, but this will take longer than the method using a modified lathe.

Similar constant-pressure tests could also be done for the drilling or sawing processes.

The degraded tool test This test consists of machining the workpiece with a softened (degraded) cutting tool. The end-point of the test is determined as either when a specified amount of flank or crater wear is reached, or at catastrophic failure. When machining the softer and more easily-cut materials, such as the various brasses, just a small variation in the softened tool steel prior to cutting will have a drastic effect on the results obtained, whereas on harder materials small variations in the hardness of the degraded cutting tool will have less effect on the results of the test.

The accelerated cutting-tool wear test As an alternative to employing deliberate tool softening in order to achieve speedy wear-rates as in the previous test, in this test the cutting conditions, and more specifically the cutting speed, are increased. If the cutting speed is increased significantly, then the tool will not behave according to Taylor's equation for tool wear (see page 189), mainly because of the higher cutting temperature generated. Therefore, it is not prudent to extrapolate tool-life data beyond that actually obtained in the tests in order to obtain quantitiative information about other ranges of conditions, with different operations and parameters. As a result, this test is classified as a ranking test rather than as an absolute test; when the results are compared against those for a standard reference material, it provides useful results.

Absolute – comprehensive data-gathering methods

The taper-turning test This test is done by turning a tapered workpiece. As a result of turning along the taper, the cutting speed will obviously increase as the diameter becomes larger (which will be in proportion to the cutting time). By knowing the original cutting speed selected, the changing rate of the cutting speed and the amount of tool flank wear for two separate tests, the values of the constants in Taylor's equation may be derived, and the tool life can be calculated for a range of future cutting speeds. As the depth of the cut must be maintained consistently throughout the test on the tapered specimen, either a taper-turning attachment must be used on a centre lathe, or a CNC program must be written (using one of the standard routines normally available, for example) on a turning centre.

The main advantages of this test are that is is a comprehensive test, producing results that are valid for a range of chosen cutting speeds, and that it is of relatively short duration but agrees closely with thorough and longer test methods. However, the results obtained may not be representative of actual cutting conditions, owing to the fact that the tool cuts at different diameters throughout the test.

The variable-rate machining test This test achieves similar results to the taper-turning test described above. In this case the increase in cutting speed is obtained by turning a parallel test sample axially, whilst simultaneously increasing the cutting speed as the tool traverses along the sample. Once again, the constants in Taylor's equation can be determined after two tests have been conducted.

The main advantages of this method over the taper-turning test are that a standard test piece can be used (e.g. a parallel length of bar), and that the results reflect truer turning conditions in that consistent diameters are being turned. However, with the advent of turning centres on which complex component forms may be machined, this latter argument becomes less relevant.

The step turning test This test was a later development which to a certain extent overcame the problems of both the taper-turning and the variable-rate tests, allowing them to be conducted on an unmodified lathe. In the step turning test, a range of discrete speeds is utilised to determine the Taylor's constants. It shows close agreement with results obtained using the taper-turning and variable-rate tests.

The HSS tool wear-rate test As the name implies, this method of assessing machinability is concerned with the flank wear produced whilst cutting free-machining steels; the main parameters are elemental additions to the steels. These tests are carried out in a similar manner to the ISO 3685 : 1977 standard for long absolute tests (the BS 5623 : 1979 test, which was its equivalent in the UK, was withdrawn in mid 1984).

6.1.2 Non-machining tests

Whenever there seems to be a need to experiment with material cutting using one of the test methods described above, it is important to establish whether any savings gained will be recouped in the actual production environment. If a company is unsure of the likely cost benefits of such testing, then a strong case can be made not to test the material at all! Fortunately, non-machining tests exist that can be used in these doubtful cases rather than simply working in the dark with no data to base the cutting conditions upon.

As their name implies, these non-machining tests were developed in order to avoid the costly test procedures and time required to obtain data by the long (absolute) or shorter (ranking) machining tests, but at the same time to give results which have a reasonable approximation to those of the actual machining tests.

Ranking tests

The chemical composition test A variety of tests have been developed by which the materials are ranked according to their primary constituents. It is obvious that the results obtained from such tests are only relevant when materials of a similar type, and with identical processing conditions and thermal history, are to be machined. Given these limitations, however, these tests have proved quite successful for screening material prior to machining.

Typical examples of this type of test are those that rank the materials using the V_{60} scale (which gives the cutting speed in m/min) and the machinability index of 100 (used in the Volvo test, Fig. 6.3c). A close correlation between the chemical composition test and absolute tests has been obtained, with accuracies claimed to be within 8%. An example of the relationship between the chemical composition and cutting speed is

$$\text{Cutting speed, } V_{60} = 161.5 - 141.4 \times \%C - 42.4 \times \%Si - 39.2 \times \%Mn - 179.4 \times \%P + 121.4 \times \%S$$

Microstructure tests These tests are principally concerned with the type of microstructure present, that is, with the shape, dispersion and type of inclusions found in a steel. The early work was primarily concerned with the microstructure of low and medium-carbon steels, and more specifically with the spacing beween pearlite laminae achieved through heat treatment. The pearlite-to-ferrite proportions clearly influenced the material's Brinell hardness value. If a set cutting speed was chosen (e.g. V_{20}), a machinability rating could be obtained for either life at a constant speed (in minutes) or relative speed for a constant tool life (in m/min). It was noted that when more than 50% pearlite was present combined with a relatively high bulk hardness, then good machining characteristics resulted.

In recent years, the commercially-available steels have tended to be produced with trace amounts of other elements to aid machinability, typified by the addition of sulphur. In fact, many of the so-called free-machining steels also contain manganese, which combines with the sulphur to produce manganese sulphide inclusions. It has been shown that it is not only the amounts of manganese and sulphur that are important, but that the shape, size and distribution of these inclusions also affect the machinability of the steel.

This test method gives a better indication of the likely machinability of the steel, but it is difficult to perform without highly-specialised laboratory equipment; even then, it is only possible to rank the materials as good, bad or indifferent.

The physical properties test Specialised equipment is also required in order to perform this test, which uses the physical properties of a material in order to determine its machinability ranking.

Researchers have produced a general machinability equation using the dimensional analysis technique, and by using conventional tests to measure thermal conductivity, Brinell hardness, and percentage reductions in area and the sample's length. This gives close agreement with the V_{60} cutting speed for a range of ferrous alloys. However, when brittle materials are assessed, the lack of a yield-point and the much smaller reductions in area may cause problems in ranking.

The Volvo test Frustrated with the general test methods available, some companies have produced their own test methods to meet their specialised machining conditions and needs. Typical of these 'customised' tests is the Volvo test, shown diagrammatically in Fig. 6.3. The test consists of fly cutting (i.e. single point milling) using an HSS cutting tool ground to a tightly controlled geometry as shown in Fig. 6.3a. The depth of cut and the feedrate are standardised, whilst the cutting speed may be varied for each specimen tested. Cuts are taken at 90° to each other across the width of the specimen, and the end-point of the test occurs when the flank wear on the tool reaches the predetermined value (shown in Fig. 6.3b).

A value of the Volvo index 'B' is awarded to the material for a given cutting speed, which is chosen to represent good through to bad cutting conditions by a numerical value; the higher the 'B' value approaches the theoretical 100, the better the material's machinability. Fig. 6.3c shows a graph of volume of metal removed against cutting speed for a number of metals; a line showing the Volvo index 'B' is also drawn. Volvo has found this test to be an excellent method of correlating materials for its own needs, and other companies have also found it useful.

Having briefly discussed machinability tests, the mechanics of the cutting action is now presented in the next section, in order that the reader can appreciate its function more fully.

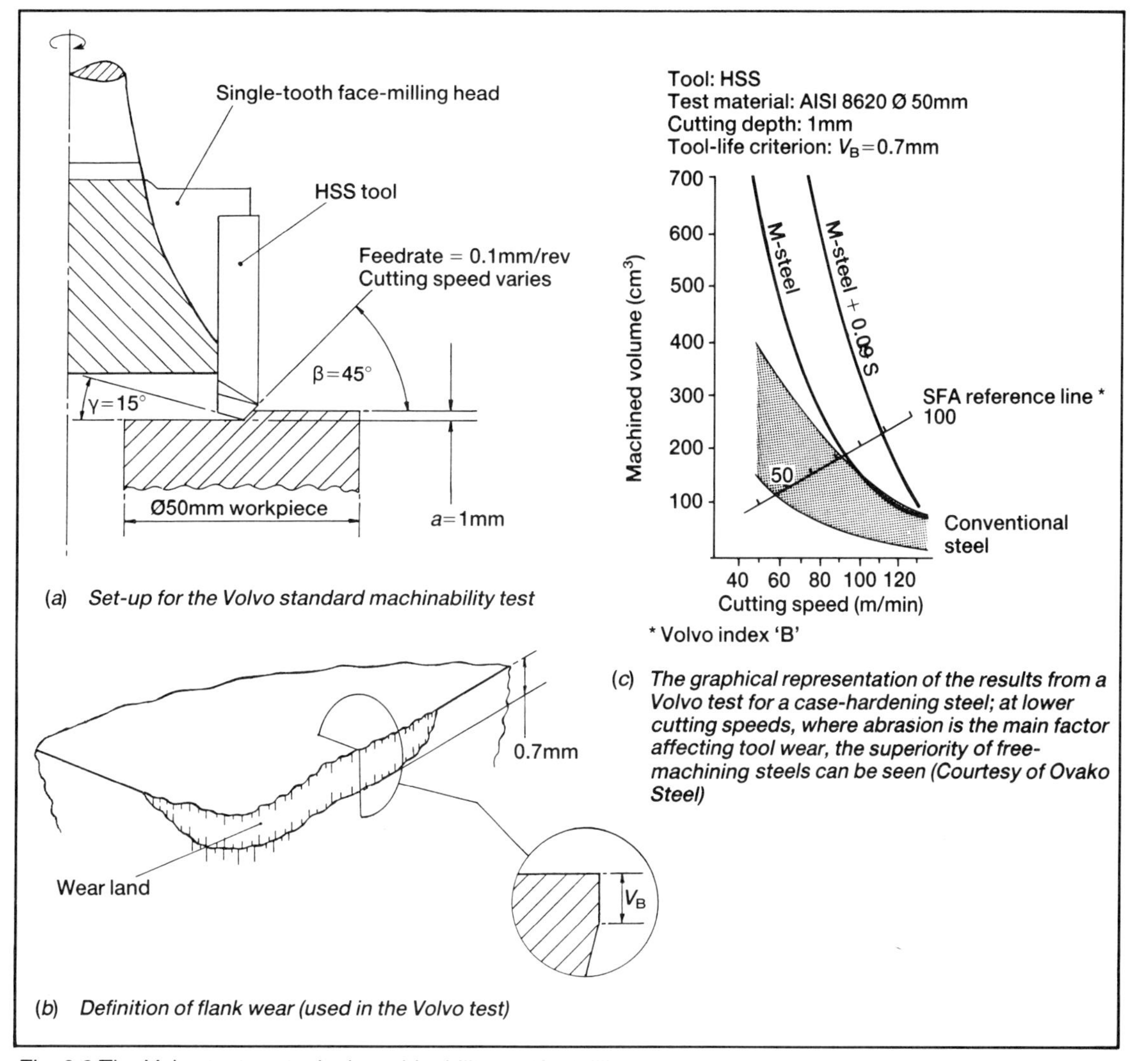

(*a*) *Set-up for the Volvo standard machinability test*

(*b*) *Definition of flank wear (used in the Volvo test)*

(*c*) *The graphical representation of the results from a Volvo test for a case-hardening steel; at lower cutting speeds, where abrasion is the main factor affecting tool wear, the superiority of free-machining steels can be seen (Courtesy of Ovako Steel)*

Fig. 6.3 The Volvo test – a typical machinability test for milling

6.2 The mechanics of metal cutting

In order to explain how a continuous chip is formed as a result of metal cutting, the simplest model, based on orthogonal cutting, will be described. Work on these models was first begun almost a hundred years ago, but in those days it was assumed that a 'splitting action' occurred in the workpiece material ahead of the cutting edge as the tool passed through it producing chips. However, this theory was soon dismissed in favour of one involving a shear plane, which, more or less has held favour ever since.

According to the shear plane model, the chip is formed during machining by deformation along the shear plane, elastically in the first instance, and then plastically, as it passes through a point of stress concentration. Thus chips are sheared to form segments, as shown in the diagram in Fig. 6.4a. This 'pack of cards' analogy, as it is usually called, is not of recent origin but was first described in 1937. The idealised diagram shows the workpiece material being cut by progressive slip relative to the tool point, at an angle that corresponds to the shear angle. It shows how each chip segment forms a small thin parallelogram, with slippage occurring along the shear plane. In reality, the deformation occurs over a zone of finite width, usually described as the primary deformation, or shear, zone. In spite of its assumptions, though, the diagram offers a good idealised model of how the shearing action occurs.

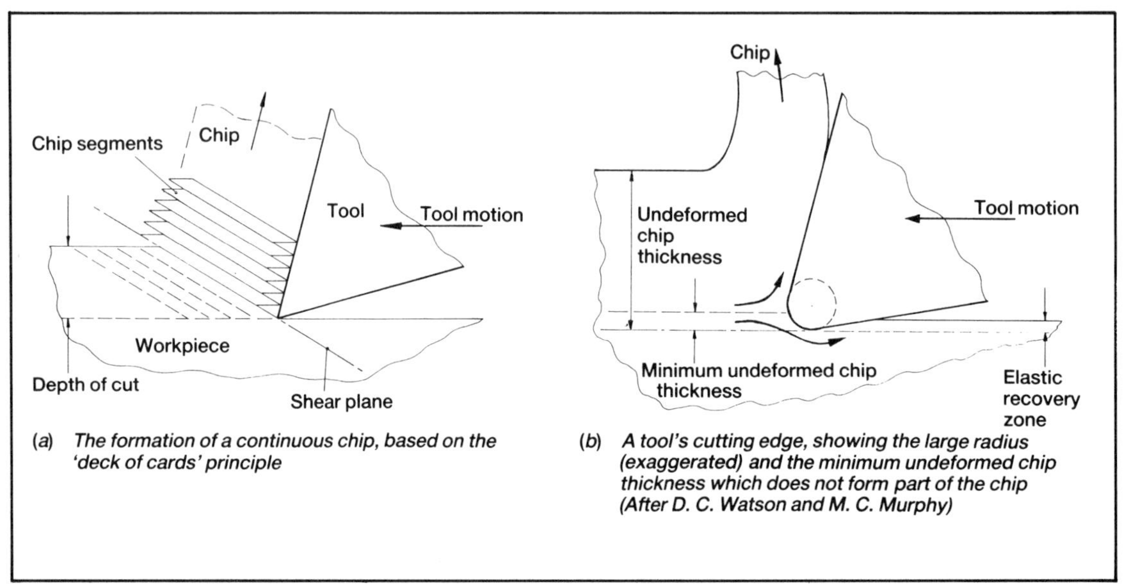

(*a*) *The formation of a continuous chip, based on the 'deck of cards' principle*

(*b*) *A tool's cutting edge, showing the large radius (exaggerated) and the minimum undeformed chip thickness which does not form part of the chip (After D. C. Watson and M. C. Murphy)*

Fig. 6.4 Formation of a chip by deformation

The problem with all these theoretical models is that they assume that the tool's point (i.e. where the clearance and rake faces converge) is sharp, whereas in reality there is always some modification to the point after the tool enters the workpiece. This effect is shown in somewhat exaggerated form in Fig. 6.4b, where the point is represented as an arc of a circle. In fact, the rake angle becomes progressively more negative the closer it approaches the clearance face, and there is a critical point on the tool above which the material is cut and forms the chip and below which it is ploughed. The minimum undeformed chip thickness is above this critical point for a specific material, and below this chips are not formed. When a tool's edge is sharp, it follows that the minimum undeformed chip thickness is small: with a cutting-edge radius of about 12μm and using a cutting speed of 210m/min, the minimum undeformed chip thickness has been found to be approximately 4μm. As a result of this minimum undeformed chip thickness and the very high pressures involved in this vicinity, material is extruded out from the interface between the tool and workpiece. This extruding action affects the surface finish and metallurgy. More will be said about this problem in section 6.8.

6.3 Chip formation

As a result of the many different cutting conditions that can occur in industrial machining operations there is an enormous variety in the shape and size of the chips that are formed (see Fig. 1.27 on page 32). This variety can be classified into three groups, examples from two of which are shown in Fig. 1.27. The three types are the continuous chip, the continuous chip with built-up edge, and the discontinuous chip. These three chip forms will be briefly reviewed in the following sections, which outline how and why they are produced.

6.3.1 The continuous chip

A steady continuous chip will be formed when cutting a ductile material under ideal conditions, perhaps using a cutting fluid. Typical examples of these continuous chip forms are those shown on the lefthand side of Fig. 1.27. Anyone familiar with machining materials such as low alloy steels, aluminium and copper will be aware of the continuous ribbon-like swarf that results. The metal shears off from the parent metal along the shear plane, remaining in a homogeneous form and not fragmenting; it behaves as if it was a rigid plastic material.

The continuous chip can cause problems in that it may curl around on itself, or envelop tools, jigs and fixtures, etc. and make their removal difficult. As a result the problems caused by this curling tendency, the normal procedure is to use chip-breakers as described in section 1.6.

Because the pressures involved whilst cutting these ductile materials are relatively low, it is unlikely that the chip will weld itself to the tool's edge. Increasing the cutting speed will have the effect of raising the temperature in the tool and workpiece accordingly but, because of this, it also reduces the shear strength required to form chips. The increase in cutting speed means that the pressure at the interface between the chip and the tool is also lower, so avoiding any welding tendency. A cautionary note about increasing the cutting speed: it will be obvious to anyone that wear on the tool will also increase as a result, and that extra heat will be generated in the workpiece. The heat conducted into the workpiece must be dissipated by extra cooling and lubrication whenever possible in order to avoid strains that would produce a warping tendency.

6.3.2 The continuous chip with built-up edge

One of the undesirable features of cutting very ductile – or, indeed, very tough – materials is the type of chip known as 'continuous chip with built-up edge'. A built-up edge is always a likely result of sterile and clean surfaces (a product of the generated heat) coupled with high pressure, as these are ideal conditions for pressure welding. This formation of built-up edge occurs particularly in situations when a low cutting speed is used and where the chip velocity is not very great across the interface between the chip and the tool. The 'sticking' phenomenon is a complex action produced by a number of interrelated functions; a quick review of them is in order at this point. A large force occurs along the interface between the chip and the tool, which is normal to the chip and to the tool's rake face. This has the effect of producing high stress conditions in the vicinity of the tool point. This means that, instead of the chip moving along the rake face at a uniform velocity, the unrestricted outer layers of the chip move at a free chip velocity whilst the innermost layers, which are in contact with the tool, move at zero velocity with respect to the tool. The region of the tool's face where this phenomenon can be observed is termed the 'sticking' region. The region beyond this, up until the point where the chip loses contact with the tool, is known as the 'sliding' region.

Fig. 6.5 shows a quick-stop photomicrograph of the build-up edge that adheres to the tool. The high deformation occurs over the sticking length or seizure zone. This is usually termed the 'secondary deformation zone', which will be discussed further later. The zone can be considered to be stationary on the tool, although the built-up edge is usually unstable. It is often called the built-up layer and is of finite thickness. This leads on to the fact that two types of built-up edge can occur, unstable and stable.

Unstable built-up edge – the worst condition

This type of built-up edge is continually destroyed and reformed at high frequency, which produces an effect where part of the built-up edge is carried away by the swarf's underside and part of it sticks to the workpiece. This adhered built-up edge gets larger and larger, until it cannot be retained on the rake face and overspills onto the flank

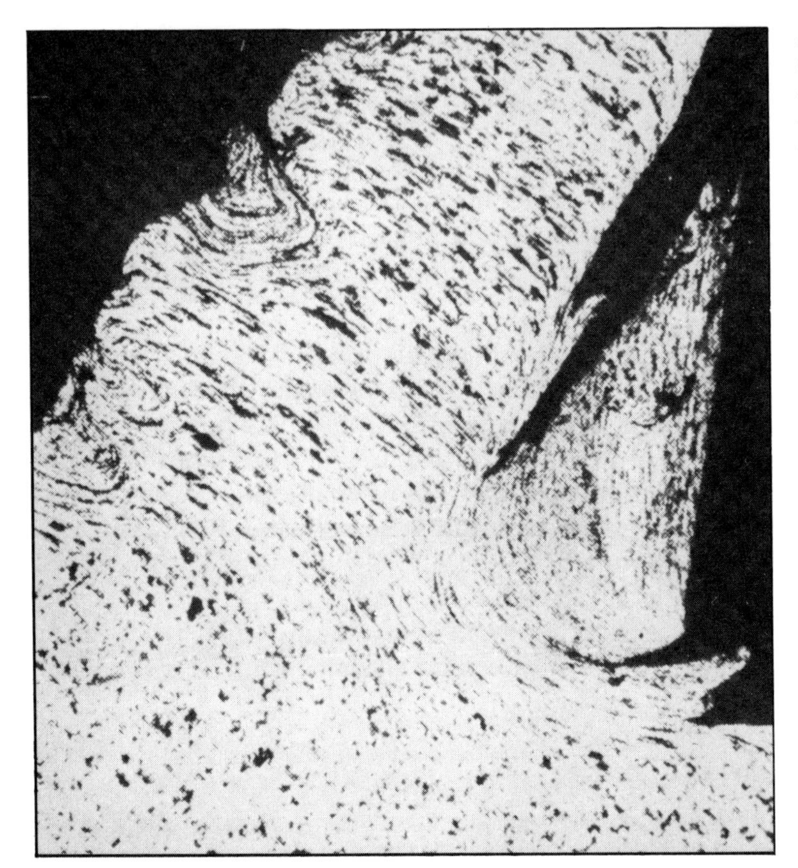

Fig. 6.5 A microsection of the cutting area, showing the built-up edge zone between the chip and the parent metal (Courtesy of Edgar Vaughan)

(clearance) face; at this point it fragments, with part (from the material on the flank face) sticking to the work and the remainder (from the rake face) being carried away by the swarf. This built-up edge material is extremely work-hardened (with hardness up to 700HV), and it alters the tool's geometry completely, causing a poor surface finish and a higher power consumption by the machine. As the unstable built-up edge is continually destroyed and reformed it takes away minute quantities of cutting-tool material from the rake face by an attrition action: this material is literally torn away by the velocity of the chip across the tool's face. The affinity between the tool and the work material play a large part in this degradation of the cutting tool by attrition. It can be partly overcome by changes of coolant or tool material – for example by using ceramic or cermets and, to a certain extent, coated carbide grades – where applicable.

Stable built-up edge

The conditions for formation of a stable built-up edge are often encouraged because it has been shown to protect the tool from high wear-rates. A stable built-up edge often occurs when cutting materials such as cast iron, and it gives a measure of protection to the tool when 'hard spots', such as sand inclusions or chilled regions, are encountered.

The condition for stable built-up edge is reached when the formation rate of the built-up edge is equal to the rate at which it is removed from the face, so that its size remains constant. Other benefits of a stable built-up edge are that it effectively causes an increase in the tool's rake angle, resulting in lower cutting forces and reduced power consumption. It also causes the tool to protrude beneath the flank face, which has the effect of reducing the amount of flank wear likely to occur.

The formation of a region of stable built-up edge is only likely if the insert geometry is modified, as it is not a natural phenomenon.

6.3.3 The discontinuous chip

A discontinuous chip is one that fractures either because the component material is inherently non-ductile, or because it produces little work-hardening during machining. In these situations, the combined shear and bending forces exceed the tensile strength of the material and discontinuous chip formation results. The formation of the chip causes the metal to undergo severe straining, and if the metal is brittle it will fracture in the shear plane's vicinity.

The shape and size of these discontinuous chips can vary considerably, ranging from needle-like swarf (e.g. with 60 : 40 brass) to rather longer segments (e.g. with SG iron), but they are all confined to a small contact area on the rake face of the tool and, as a result, draw less power from the spindle than continuous chip forms. Fig. 1.27 shows the effects of increasing the depth of cut and the feedrate on the tool's chip-breaking ability (irrespective of whether the material is ductile or not), and a range of types of discontinuous chips are illustrated.

It is worth mentioning that discontinuous chip formation does not occur solely with brittle materials; if ductile materials (e.g. medium alloy steels) are machined at low cutting speeds or with a very low rake, discontinuous chips can be produced. The stainless steels and nickel alloys are also susceptible to discontinuous chip formation, as a result of work-hardening of the chip. Artificially inducing ductile materials to form discontinuous chips by the use of chip-breakers has the problem that it may set up high-frequency vibration in the tool and thus lower the tool life or affect the surface finish. This problem can be lessened by identifying and using a cutting fluid of the correct grade.

Generally, discontinuous chips are desirable, in the sense that the swarf can be removed more easily from the cutting area and, at the same time, the so-called 'candyfloss' effect around the tool, work, etc. can be avoided. The swarf may, however, prove to cause problems by sticking to the moving parts of the machine or clogging up the coolant pump. Cast iron is particularly damaging in this respect, as the hard abrasive particles can increase coolant pump wear or, at the very least, give intermittent delivery of coolant – which is the 'life blood' of any high-performance cutting tool nowadays.

6.4 Tool chatter and its effect on chip morphology

Tool chatter is the result of forced vibration set up between the workpiece and the tool. The vibration is usually attributed to such causes as lack of stiffness in the machine tool, poor workpiece rigidity, insecure tool clamping, and use of an incorrect or poor lubricant.

The effects of chatter on chip morphology are shown in the photomicrographs in Fig. 6.6, which were produced on a scanning electron microscope. The micrographs illustrate the complex shearing that occurs at two cutting speeds, one very low (12m/min) and the other reasonably high (120m/min). The low cutting-speed sample is shown in Fig. 6.6a and b, and the high cutting speed one in Fig. 6.6c and d.

The low cutting-speed micrographs show longitudinal views of the lamellar constuction of the chip's surface; this is a common type of continuous chip formation. With this type of chip, the shearing occurs by a progressive build-up of compressive deformation, followed by the explosive propagation of shear completely through the chip's thickness.

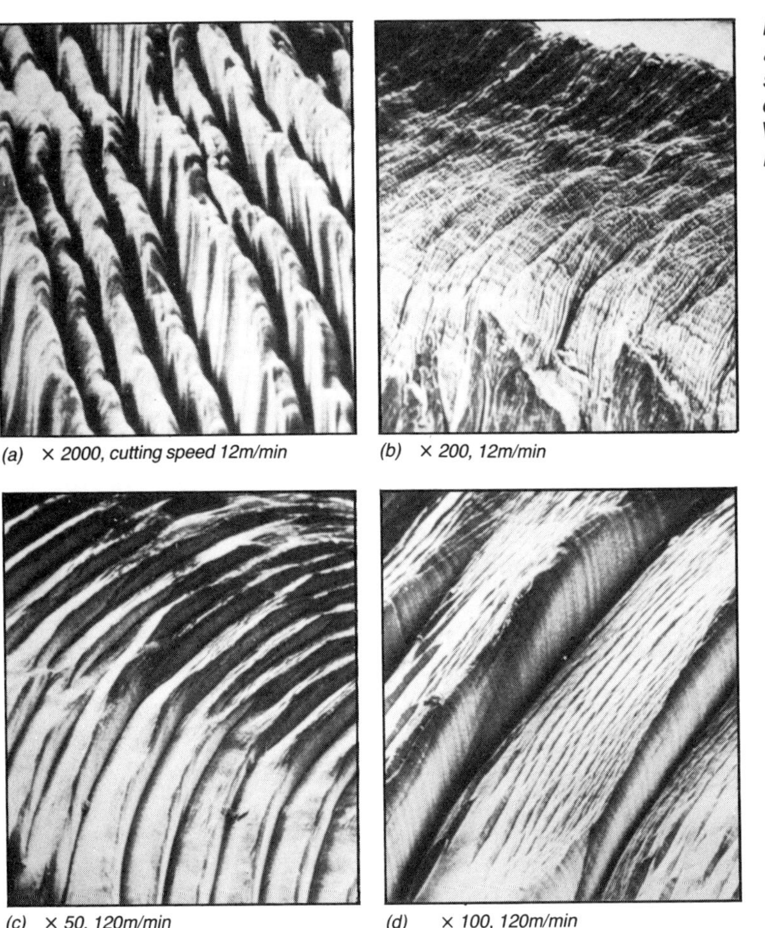

(a) × 2000, cutting speed 12m/min (b) × 200, 12m/min

(c) × 50, 120m/min (d) × 100, 120m/min

Fig. 6.6 Variations in chip surfaces at different cutting speeds, showing irregular shearing of the chip (Courtesy of Edgar Vaughan/D. C. Watson and M. C. Murphy)

In addition, the sticking that occurs at the interface between the chip and the tool produces a subsequent bending of the primary shear zone and a region of intense deformation, called the secondary deformation zone (See Fig. 6.17 on page 193).

The micrographs illustrating the chip morphology produced by the higher cutting speeds show that in this case a pronounced segmental morphology results. These micrographs (Fig 6.6c and d) show what is normally referred to as a saw-toothed, serrated or semi-continuous chip form. The noticeable periodic variations in the chip's geometry are not thought to be caused by any vibrations in the machine tool, but are probably the result of the stress becoming unstable; any such instability has the effect of causing the shear zone plane to oscillate backwards and forwards whilst cutting takes place (Fig. 6.7a). The differences in the shapes of the segments and in their frequency that occur with different cutting parameters are thought to be dependent upon the frequency of the shear plane's oscillation relative to the cutting speed.

Tool chatter can have disastrous effects on the tool's life: in extreme cases it could be reduced to a mere 1% of its expected life! Other consequences of tool chatter include side effects that range from premature failure of the machine tool's bearings to a high noise level that creates considerable discomfort to the operator. It also causes degeneration of the surface finish (Fig. 6.7a) and dimensional inaccuracies on the workpiece which are probably unacceptable, causing the workpiece to be scrapped. Once again, the use of the correct cutting fluid will minimise the consequences of tool chatter by providing the lubricating and cooling properties necessary for efficient cutting to take place.

Fig. 6.7 The effect of chatter on the workpiece

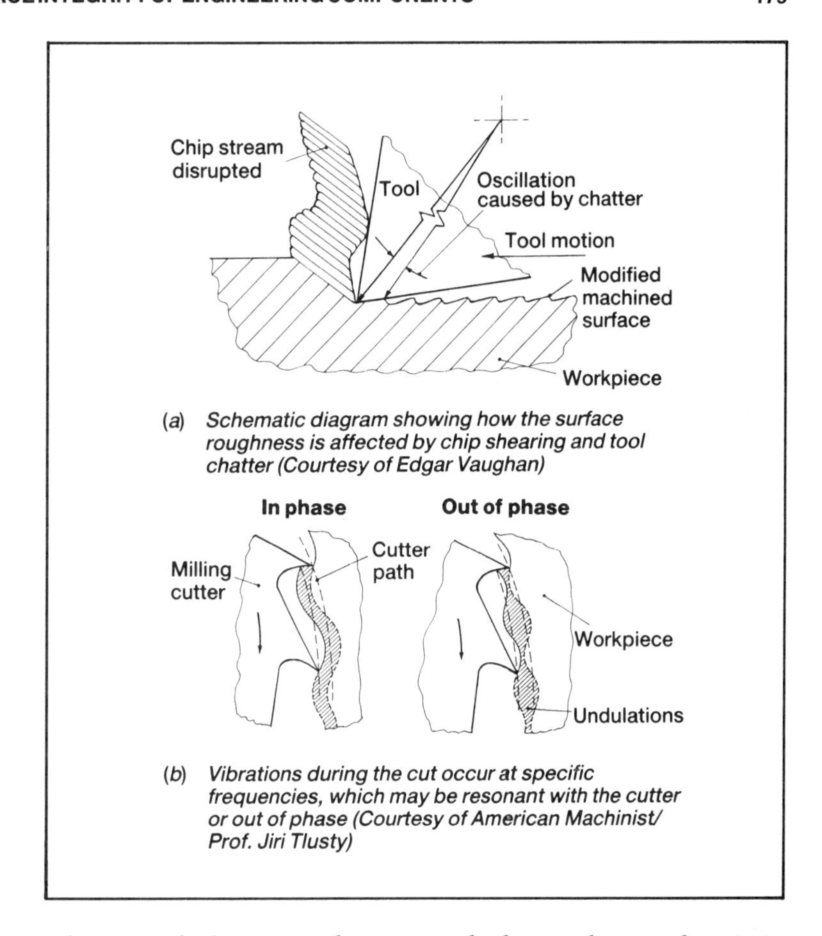

(*a*) *Schematic diagram showing how the surface roughness is affected by chip shearing and tool chatter (Courtesy of Edgar Vaughan)*

(*b*) *Vibrations during the cut occur at specific frequencies, which may be resonant with the cutter or out of phase (Courtesy of American Machinist/ Prof. Jiri Tlusty)*

To complete the discussion about tool chatter, it has recently been shown that it is possible to utilise chatter to advantage when milling at high speed using multi-toothed cutters. This process (Fig. 6.7b) is known as high-speed high-power (HSHP) milling; it utilises the fact that the frequency of the teeth on multi-toothed milling cutter may approach the natural frequency of the system, namely the machine tool. According to one definition, chatter is the regeneration of waviness on the cut surface. Normally, every tooth will cut into the undulating surface left by the previous tooth, and at low speeds there will be many waves between two subsequent teeth. However, at higher speeds there are only a few undulations (Fig. 6.7b), and the actual phase at which a tooth meets the preceeding waviness will have a greater influence on chatter. So the effect of changing the spindle speed is to alter the limit of stability considerably, the greatest changes occurring around the speed at which the tooth frequency equals the frequency of the dominant mode.

A mixture of forced vibrations is always present in milling, owing to the periodicity of the cutting force, and at high depths of cut this results in chatter, but this is not the case with turning operations. In a well-developed chatter situation, the amplitude of the relative vibrations will considerably exceed the average chip thickness at the maximum chip load. Therefore periodic force and vibration are not directly related at this point. If, as a result of the frequency and speed used, the undulations generated by subsequent teeth meet in phase (Fig. 6.7b *left*), there is no variation in chip thickness and, as a result, no periodic force component due to vibration.* This cutting condition is known as the 'resonance case'. However, the regeneration of waviness usually dampens the resonant forced vibration. The out-of-phase condition (Fig. 6.7b *right*) has the same amplitude of

*Because of the intermittent nature of the cut there will still be a periodic once-per-tooth force component in all cuts except for those at full depth, but this component is minute.

vibration, but there is a phase difference of 180° between successive teeth. The variation in chip thickness and the corresponding periodic force are therefore at their maximum levels. The relationship between vibration and surface condition is of still greater complexity than that between vibration and force, and the consequence is that the machined surface has virtually no chatter marks on it.

The advantages of multi-toothed milling using the out-of-phase condition rely on the fact that the cutter can be tuned to the 'miracle speed', as it has been called, at which the tooth frequency of the cutter equals the natural frequency of the dominant mode. Therefore research is continuing into developing a system that will monitor force or vibration, and then control the spindle speed to the most stable condition, i.e. a method of adaptive control.

6.5 Tool wear – a tribological condition

The selection of the cutting-tool material is fundamental in determining the extent and type of wear to be expected. Fig. 6.8 shows the development of just some of the cutting-tool materials developed over the last century. The graph clearly shows that the prime objective of developing a material for a cutting tool is to increase the cutting speed so that shorter cutting times will result, leading to greater economic savings. The problem with increasing the cutting speed is that the net result is high temperature in the contact zone between the tool and the workpiece (Fig. 6.9), and that this leads to thermal softening when the cutting-tool material's specific temperature threshold is reached, in the case of metallic-based tools. An often-quoted example of this phenomenon is that high-carbon steel derives its cutting ability from the heat treatment it receives but loses its edge hardness at relatively low temperatures, typically 200°C. High-speed steel, on the other hand, retains its hardness at temperatures approaching dull-red heat, 600°C. This means that high-speed steel can be used as a cutting tool material at a temperature three times

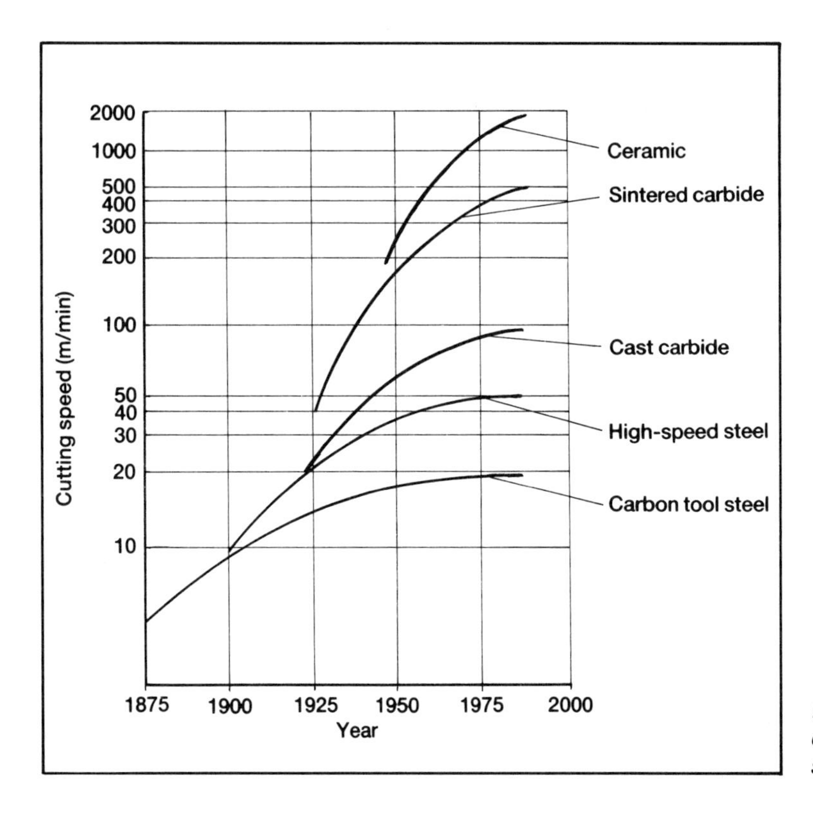

Fig. 6.8 The development of cutting-tool materials – cutting speeds (Courtesy of Krupp Widia)

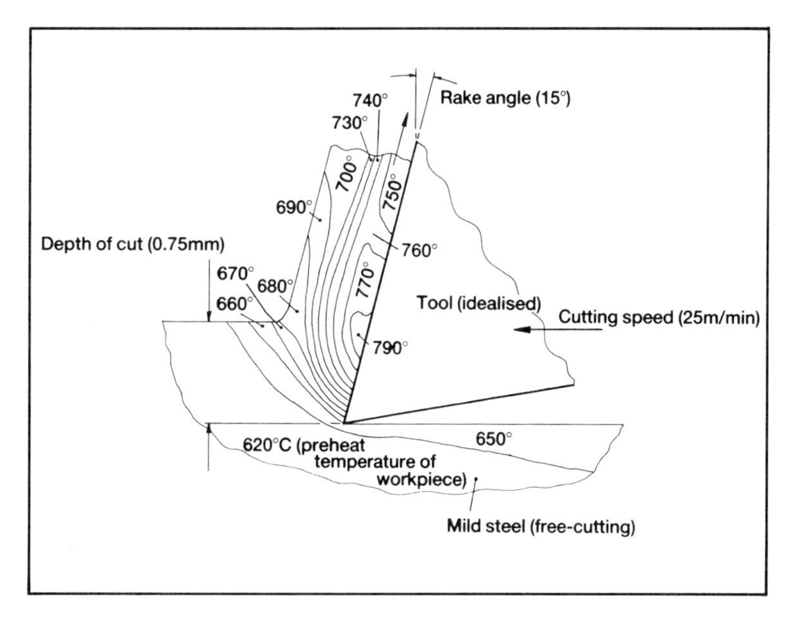

Fig. 6.9 Temperatures typically found in the workpiece and the chip

higher than high-carbon steel can, showing real savings in cycle times, despite the fact that at room temperature high-carbon steel is the harder. However, both these cutting tool-materials are now considered to have been superseded by the various carbides, ceramics, cermets and superhard materials, such as polycrystalline diamond and cubic boron nitride, for CNC tooling applications.

6.5.1 The effect of tool temperature on tool wear

When metal-cutting operations occur, heat is generated in the primary and secondary deformation zones, resulting in a temperature distribution which differs widely throughout the workpiece, chip and tool. The influence that heat plays on the life of a tool was described by F. W. Taylor in 1907 in a paper entitled *On the Art of Cutting Metals.* He identified the necessary path forward towards developing high-speed steel. Even earlier than this, it was clear to J. T. Nicolson doing some experimental work in Manchester in 1900, the results of which he later published in *The Engineer* (1904), that the tool temperature varied according to the geometry of the tool. The greatest problem in those early days was a means for recording the temperature and observing its distribution throughout the tool; even today, realistic methods of determining the isotherms within the tool are difficult. This meant that progress in formulating the reasons why wear occurred at specific regions on the tool was slow.

The conversion of work done into heat is the result of deforming a workpiece to produce a chip, then moving the chip whilst passing the tool over a freshly-cut surface. As this is a plastic strain situation, only about 1% of the energy is stored as elastic energy, the remaining 99% being conducted into the chip, tool and work material in the form of heat. It has been shown that under most normal cutting conditions, nearly all the work done is in chip forming at the shear plane, and that most of this generated heat is passed into the chip (Fig. 6.9), with a proportion being conducted into the workpiece and tool.

The heat conducted into the chip is greatest some way up the rake face (Fig. 6.9), and this results in eventual crater wear occurring in the tool at this position (Fig. 6.10). This is not the only form of wear that is produced: just as serious are the flank wear that occurs on the clearance face of the tool and the other wear mechanisms (Figs. 6.10 and 6.11). The elevated tool temperature causes some effects of crater wear, and this was traditionally attributed to the heat generated at the interface between the tool and workpiece by

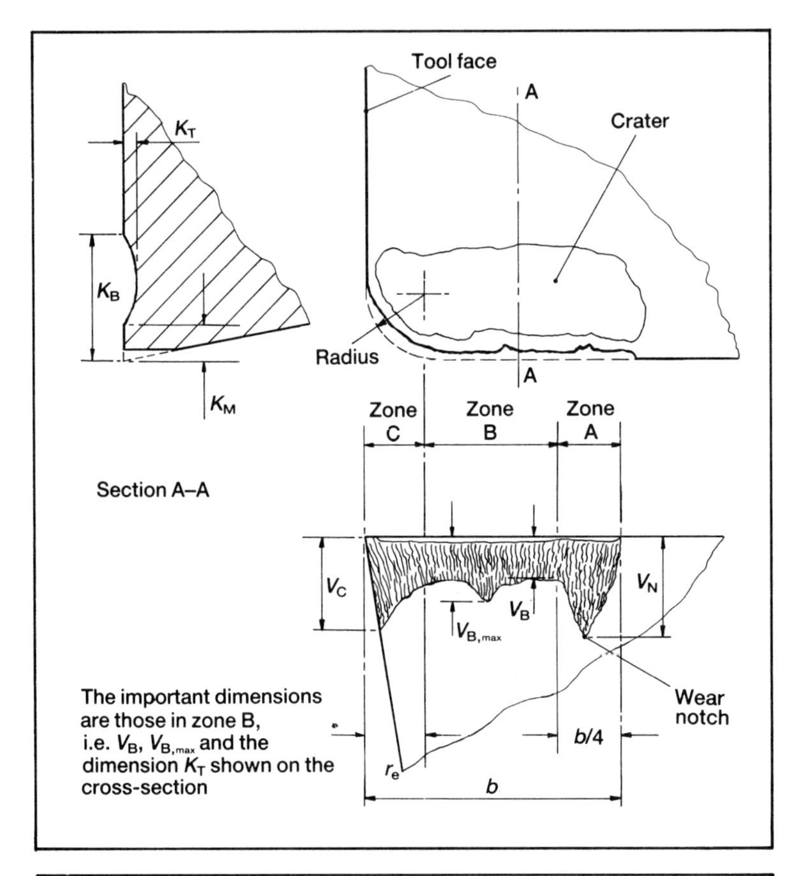

Fig. 6.10 A typical wear pattern that might occur on a carbide insert under normal operating conditions (Courtesy of Seco)

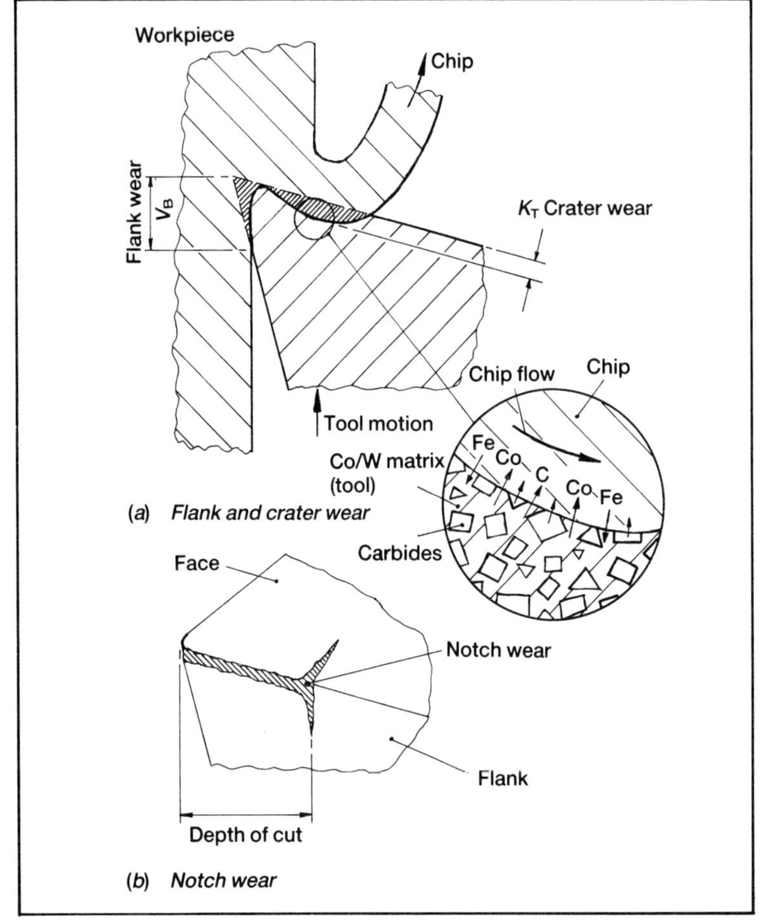

Fig. 6.11 Types of wear on inserts

means of friction and its conversion to heat energy. The current view is that the normal condition is one in which seizure through strain effects in the flow zone is the main contributor in raising the tool's temperature, but the reasons for this are beyond the scope of this book.

So far a brief mention has been made of only two of the types of wear which may affect the tool whilst cutting, namely crater and flank wear. These tool-wear conditions need more clarification and other wear mechanisms must be discussed.

6.5.2 The mechanisms of wear on the tool

The tribological (from the Greek *tribos* meaning 'to rub') nature of cutting-tool wear may be divided into six categories, which are discussed in the sections that follow.

Diffusion wear

This is a particular problem resulting from the interaction that occurs between metallic workpieces and cutting-tool materials during machining; the wear rate depends heavily on the metallurgical relationship between the materials. An example of this effect is the diffusion that occurs between conventional steel workpieces and carbide cutting tools. The diffusion process is highly temperature-dependent, and when machining at high cutting speeds, and therefore at high temperatures, there is actual atomic transfer across the interface between the tool and the chip (Fig. 6.11a). This diffusion is in accordance with Fick's first law which states that the flow will be proportional to the concentration gradient; in this example, it is a two-way atomic transference. As a result of the high interface temperature, ferrite diffuses from the chip into the tool. This weakens the atomic bonding by increasing the solubility of free carbon in the tool. The carbon diffuses to the chip, simultaneously increasing the solubility of cobalt in the tool. Once the cobalt has diffused to the chip's underside, the remaining tungsten carbide in the tool is left unsupported and is rapidly swept away by the velocity of the chip across the interface. This action is continuous, and eventually a crater forms in the region of highest temperature on the rake face, as shown in Figs. 6.10 and 6.11a.

The fact that cobalt is the binding element that holds the tungsten carbide in solution means that the tool now has a low shear strength at its exposed surface and the carbide is rapidly transported away from the tool's surface. Additions of tantalum carbide are often admixed and sintered into the insert to lower the diffusion rate which causes crater wear. Their addition modifies the structure of the tungsten carbide grains and has the effect of lowering their solubility in the workpiece. Most carbide cutting tools now have a thin coating applied to them (see section 1.2) to interrupt the diffusion mechanism by adding an inert barrier at the interface between the workpiece and the tool.

Note: The importance of diffusion as a tool-wear mechanism is the subject of debate between some researchers in the field.

Attrition wear

This type of wear mechanism is found principally on the tool's face; it is normally associated with the relatively low cutting speeds that lead to the formation of an unstable built-up edge. The low temperature means that wear by diffusion or plastic shear does not occur, and chip flow across the interface between workpiece and tool tends to be of an irregular or laminar nature, resulting in an unstable built-up edge. When the built-up edge breaks down, it tears away microscopic fragments of the tool with it, and this mechanism may be termed attrition. Small fragments of the tool's edge are continually and progressively removed as the built-up edge breaks down.

Attrition is normally associated with a slow progressive type of wear, but this may be accelerated if an intermittent cutting condition occurs, or if vibrations are present during machining owing to a lack of rigidity of the tool or workpiece. This type of wear is not influenced by elevated temperatures and, indeed, it is likely to disappear completely as the chip flow becomes more laminar. It is possible to reduce attrition wear by using tools with a grain size less than 2.5μm.

Fatigue wear

Under conditions where the wear by abrasion and attrition is insignificant, the dominant wear mechanism will be by fatigue. Fatigue conditions occur when the tool surface is repeatedly subjected to a loading and unloading cycle. This leads to small portions of the tool material becoming detached from the tool's surface. The fatigue effect may be a result of an intermittent cutting action (e.g. eccentric turning or milling), and it may lead to the edge of the tool becoming chipped. Fatigue wear and edge chipping may also be initiated by such sub-surface features as non-metallic inclusions near the tool's surface or even fatigue cracks just below the surface.

If the stress on the tool is lower than a certain limit, it is unlikely that cracking of the tool by fatigue will occur, as the contact pressures are determined by the workpiece material's yield properties. Any fatigue problems can always be lessened by changing to cutting-tool materials that are appreciably harder than the workpiece.

Abrasion wear

Of all the wear mechanisms likely to occur on the cutting tool, the most common is abrasion. Abrasion of the tool is caused by the work material containing high surface concentrations of hard particles (e.g. sand pockets, 'scabs' from the casting process or hard oxide coatings on hot-rolled bar stock). Under these conditions, the hard particles abrade the tool's surface in a similar manner to a grinding wheel. Obviously, when such conditions are expected, the surface layer can be removed by a variety of treatments, such as bead blasting, etc.

The material to be cut may in fact be very soft, but there could still be precipitates or hard inclusions present within the structure, resulting from the heat treatment and the manufacturing process respectively. The nature, shape and size of the hard inclusions will affect how they abrade the tool's surface: those with angular sharp edges are much more effective at increasing the wear-rate than the hard inclusions that are smooth and spherical. Typically, the hard angular inclusions produce a 'micro-cutting' effect on the tool, whereas the more spherical types have a tendency to deform the surface plastically, with a grooving action.

Abrasive wear is not confined to any one face of the cutting tool: it may be seen on both the rake and the flank faces simultaneously. The abrasive action tends to produce a flat surface on the tool and causes such conditions as flank and notch wear (Figs. 6.10 and 6.11b).

Notch wear in particular is the result of either machining surfaces that have been extensively work-hardened or machining materials with hard oxide films. The position of the notch on the cutting tool (Fig. 6.11b) is related to the depth of cut used. This type of wear tends to be much more excessive than flank wear caused by the action of the abrasive surface particles or work-hardened material on the tool's surface.

Electrochemical action

If a cutting fluid is used, it is possible to cause an electrochemical reaction between the workpiece and the tool under appropriate conditions. This reaction results in the formation on the tool's face of a rather weak layer of low shear strength. This effect is normally considered desirable as it lowers the frictional force on the tool, which in turn reduces the forces required for cutting whilst simultaneously lowering the tool's temperature. However, it may also cause small amounts of tool material to be removed by the action of the chip flow. The benefits gained from using a cutting fluid may be more than cancelled out by the tool wear caused by electrochemical action under certain specific conditions.

Other factors that increase tool wear

These factors include plastic deformation, brittle fracture and edge chipping; in fact, none of these is considered to be a true wear process, but they do contribute to the actual wear on the tool.

A major problem that occurs when using some carbide cutting tools at high speeds or feeds, or when machining relatively hard workpiece materials, is a tendency for the insert to deform plastically under compression (Fig. 6.13, no. 5).

When a high cutting speed is chosen, the insert may deform plastically as a result of the high tangential force (see Fig. 1.19 on page 21). This 'bulging' is usually caused by several interrelated factors. For example, the high cutting speed causes the temperature of the tool to increase locally, a factor which is further exacerbated when a small nose radius is used. This local increase in temperature, coupled with the high speed, causes the tool geometry to change. So the chip flow pattern is modified accordingly, leading to tool wear. By using a larger nose radius on the insert, (when appropriate), the heat can be more easily dissipated. The heat conductivity of the tool may also be improved by changing the composition of the insert.

When high feedrates are utilised in conjunction with the other wear-promoting factors already mentioned, namely a small nose radius and a low conductivity insert, the axial force (see Fig. 1.19) will also plastically deform the insert. In this case, the clearances and plan approach angles will change, rather than the rake angle as in the previous case. Once again tool wear will be increased, leading to a situation where premature tool failure results. In fact, very little deformation is necessary on carbide inserts before some form of premature failure occurs. If a carbide insert is deformed plastically, cracks are formed which lead very shortly to sudden tool fracture. Inserts with a lower cobalt content offer increased resistance to deformation and, of course, to cracking.

Brittle fracture is associated with large portions of the tool becoming detached, causing sudden failure. This type of failure is associated with very high feedrates and depths of cut; it may be caused by a complex distribution of stresses produced within the tool when specific machining conditions are adopted. If adaptive control is used, the likelihood of high feedrates and depths of cut affecting the tool can be minimised by using the 'over-ride' facility within the software to protect the cutting insert. Normally, the conditions that cause complex stress distributions in the tool are associated with machining high-strength materials using carbide tooling. The cutting is controllable until a quite small amount of flank wear has been experienced by the insert, which results in a high tensile stress developing. Because carbide inserts tend to be weak in this aspect, premature tool failure often results.

Whenever there is an intermittent cutting action, the edge-chipping effect is a common wear phenomenon. Intermittent cutting operations promote cyclical thermal and mechanical stresses in the cutting insert; the outcome is that fine cracks develop near to the cutting edge, with a degree of insert flaking. This problem can be reduced by changing the insert geometry or its toughness*, or by using adaptive control techniques.

Fig 6.12 shows an insert that is well past its best! Heavy flank wear can be clearly seen on the front clearance face of the insert, and a large crater has developed along the depth-of-cut contact region. Notice how the crater has been formed some way in from the tool's edge, on the rake face. This illustrates the point that its position on the rake face coincides with the highest-temperature region of the chip and shows the effects of diffusion and tribological action on the tool.

Fig. 6.13 summarises the criteria for wear and failure of a hard-metal insert. As a rule, several wear mechanisms act simultaneously. For a specific combination of cutting tool and workpiece material, the wear is determined primarily by the cutting speed, V_c. A rise in cutting speed produces an increase in temperature leading to temperature-induced wear. The 'Taylor's curve' (Fig. 6.13 *top*) illustrates the combinations of time and cutting speed at which the different wear mechanisms are likely to occur. When low cutting speeds are utilised, there is a greater chance that a built-up edge condition will occur; at the other extreme, using high cutting speeds, there is a high probability that the insert will be plastically deformed. At intermediate positions on the graph, where the 'optimum' cutting speed is likely to be selected, several wear mechanisms can occur simultaneously.

Cutting conditions, feedrates and cutting speeds, compatibility of tool and workpiece, tool geometry, lubricants, etc. – all these factors have a bearing on the likely life of a tool and are reflected in the number of parts machined per insert edge. It is worth taking a closer look at these factors that affect tool life in its broadest sense.

6.5.3 Tool life

An insert edge's life is assessed according to three criteria, namely its ability to hold workpiece tolerances and to maintain specific requirements for surface finish and an

Fig. 6.12 Flank and crater wear in P15 hard-metal insert after 26.5 minutes of turning hardened and tempered AISI 4340 at a cutting speed of 200m/min (Courtesy of Ovako Steel/University of Hanover)

*The appendixes list possible causes of cutting-insert problems and suggests remedies, and can be used as a starting point in deciding on changes of the insert grades.

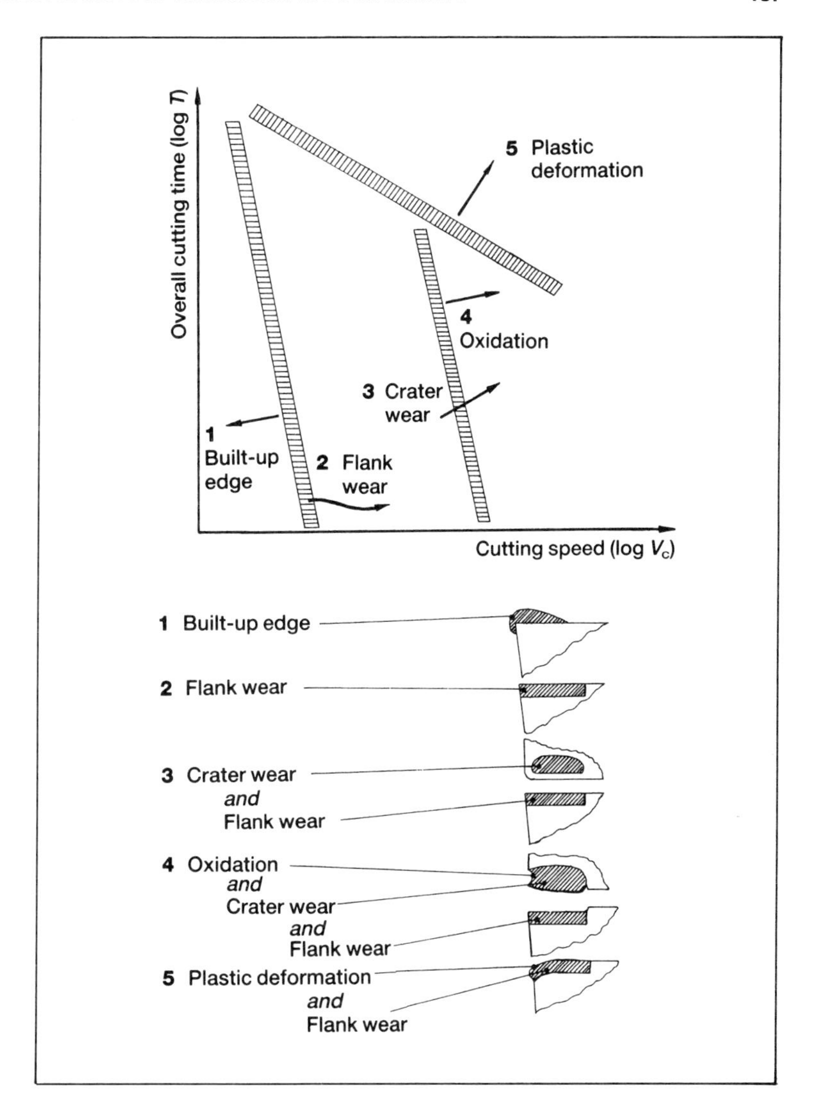

Fig. 6.13 The wear and failure criteria for hard-metal inserts (Courtesy of Krupp Widia)

efficient chip-breaking ability. If the insert no longer satisfies these criteria, its useful life is ended and it should be discarded. The tool's life is the key factor in an estimation of the productivity level expected for specific cutting conditions. Seen from another viewpoint, a programmer may deliberately choose an insert they are familiar with, that wears out in a progressive manner, rather than have the unpredictability associated with an insert of uncertain machining capability that may break down at any time.

Before discussing criteria for determining when a cutting insert is 'worn out', it is necessary to establish exactly what this means. Does it mean when the dimensional accuracy becomes unpredictable or when the surface finish deteriorates, for example? Often it is only by experience that one can judge how much wear on the cutting edge can be allowed before machining is discontinued, assessing the degree of wear by the flank wear on the insert. As a rule, flank wear can be considered a reasonably dependable criterion for assessing when a cutting edge is worn out, but there are many other factors that can also cause a shortened tool life; in this respect, the degree of cratering is much more significant.

Tool wear can be determined by several means, but the usual method is to observe and measure the actual wear as it progressively develops. The effective cutting time or tool life, T, is specified as the time elapsed before a predetermined degree of wear is reached. The extent of tool wear can be observed and measured as follows (see Fig. 6.10):

- The extent of flank wear from the original edge, if it is of a relatively uniform nature, may be found by spreading assessment over the three zones, A, B and C. The mean flank wear $V_{B,C-A}$ is measured over the cutting part of the edge across these zones. If excessive wear develops at one position on the cutting edge, such as a wear notch V_N, this zone is ignored when determining the mean wear; under these conditions the maximum flank wear V_{Bmax} is reported instead, together with the zone where it occurs.
- The extent of cratering is specified by the maximum depth of the crater from the original rake face K_T, and in some cases by its size as measured by the K_M and K_B dimensions.

These are the criteria used for estimating the extents of flank and crater wear. It has been found that progressive flank wear develops according to a fixed pattern, and that there are three stages of flank wear that occur during the life of a tool. These are the initial (primary) wear, progressive (secondary) wear and catastrophic (tertiary) wear (Fig. 6.14a).

Initial or primary wear When a new tool is first used to machine a component, there is a rapid breakdown of the edge, shown by the initial high wear-rate in the graph of wear against time (Fig. 6.14a). This wear-rate is dependent upon the cutting conditions and the workpiece material; it will decrease as the cutting speed is increased. It is often stated that the high wear-rate in this region is caused by crumbling of the edge, rather than the insert being worn in.

Progressive or secondary wear Once the initial wear has occurred, there is a steady progressive stage of insert wear, with a much less dramatic increase than in the initial stage, when realistic cutting speeds are applied. Towards the end of the progressive wear zone, which in this case is when the flank wear V_B, is about 0.8mm, it is usual to replace the tool with a 'sister tool'. Once this extent of flank wear is reached, then to all practicable purposes the tool's life is ended.

Catastrophic or tertiary wear The final stage of wear is considerable and rapid, leading to catastrophic failure of the insert. This failure is caused by a combination of high flank wear and a large crater formation, which reach a point where the tool is sufficiently weakened for the tool forces to cause it to fracture. Inevitably, if such a rapid edge breakdown occurs during the final pass along the surface of the work, it is likely that the workpiece will have to be scrapped. When the workpiece has a high value, which has been increased as a result of the machining, any savings made by using tools beyond the end of the progressive wear zone to produce extra components will be more than cancelled out by the value of the component scrapped!

Under the conditions of medium or rough turning operations, the limiting value of flank wear normally lies between 0.3mm and 1.0mm (Fig. 6.14b). Experience shows that as the flank wear approaches the upper value (1.0mm), the wear rate increases rapidly as wear is now in the catastrophic zone. If the tool is not replaced, the extent of the eventual damage may vary in degrees from simply the cutting edge to the complete tool, the component or, under extreme conditions, the machine as well. As a practical rule of thumb, 0.5mm is often given as an acceptable upper limit for the wear. Using this value for flank wear and keeping all the other variables constant – apart from the cutting speed, a relationship can be established between the cutting speed and the insert's life. Fig. 6.14b shows the relationship between the cutting time and flank wear for five different cutting speeds, ranging from 170 to 260m/min.* The flank-wear limit of 0.5mm was reached after periods of between 4.8 and 33.3 minutes, depending upon the cutting speed used. By simply increasing the cutting speed from 170 to 260m/min, the useful life of the cutting edge was reduced to about one-sixth.

What would be the correct speed from the point of view of economics? To assess this, it is necessary to develop the relationship between cutting speed and wear two steps further. If the five tool lifes are plotted against their cutting speeds on logarithmic paper, a curve such as the one shown in Fig. 6.14c is produced. This curve, which is more or less a straight line, is the 'Taylor's curve', which is often called the '*VT* curve' or a 'tool-life diagram'. It indicates succinctly the relationship between cutting speed and edge life, for one workpiece material and one specific machining operation. The features that characterise it are its position and its gradient; these are expressed by the general formula†

$$VT^{\alpha} = C$$

The value of α indicates the slope of the curve, whilst the value of C determines

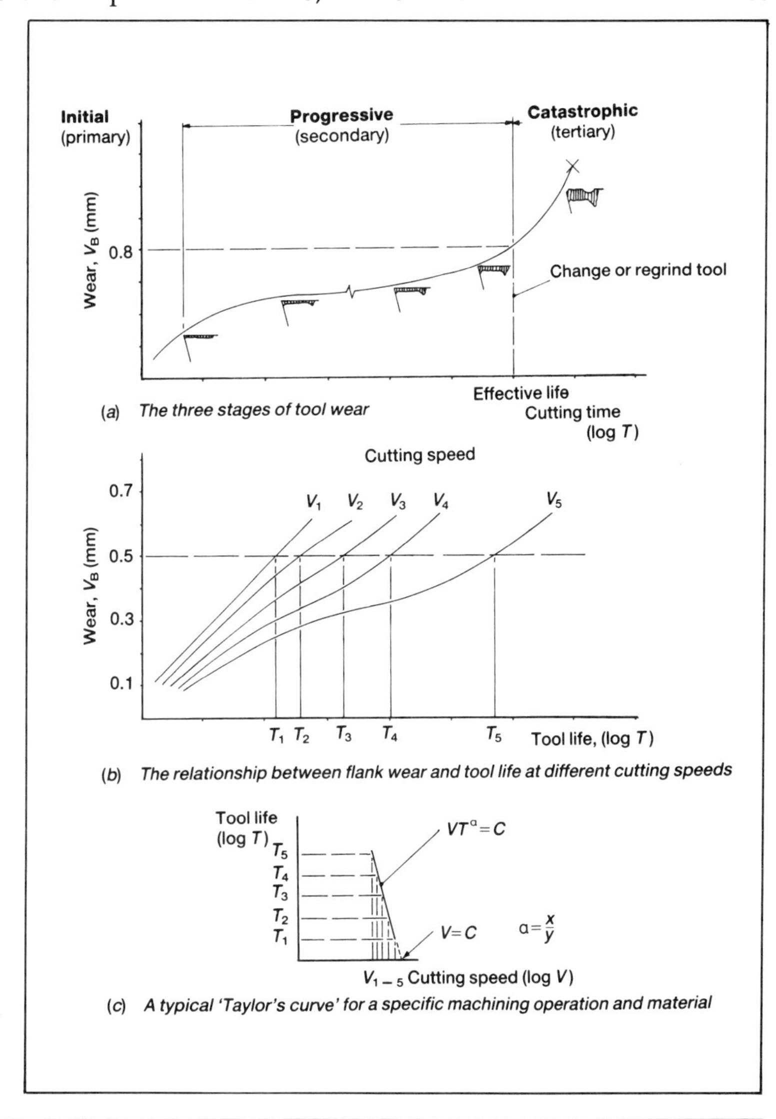

Fig. 6.14 Tool-life diagrams (Courtesy of Sandvik (UK) Ltd.)

*Similar curves would be obtained if crater wear was the parameter being measured.

† This forumula is for the general case; from this equation others have been developed which define the machining characteristics with more mathematical rigour, but it is not the intention to deal with these here, as they are fully covered elsewhere in the available literature.

its position. The value of α, which is a constant, may be obtained graphically from values of the x and y-coordinates; the value of the constant C is found by extrapolating down to the cutting-speed axis.

6.6 The use of cutting fluids during machining operations

It is apparent to anyone who has machined a component that cutting operations basically use the tool as a wedge which is thrust into the body of the material, thus prizing out a chip by a shearing action. This operation is accompanied by the generation of large amounts of heat from the following sources:

- Internal friction caused by shearing the chip from the workpiece.
- Internal friction due to chip deformation.
- External friction between the clearance face and the workpiece.
- External friction between the rake face and the chip.

The internal friction will depend upon the nature of the workpiece material and the prevailing cutting conditions. This friction can never be entirely eliminated but use of a cutting fluid will reduce the amount of unwanted heat generated by the cutting action. A reduction in the external friction can be achieved by a boundary layer of a suitable lubricating oil between the rubbing surfaces. Despite the presence of such a lubricating film, there is still likely to be some unavoidable metal-to-metal contact between the tool and the work under most normal cutting conditions. When extremely high pressures are encountered at the interface, the oil film will occasionally break down, resulting in an immediate rise in the local temperatures and causing pick-up or spot seizure. Many cutting fluids now contain extreme-pressure (EP) additives to minimise this effect when working under these arduous cutting conditions.

From these observations and from those in Chapter 5, it is evident that a lubricant fulfils many other functions besides simply reducing wear: it also lowers the temperature and therefore reduces the extent of temperature-induced wear, and whenever machining with high-speed steels (HSS), an adequate emulsion-type coolant or a lubricant (oil) should be used (Fig. 6.15a). Fig. 6.15a and b show how cooling lubricants increase the tool's life compared with cutting in the dry condition.

In general, most turning and machining-centre applications use the 'water-soluble' oil emulsions or coolants. These contain a microscopic dispersion of the concentrate in water (Fig. 6.16). * The microscopic oil globules are homogeneously dispersed throughout the coolant. The basic ingredients of these emulsions are water, oil and wetting agents, but other additives are necessary in order to achieve the desired properties of a cutting fluid:

- Water acts as the carrying medium for the oil. Its other major function is to remove the heat from the cutting region.
- Oil acts as the lubricant and reduces the friction produced at the interface by the cutting action.
- Wetting agents are required to break down the surface tension between the oil and water, so that a dispersion of the concentrate (oil) throughout the water will form.

*In the early (unstable) emulsions, the oil droplets were typically up to 20μm in diameter, but most of the recent emulsions, (e.g. the one in Fig. 6.16) have diameters of less than 1μm. By modern standards, an oil droplet size of 5μm is considered too large for a stable product.

- Other additives, such as extreme-pressure (EP) additives and individual manufacturer's ingredients, are used to extend the range of the emulsions to cover a variety of cutting conditions.

The influence of the cutting fluid on the machine process depends on a complex interrelationship between many factors, including the machine, the tool, the metal, the component geometry and, not least, the cutting fluid itself. Once the balance is correct, a truly productive process will result. Unfortunately, it is no exaggeration to say that machine tools are underutilised in many companies by as much as 50% because of an unbalanced cutting process.

The classical cutting process, in which the tool moves into the workpiece, causing a chip to form by the combined actions of shearing and bending, is illustrated in Fig. 6.17. This cutting action produces two deformation areas, which affect the total mechanism of cutting and are known as the primary and secondary deformation, or shear zones. The primary shear zone and its associated shear plane absorb the greater part of the energy associated with the cutting process, whereas the secondary shear zone is influenced by the interface friction, amongst other things. The effects on the deformation zones of using an efficient lubricant can be seen by comparing Fig. 6.17a (efficient lubrication)

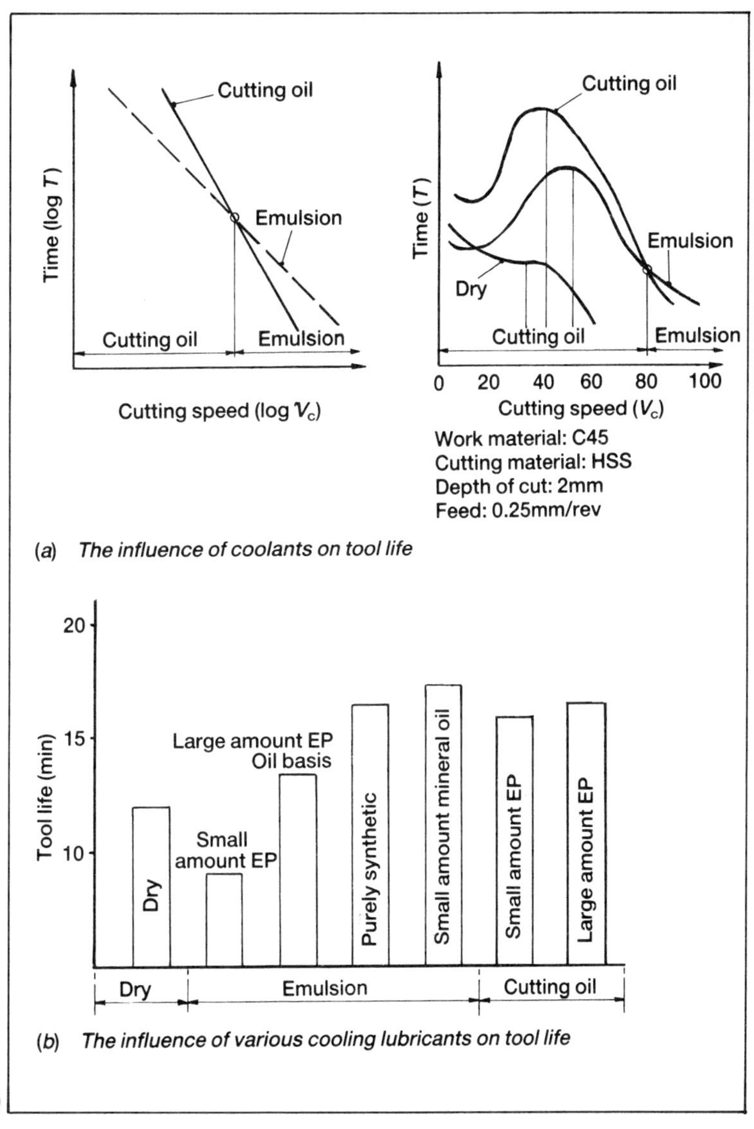

Fig. 6.15 Extending tool life with lubricants (Courtesy of Krupp Widia)

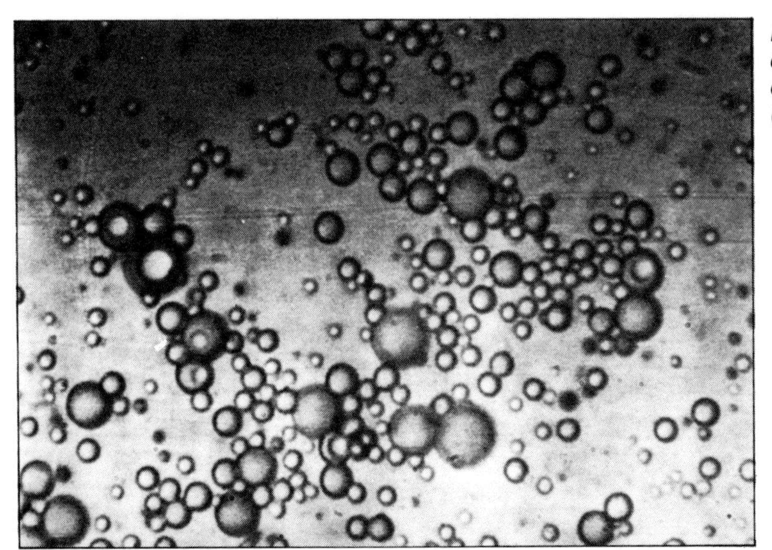

Fig. 6.16 An emulsion, showing dispersed oil particles in the continuous water background (Courtesy of Edgar Vaughan)

with Fig. 6.17b (inefficient lubrication); they are considerably smaller when there is efficient lubrication. In the case of the inadequately-lubricated tool, the shear angle is less and the primary deformation zone is pushed further ahead of the tool; therefore the deformation area is larger and absorbs more energy than in the lubricated condition. Because the deformation zones are larger in the poorly lubricated state, the cutting action produces greater residual stresses in the workpiece. The larger primary shear zone influences the size of the dependent secondary shear zone by making the metal more susceptible to work-hardening.

Conditions of poor lubrication produce high frictional forces (drag) at the interface between the chip and the tool, resulting in pressure welding on the tool. Further tool wear and heat generation will surely occur, because of the increased length and depth of the secondary deformation zone. How does the use of an appropriate lubricant assist in improving these harmful cutting conditions? Very simply, a lubricant aids by producing a layer of low shear strength at the interface between the chip and the tool, thus minimising the areas of primary and secondary deformation (Fig. 6.17a). Other effects of the increased temperatures in the deformation zones will be discussed in section 6.8.

Because friction is an 'invisible entity', there is often a tendency to underestimate its true effect on the machining process. When very high loads are imparted to the tool, the friction between the tool and the chip may be so severe that welding is induced on minute areas, giving rise to the built-up edge phenomenon previously mentioned, with its consequences of rapid tool wear and degraded surface finish. As would be expected, considerable energy, intended for cutting, is dissipated in the form of heat. Many factors affect the amount of heat generated during cutting, including the cutting speed, the feedrate, the material characteristics, the tool geometry, and – by no means the least – the heat-retention properties and cooling efficiency of the cutting fluid.

The correct use of cutting fluids is industry's most useful ally against friction. It reduces friction, minimises wear, cools the tool and workpiece alike, and flushes chips and swarf away from the cutting region. The use of coolants as a means of improving chip control and disposal will be discussed in the next section.

6.6.1 Using a high-velocity coolant system for improved chip control and disposal

If chips are uncontrolled during metal-cutting operations, they can be counterproductive and even dangerous. The damage that can result from continuous chip formation is

Fig. 6.17 The effect of lubrication on the magnitudes and positions of the deformation zones and the resulting shear plane

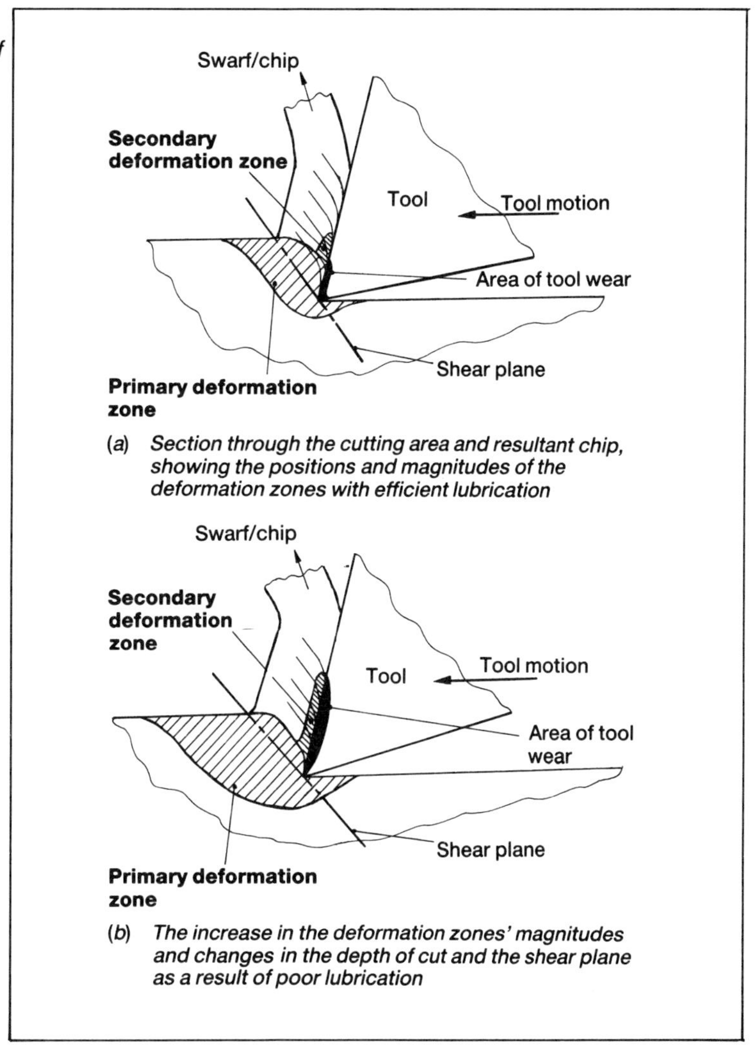

known to every machine-tool operator. It is also known that the workpiece can also suffer damaging effects. In the era of minimally-manned or untended machining, the cutting operation may not be economically or safely run without some form of provision for operator intervention in circumstances where uncontrolled chips occur. Continous chips also pose swarf-disposal problems. Recent advances in chip breaking, discussed in Chapter 1, have solved many of the problems associated with chip formation, but some difficulties still remain. For example, under certain machining conditions, such as the turning of 'superalloys' (e.g. titanium alloys) or the machining of threads on alloy steels, consistent chip control by traditional means is almost impossible. One solution is to provide a continuous force at the interface between the tool and workpiece, in the form of a high-velocity high-pressure flow of coolant, as shown in Fig. 6.18.

One such system (Fig. 6.18) supplies coolant to a nozzle on the tool-holder approximately 2.5mm to 4mm away from the interface between tool and chip, at a pressure of 110 bar and a velocity of 122m/s. This system consists of three basic components; a pumping unit, the piping necessary to transport the fluid, and a tool-holder through which the coolant can pass and which can focus the high-pressure flow of coolant on to the interface. As well as the sophisticated pumping unit (shown schematically in Fig. 6.18), there is also a programmable control unit which is adapted to the needs of the user and the machine tool, with sensors that monitor for loss of pressure, low coolant flow, low sump level or a dirty coolant filter and signal the information to the operator.

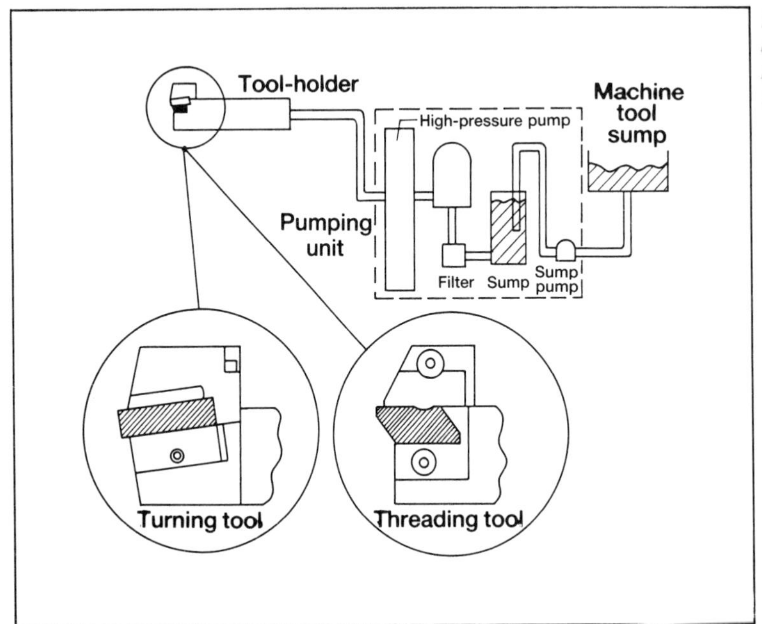

Fig. 6.18 A high-velocity coolant-distribution system for improved chip control and disposal (Courtesy of Kennametal (UK) Ltd.)

For this system to work efficiently, several critical variables need to be established and maintained. Firstly, the coolant pressure must be great enough to achieve an exit velocity of 122m/s and a flow rate of at least 19 litres per minute, in order to control the chips consistently over a wide range of operating parameters. Another key factor is the diameter of the nozzle which affects the exit pressure and the velocity; the nozzle's position is in a 'known window' in relation to the interface between the chip and the tool. Finally, the configuration of the nozzle will dictate the maximum feedrate and depth of cut obtainable.

As well as its chip-breaking and swarf-disposal ability, such a system has also been shown to improve tool life over other coolant-delivery systems by a factor of between 30% and 100% over a range of materials with very poor machinability. In action (Fig. 6.19), these high-velocity coolant systems show positive pay-backs in very little time,

Fig. 6.19 Using the Hyperson high-velocity coolant system (Courtesy of Kennametal (UK) Ltd.)

even if one just considers the increased tool life and discounts their ability to machine the almost un-machinable materials. The consistent high-pressure high-velocity coolant-delivery method of controlling chip-breaking is very impressive even under 'normal' machining conditions, let alone in those severe cutting applications.

So far, little has been said about the economics of machining, which are very dear to the heart of any company engaged in making a profit! Numerous methods are used to determine economic machining conditions. The following section deals with just one of these techniques.

6.7 The economics of machining

6.7.1 Machining costs

A variety of methods are used in industry for costing out a job. One of these, based upon the metal-cutting productivity, utilises the machining cost per component. This cost is a combination of the fixed costs and two variable costs, namely the machine costs C_m and the tool cost C_t(Fig. 6.20a). It is normal procedure to establish these variable costs from calculations based upon machine trials. Once the depth of cut and feedrate have been selected for a particular operation, the other variable is the cutting speed. Fig. 6.20a shows the effect of different cutting speeds upon the machining cost per component.

An accurate picture of the costing of a job depends on an understanding of several factors, namely fixed costs, machine cost and tool costs; these are described below.

Fixed costs The fixed cost C_f is independent of the cutting speed. It is a constant charge for each component, related to handling operations such as loading and unloading of the component in the machine tool and so on. The fixed charge will be affected by such factors as the tool and work-holding equipment used – fixtures, vice, chuck, mandrel, etc. It is obtained by dividing the hourly charge by the number of components machined per hour.

Machine costs The machine cost C_m is calculated differently from company to company. It generally includes such costs as the rate of machine depreciation, labour costs, the distribution of direct and indirect costs, etc.

In recent years there have been sharp increases in a company's costs; this is particularly apparent in labour costs, which have more than doubled recently, and in the capital costs of machines, which have also risen by similar proportions. The machine costs per component, which also depends on such factors as down-time and handling, can be calculated as follows.

Machine charges per hour	=	(A)
Cutting time per component (minutes)	=	(B)
Machine cost per component	= $A \times B \div 60$	
	=	(C_m)

Obviously, the machine cost per component will decrease as the cutting speed is increased. Therefore, this variable cost will be less for companies utilising the latest tooling materials and techniques, which have considerably higher cutting speeds and result in greater stock-removal rates.

Tool costs The tool cost C_t will consist of the sum of the purchase cost, grinding cost (where applicable) and the tool-changing cost for each machined component. The following example shows how tool costs can be calculated for a turning operation; the method can be modified for machining-centre or other metal-cutting operations.

Initial cost of tool	=	(A)
Number of cutting edges per tool	=	(B)
Tool cost per cutting edge	$= A \div B$	
	=	(C)
Cost of insert	=	(D)
Number of cutting edges per insert	=	(E)
Insert cost per cutting edge	$= D \div E$	
	=	(F)
Machine charges per hour	=	(G)
Tool changing time (minutes)	=	(H)
Cost per tool change	$= G \times H \div 60$	
	=	(I)
Regrinding charges per hour	=	(J)
Regrinding time (minutes)	=	(K)
Regrinding cost	$= J \times K \div 60$	
	=	(L)
Tooling cost per edge	$= C + F + I + L$	
	=	(M)
Number of components per cutting edge	=	(N)
Tooling cost per component	$= M \div N$	
	=	(C_t)

Obviously, the cutting speed has a major impact on the value of C_t, because an increase in cutting speed normally results in faster tool wear, and the tool charge per component will increase as a result. Tool costs now account for only a small proportion of the total costs of production, owing to the fact that the latest tooling can operate at higher feeds and speeds than the tooling previously used, and lasts longer. When the tool costs actually rise (as a result of higher tool wear-rates) as cutting speeds are increased, it naturally follows that tool performance will also increase.

Having found the three costs, C_f, C_m, and C_t, they may be added together to obtain the total machining cost per component C. If the values of C are plotted against cutting speeds, the curve shown in Fig. 6.20a will be generated. It has a characteristic shape with a dip along its length, the lowest point of which indicates the minimum cost per component. This low region of the curve can be regarded as the 'economic machining zone', in which all machining operations should be carried out by optimising the various parameters that must be set when machining workpieces.

6.7.2 Machining components economically

As explained above, the lowest point on the graph of total machining costs against cutting speed represents the minimum machining cost per component; it is found at a cutting speed which is normally known as the economic cutting speed, V_e. The value of V_e can be obtained from the '*VT* graph' or 'Taylor's curve' shown in Fig. 6.14c once the economical cutting-edge life has been determined. Alternatively, the economical

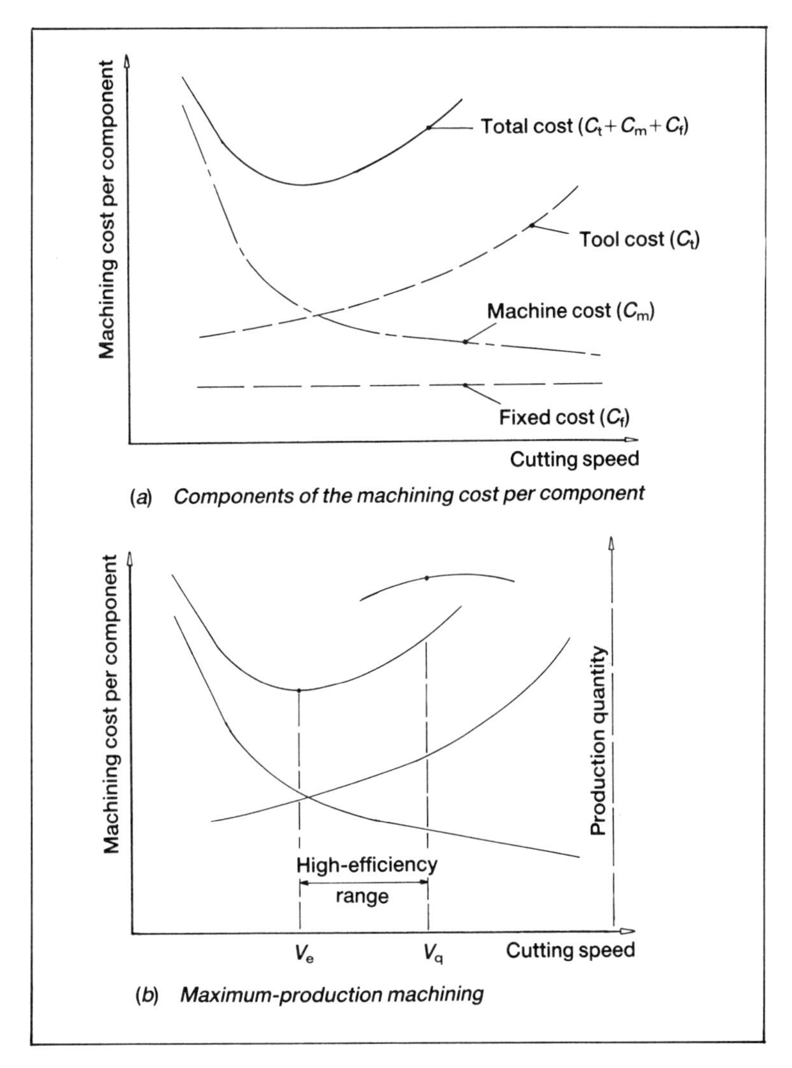

Fig. 6.20 Machining economics (Courtesy of Sandvik (UK) Ltd.)

(*a*) *Components of the machining cost per component*

(*b*) *Maximum-production machining*

cutting-edge life can be calculated for a specified feed by means of a modified version of the Taylor formula:

$$T_e = \left(\frac{1}{\alpha} - 1\right)\left(\frac{C'_t}{C'_m} + t_c\right)$$

where T_e = economical tool life,
α = slope of the VT curve (measured from the graph),
C'_t = cutting-tool cost per cutting edge (obtained as described above),
C'_m = machine charge per minute (normally established by the machine-shop management), and
t_c = tool-changing time for the cutting operation; this will vary according to whether the tooling is of the conventional or quick-change variety.

When establishing the values of C'_m and C'_t , it is important to consider the distribution of the various overhead costs related to the machine and to the tool.

6.7.3 Maximum-production machining

The most economic cutting speed is that which produces the lowest production cost per workpiece but, for a variety of reasons, it may be necessary to exceed this speed occassionally. The additional costs incurred can be justified in special cases, for example when higher production rate is required, or if the company invests in new machine tools.

When the number of components produced per hour is plotted against cutting speeds, a typical curve results, which is shown together with the cost curves in Fig. 6.20b. The highest point on the curve represents the maximum production rate; it is found at the cutting speed V_q, which is normally higher than the economical value V_e. The range of cutting speeds between these two values can be described as the high-efficiency range for a particular operation. The production rate – the number of components produced per hour – can be calculated according to the following formula:

$$P = \frac{60}{t_p}\left(1-\frac{t_c}{T}\right)$$

where t_c = tool-changing time,
T = tool life, and
t_p = total time per component (including machining, handling and down-time)

6.7.4 The improved return on the investment

An alternative way of looking at what cutting tools can do for economical production is to relate any improvements in productivity to the actual machinery cost and the total invested capital.

The results of machinability tests such as those described in section 6.1 are rarely related to actual increases in the rates of production, and if they are, the figures seldom mention their influence on the total economics of production. A significant economic argument is that any figures obtained from testing should highlight the improved return on investment which can be obtained by any company utilising the latest tooling in conjunction with modern cutting conditions.

The following formula can be used to calculate the improved return on investment for any particular operation, machine or machine group:

$$\text{Improved return on investment} = \frac{\text{TS} \times \text{MC}}{\text{MTI}}$$

where MTI = machine-tool investment,
MC = machine charge per hour, and
TS = 42 time savings per year (in hours).

This section has only briefly described some of the techniques which a company can use to determine the economic data required to establish an efficient metal-cutting capability; more intricate economic models are given in the publications listed in the Bibliography.

6.8 Surface integrity

Until the middle to late 1960s, measurement and analysis of machined surfaces consisted basically of surface-texture measurement, using a stylus-type of instrument to 'track' the surface, amplifying and recording its texture. Some mechanical and metallurgical tests might also be carried out, but these tended to be done on the component as a whole, and not on the specific machined surface in question.

Two researchers (M. Field and J. F. Kahles) did important work on critical components such as the hydraulic piston and cylinder assemblies used in the undercarriages of aircraft. These highly-stressed parts require stringent inspection procedures to ensure that failures are minimised. Early failure of these assemblies was found to be caused by fatigue initiated at, or near to, the machined surface, even though it had previously seemed to be satisfactory when inspected and tested by the techniques described above. At first, the reasons for these premature failures were not readily apparent, but on closer investigation several effects were noted:

- Plastic deformation due to hot or cold work.
- Laps, tears and crevice-like defects associated with the tool's built-up edge.
- Phase transformation.
- Intergranular attack.
- Microscopic and large-scale cracking.
- Residual stresses in the surface layer.
- Hydrogen embrittlement caused by chemical absorption.

These principal causes of the defects were first noted by Field and Kahles. They also showed that any alterations of the machined surface were due to one or more of the following:

- High temperatures or temperature gradients developing in the machining process.
- Plastic deformation.
- Chemical reactions and subsequent absorption into the surface.

Clearly, these effects of the machining process are important, so how can they be assessed? The answer to this question lies in the tests that Field and Kahles devised, which are now widely used and recognised to be realistic methods for assessing surface integrity.

6.8.1 Methods used for assessing surface integrity

Field and Kahles recommended standard tests that can be used to assess a machined surface and indicate its likely surface condition. The tests fall into three 'sets' or groups: the minimum surface-integrity data set, the standard surface-integrity data set and the extended surface-integrity data set. Each of these is sub-divided into discrete testing procedures. All of them are based on the procedures in the minimum surface-integrity data set, with additional more complex tests in each successive group.

The minimum surface-integrity data set consists of tests for the following surface properties:

- Surface finish.
- Macrostructure (i.e. the structure seen at magnifications of 10 × or less), which sub-divides into tests for macrocracks and macroetch indications.
- Microstructure, which is sub-divided into microcracks; plastic deformation; phase transformations; intergranular attack; pits, tears, laps and protrusions; built-up edge; melted and redeposited layers; and selective etching.
- Microhardness.

The standard surface-integrity data set consists of all the previous tests, plus more in-depth appraisal of the following aspects, for applications that are more critically influenced by the surface integrity:

- Testing for fatigue properties.
- Stress corrosion tests.
- Assessment of residual stress and distortion.

The extended surface-integrity data set consists of all the tests mentioned above, plus the following:

- Extended fatigue tests to obtain design data.
- Extra mechanical tests, such as for tensile strength, stress rupture and creep, and other specific tests when appropriate, such as bearing performance, frictional tests and the sealing properties of surfaces.

It is important to note that these surface-integrity tests were principally designed for the assessment of wrought products. It follows, that their application to materials processed by other methods (e.g. powder metallurgy, squeeze casting or metal-matrix composites) may require the test procedures to be modified.

6.8.2 Surfaces produced by milling

Milling operations produce different surfaces from turning, in which there is usually only one cutting edge. Milled surfaces are almost always generated by more than one edge, the exception being when a single-point fly-cutting action is used. Because the numerous cutting edges on the milling cutter are never absolutely identical in their positioning or geometry, particularly when cutters with indexable inserts are used, milled surfaces tend to contain a mixture of several surface variations. Typical surfaces produced by milling are shown in Fig. 6.21, which illustrates that the texture of a

Category (enlarged profile view)	**Examples of kind of variation**	**Possible reasons for variation**
1 Variations in shape 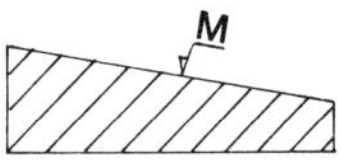 (M)	Unevenness Out of round	Errors in the guideways of the machine tool or of the component Incorrect clamping of the component Distortion due to hardening Wear
2 Waviness 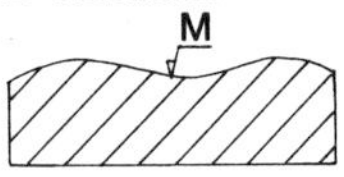 (M)	Waves	Eccentric clamping or errors in shape of a cutter Vibrations of the machine, tool or component
3 Grooving 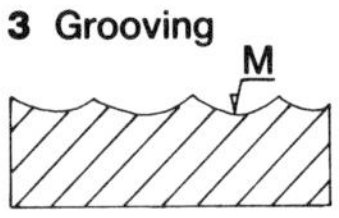 (M)	Grooves	Shape of the cutting edge, feed or supply of the tool
4 Minor imperfections 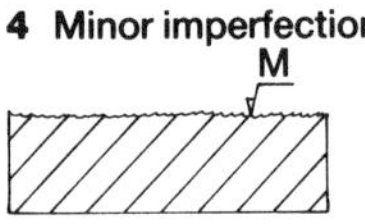 (M)	Scores Flakes Arches	How chips are produced (tear chip, shearing chip, built-up edge), deformation of the component at the sand-blasting, serration (after galvanic treatment)
5 (Cannot be shown on a picture)	Structure	Crystallising processes, alterations of surface by chemical influence (e.g. corrosion)
6 (Cannot be shown on a picture)	Lattice structure of the material	Physical and chemical reactions, in the structure of the material, strains and shearing strains in the crystal lattice
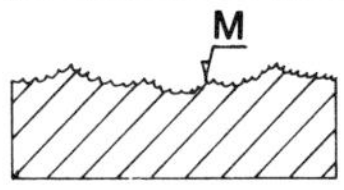 (M)	There is a general overlapping of categories 1 to 4	

Fig. 6.21 Examples of variations in milled surfaces (Courtesy of Hertel)

machined surface is a combination of waviness (long wavelengths) and surface roughness (short wavelengths). Over an even longer length of the surface, there could be superimposed an 'unevenness' produced by such factors as guideway errors, heat-treatment distortions, etc.

These long-wavelength variations on a milled surface are due to positioning and geometry faults and can be overlapped by a further variation related to the concentricity, or plane-running errors of the milling cutter or spindle (tumbling). Any waviness resulting from these errors may be identified because it depends on the feed, which will be reflected in the frequency of the cutter revolutions.

When assessing the surface generated by a milling cutter, it is necessary to look at the depth of the surface profile, which contains elements of both waviness and roughness (Fig. 6.21). In order to obtain a clear and unbiased result for the machined surface, a reference line of the same trace length is usually measured and the results stated. In any case the trace length should be larger than the feed rate per revolution of the cutter.

Note: The surface-texture parameters R_a and R_z mentioned in section 6.9 do not measure waviness and, as a result, they are not truly applicable for assessing milled surfaces.

Surfaces obtained using radiused milling cutter inserts

The cutting edge will leave its profile on the surface of the workpiece being machined, both when milling (Fig. 6.22) and when turning (Fig. 6.24). In machining using a milling

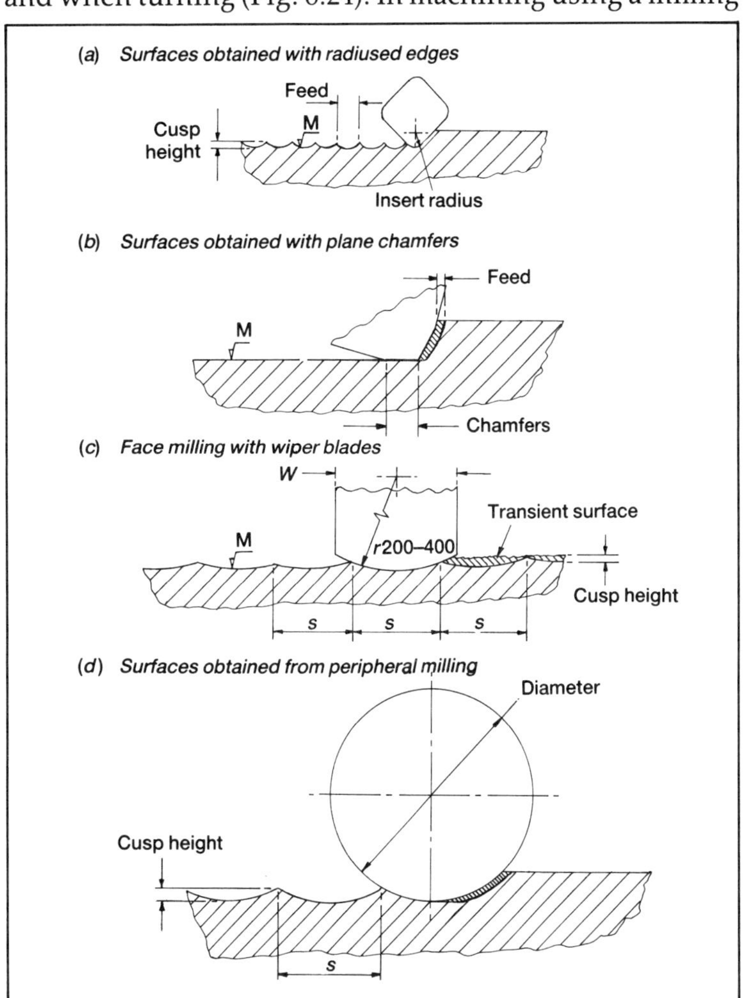

Fig. 6.22 Typical surfaces obtained by milling (Courtesy of Hertel)

insert with a radiused cutting edge, a groove will be formed (Fig. 6.22a) whose depth, which is often known as the 'cusp height', depends upon the feed rate and the insert radius. In order to obtain a good surface, with a small peak-to-valley height and minimal waviness, a large radius is required on the insert or tooth, and a small feedrate must be programmed. This method of obtaining a good surface from radiused cutting edges is rather uneconomic, and these cutters are therefore usually reserved for roughing operations, where the surface quality is less important.

Surfaces obtained using plane-chamfered milling cutter inserts

Whenever a high-quality finish is required, it is recommended that the milling insert or tool is ground with a corner chamfer (Fig. 6.22b). It is important that this corner chamfer runs parallel to the component's surface. If the feed per revolution is less than the chamfer length, only one of the cutting edges – the one which protrudes the most – will be responsible for the quality of the surface generated. In this manner, the peak-to-valley heights on the resulting machined surface can be reduced to less than 12μm. However, if the feedrate per revolution of the milling cutter is greater than the chamfer length, then more than one cutting edge is effectively operating. In this case, the waviness of the surface produced relates to the inaccuracy in plane running of those cutting edges which have generated the surface. Generally, the most favourable peak-to-valley heights are obtained when the cutting-edge protrusions are equally distributed over the cutter periphery, producing a mutual overlapping effect.

From these statements, it is evident that the inserts in multi-toothed cutters must be true to size and angle, and that their positioning in the cutter body must be guaranteed. When fitting new inserts into the cutter body, therefore, it is important that a tool presetter is used so that any cutting edge that is incorrectly seated can be easily detected. This can be done by simply rotating the complete cutter whilst it is in the presetting device and noting the inserts which do not conform to the X and Z coordinates; these inserts may then be re-seated or changed.

Surfaces obtained in face milling, using wiper blades

If it is impossible to obtain a satisfactory machined surface by milling with high feedrates using corner chamfers (i.e. using a feed per revolution greater than the corner chamfer), it is possible to improve the surface quality, in terms of waviness and peak-to-valley height, by using cutters which incorporate wiper blades. These additional blades are situated in the gaps immediately behind the other milling inserts. They have a slightly crowned shape, with a radius of between 200mm and 400mm (Fig. 6.22c). Slight inaccuracies in parallelism in their initial tool presetting may be tolerated; even with a slight misalignment the regular saw-toothed profile effect is minimised.

The wiper blades' cutting depth should be between 0.05mm and 0.1mm, and the feedrate per revolution of the cutter must be less than their length. A general rule is that the feed per revolution should be approximately half the length of the wiper blade.

If large cutters are used, a greater number of wiper blades can be incorporated, but attention must be paid to ensure that the wiper blades 'run true' to one another.

Wiper blades are normally incorporated into the cutter body when machining short-chipping materials, such as grey cast iron, brass, etc. They are less likely to be incorporated when cutting steel, as they restrict the chip flow, resulting in greater passive forces which lead to a vibration tendency.

Surfaces obtained from peripheral milling

In peripheral milling, the machined surface is generated by the main cutting edge and this results in a bicurve profile (Fig. 6.22d). The cusp height or wave depth is geometrically dependent upon the diameter of the cutter and the feedrate per revolution. If a condition arises in which the machined surface is produced by just one cutting edge, as a result of untrue running, the theoretical cusp height (in μm) is given by the following relationship:

$$\text{Cusp height} = \frac{250\,(\text{feed per revolution})^2}{\text{cutter diameter}}$$

Where the feed per revolution is measured in millimetres per revolution, and the cutter diameter is measured in millimetres.

However, there should normally be more than one cutting edge participating in the production of the machined surface (as is the case when the cutter is 'running true') as this gives smaller cusp heights on the surface profile, which is normally to be encouraged.

The influence of spindle camber on the machined surface

The term 'spindle camber' describes a slight inclination of the milling spindle axis compared with its zero position, which is related to the feed of the machine (Fig. 6.23). This technique is used to avoid the so-called 'recutting effect' that occurs when face milling using a large cutter; it also avoids the additional insert wear at the cutting edges.

In reality, the spindle camber is very slight, generally only amounting to between 0.1mm and 0.3mm over a length of 1000mm. Converted to angular measurements, this is a value of 20 to 60 seconds of arc; this is greatly exaggerated in the upper diagram of Fig. 6.23.

When machining with a face-milling cutter using a spindle camber, a concave surface is produced: it is not possible to produce a flat plane surface. The surface concavity generated by the spindle camber depends upon the relationship between the cutter diameter and the width of the surface, and the depth (in mm) of the concavity, f, can be calculated using the Kirchner–Schulz formula:

$$f = \frac{q}{1000}\left[\frac{D_e}{2} - \left(\frac{D_e^2}{4} - \frac{e^2}{4}\right)^{1/2}\right]$$

where D_e = effective diameter of the cutting circle,
e = width of the surface, and
q = 1000tan θ , where θ is the spindle camber.

Alternatively, a reasonable estimate of the concavity f can be obtained from the graph in Fig. 6.23 that shows the variation in shape for a variety of spindle cambers and face-milling cutter diameters. These concave surface modifications produced by the spindle camber are never large deviations from the 'true' plane surface; even under the extreme conditions of a small cutter diameter (100mm) and a large q value (0.05mm), the deviation only amounts to 25μm over a workpiece width of 100mm – which will be within the allowed tolerances in most commercial situations.

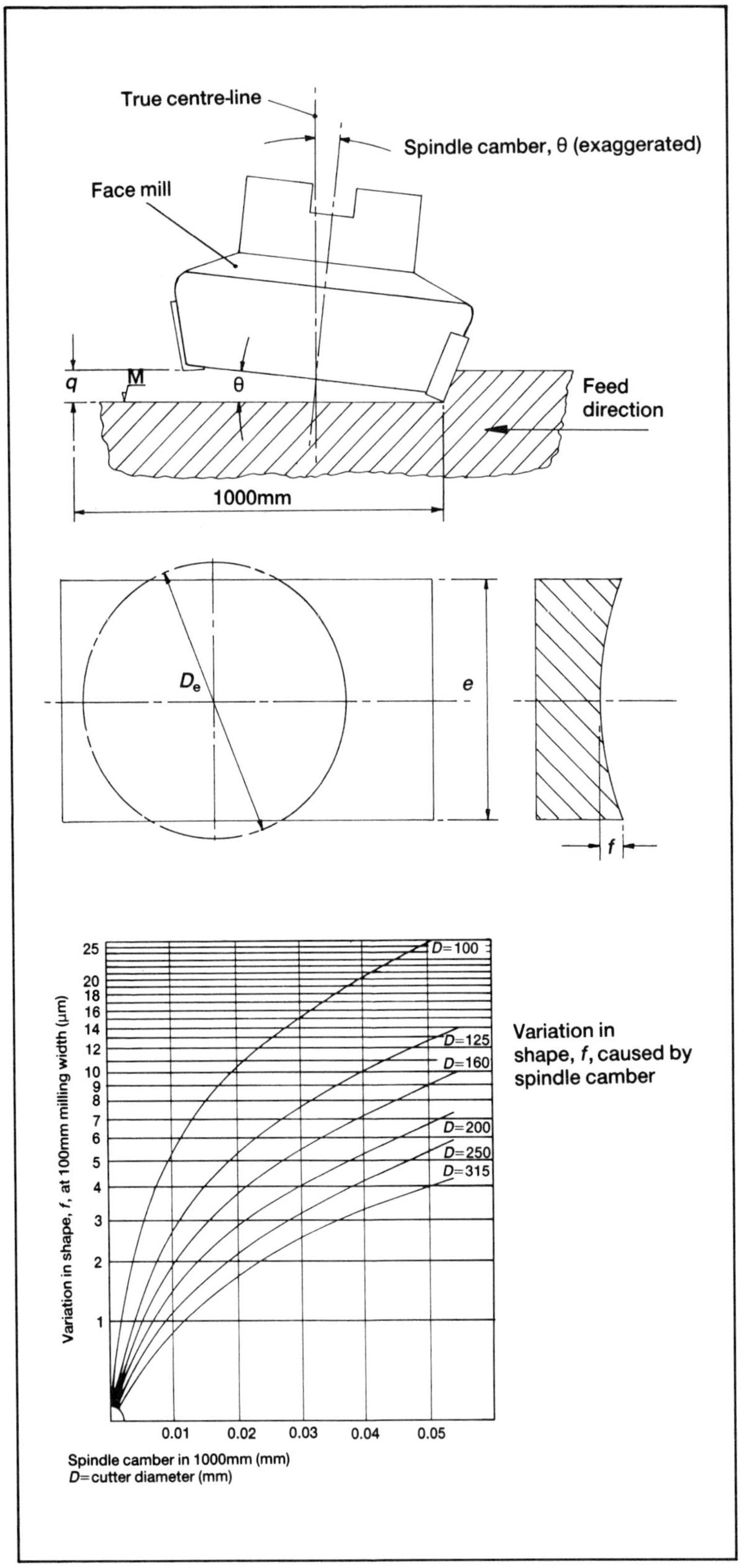

Fig. 6.23 The influence of the spindle on the workpiece (Courtesy of Hertel)

6.8.3 Surfaces produced by turning

In most metal-cutting operations the accuracy and surface finish can be improved by using higher cutting speeds – if the cutter wear-rate permits it. The factors of greatest importance in most turning operations are the nose radius of the tool and the feed-rate.

Generally speaking, the higher the feedrate, the greater the peak-to-valley or cusp height R_t (Fig. 6.24). Conversely, the lower the feed rate, the more peaks there are. The peak-to-valley height may be found from the equation

$$R_t = \frac{s^2}{8r}$$

where s is the feedrate and r the nose radius.

The operation shown in Fig. 6.24 is an example of a sliding action (turning diameters) but similar conditions apply in surfacing operations (facing off 'to length').

Surfaces produced by facing off

In the most basic form of turning operation, facing off a bar, the turning tool generates a flat surface by a spiralling action towards the bar's centre. If the face produced is either dished (concave) or bevelled (convex), a number of factors may be responsible.*

Possible causes of a convex face are:

- Too great an overhang of the workpiece.
- Jaws strained in the chuck.
- Poorly-supported tool or a lack of rigidity.
- Large nose radius and large plan approach angle on the tool (see Fig. 1.18 on page 20).
- A high infeed, pushing the workpiece off its centreline.
- Misalignment of the X- axis slideway.

Possible causes of a concave face are:

- Poorly-supported tool or a lack of rigidity.
- Small nose radius and large negative plan approach angle on the tool, creating a tendency for the tool to be dragged into the workpiece.

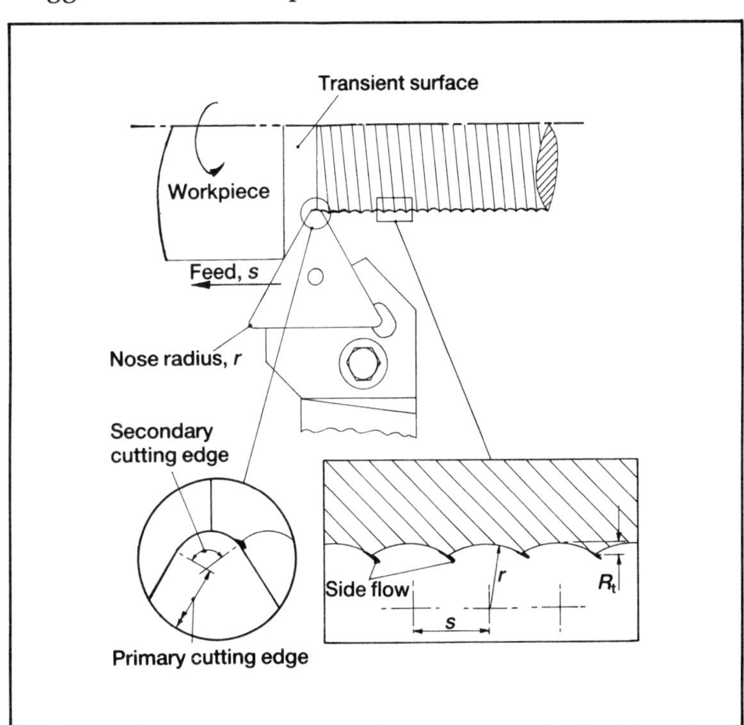

Fig. 6.24 A typical surface produced by turning (Courtesy of Sandvik (UK) Ltd.)

* Extremely convex or concave turned surfaces will result from combinations of these factors.

Surfaces produced during sliding

The sliding action is more complicated: the component is machined to a consistent diameter along its turned length as a result of the tool's helical action generated by the rotation of the workpiece and the feed of the tool. However, a number of factors may cause the workpiece's diameter to vary along its length.

Possible causes of taper on the workpiece are:

- Poorly-supported workpiece with too great an overhang.
- Incorrect alignment of the programmable tailstock (used to support the workpiece) when supposedly positioned on the centreline of the Z-axis (causes long tapers).
- Jaws strained in the chuck (causes short tapers).
- Tool poorly supported and pushing away from the workpiece.
- Worn headstock bearings.

Possible causes of the workpiece being larger at the centre than at the ends (i.e. a 'barrel' shape) are:

- The workpiece pushing away from the tool owing to lack of support from the chuck to the tailstock (this is particularly relevant in the case of long thin bars).
- Large nose radius and large plan approach angle on the tool (when turning long thin bars).
- Not using the programmable steady (if one was supplied with the turning centre).
- Poorly-supported tool.

Possible causes for the workpiece being thinner at the centre than at the ends (i.e. a 'candlestick' shape) are:

- Small nose radius and large negative plan approach angle on the tool, dragging the workpiece into the tool in the central region (this is particularly relevant in the case of long thin diameters).
- Poorly-supported workpiece.
- Not using the programmable steady when supplied.

The machining conditions causing 'barrel' and 'candlestick' shapes can be minimised if a 'balanced turning' operation is carried out. This requires two independently-programmable turrets, which reduce and, to a certain extent, negate the generated cutting forces that cause deflection of the workpiece on long slender components.

The conditions listed above concern the generation of a product by metal cutting on a turning centre. Forming techniques, on the other hand, are not too badly affected by such geometrical inaccuracies, as they are usually confined to much smaller machining areas and rely on the tool shape to produce the desired form on the workpiece. This tool form will be reproduced on the surface of a slightly-inaccurate shape caused by the combination of feedrate and cusp height on the 'true' diameter. Other machining processes, such as boring and screw cutting, will suffer from form inaccuracies to a lesser degree. Strictly speaking these geometrical inaccuracies are not surface-integrity problems, but they have been included in this section to highlight the point that these form errors are not only to be found with conventional machine tools, on which they are well-known.

6.8.4 Some aspects of the drilling process

Machining operations that are carried out with normal speeds and feeds, and using cutting fluids when appropriate, can be considered as 'gentle' cutting actions or regimes; typical examples are turning, milling and some drilling operations, using drills with through-the-nose coolant supplies, typified by the U-drill. However, if the feed is increased or cutting conditions deteriorate considerably (as a result of poor machinability owing to the combination of tool and workpiece, or the feed and speed selected), drilling processes are considered as 'abusive' cutting regimes. Twist drilling is a relatively inefficient metal-cutting process. This is particularly troublesome because the drilling of holes is probably the most common machining operation performed on materials, and below about 8mm diameter (which includes the majority of drilled holes) the U-drill types cannot be used, as insert positioning and locking arrangements cannot be accommodated in such small tool bodies at present.

Before discussing the problems encountered in drilling, we shall review the mechanism involved in generating a hole by twist drilling.

The geometry and cutting action of the twist drill

A conventional twist drill has two cutting regions: the cutting edges (or lips), which cut, and the chisel edge, which also cuts but which also rubs, extrudes and indents. Therefore, a twist drill produces a complex cutting and extruding action.

The formation of chips along the lips of the drill is fairly conventional, in that a positive rake angle is present along the lip at the periphery of the drill. This is complicated by the fact that the lips are offset – parallel to a radial line ahead of the centre – by an amount equal to approximately half the web thickness at the drill point. This offset means that the cutting edges are at an oblique angle to the direction of motion, so that the chip flow is inclined in a direction normal to the cutting edge. However, the angle increases somewhat as the chip flow approaches the chisel edge from the periphery along the lip. The oblique cutting action tends to increase the effective rake angle, so that, at the intersection between the lip and the chisel, the effective rake angle is highly positive.

An extremely complex metal-removal mechanism occurs along the chisel edge. At radii near the flute bottom, the chisel edge and the clearance surfaces form a cutting edge with a high negative rake, but as the drill centre is approached, the action can be thought of as that of a blunt wedge-shaped indentor. The severely deformed metal produced directly under the chisel edge and normal to it, shows that the metal-removal process is relatively inefficient. The deformed products produced in this region must be ejected if the drill is to produce a hole. It has been shown that they are wiped, or extruded, into the drill flutes and usually intermingle with the chips produced by the main cutting edge. Many variations of point shape have been used to overcome the extrusion effect caused by the chisel point. Some of these will be considered in the next section.

Twist-drilling forces and the methods used to improve cutting efficiency

The two principal forces on the drill are the thrust and the torque. The thrust results from the axial pressure produced by the feedrate and the extruding effect of the drill point: it is by far the largest contributor to the total force. The torque is the direct result of the rotational motion and the drill's diameter. Both of these forces are obviously highly dependent on the hardness of the material they are drilling. In fact, the extrusion effect caused by the chisel point produces a relatively large force component, being highly inefficient. There is also a small force contribution from some rubbing of the drill's margin, and finally the cutting force, which is reasonably large and is the only efficient cutting force in the whole hole-generating action.

Many methods have been devised to try and improve the hole-generating efficiency. They range from various web-thinning methods, for example the split point and the four-facet point, through to changing the point angles and the actual cross-sections of the drills. Web-thinning techniques help to reduce the extruding action of the drill, which decreases the overall forces generated; at the same time they improve the drill's self-centring action and cutting efficiency.

The so-called 'white-layer' effect in drilling steels

When twist drilling steels under abusive conditions, the drill's cutting action produces a 'white-layer' effect as it passes through the steel component. The white layer is caused by the point of the drill and forms in the regions of the drill's lip and margin and on the transient surface (Fig. 6.25). The white layer on the transient surface is removed with each successive rotation of the drill, but occasionally some of it is left behind in the drill margin area and remains at and near the hole's surface, and along its length.

The white layer is the result of a number of factors, including a reasonably hard component material which generates high temperatures that are not fully dissipated by the coolant (since its supply is, at best, erratic) so that abusive cutting conditions arise. This is compounded when excessive feeds and speeds are chosen.

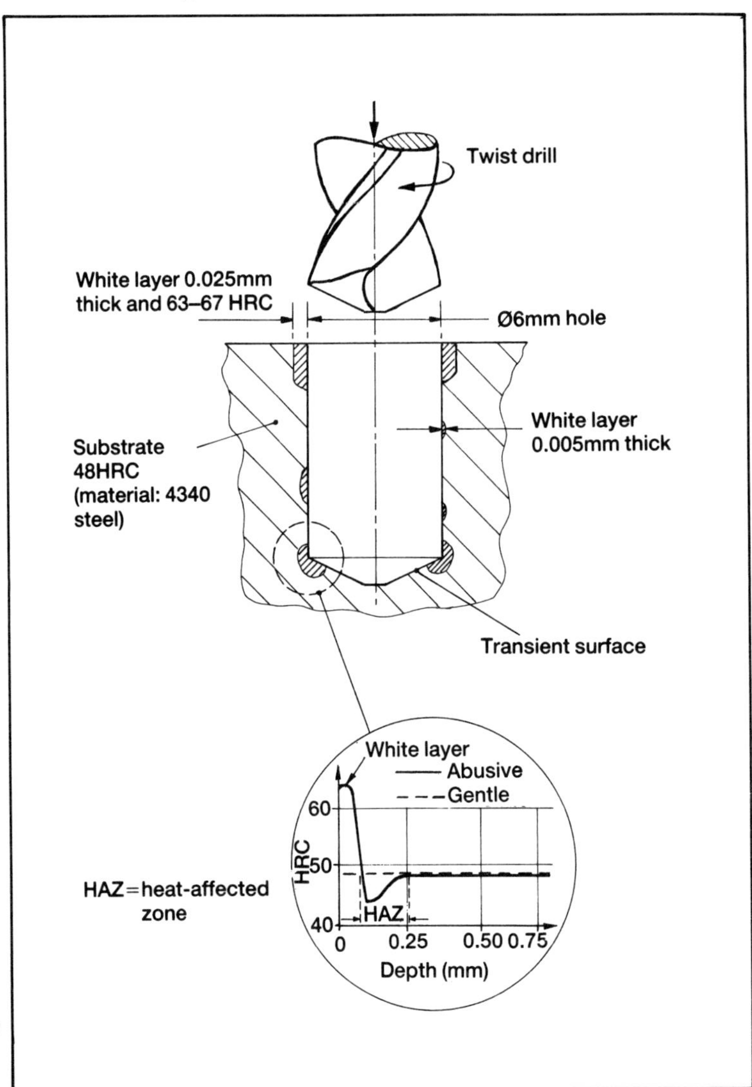

Fig. 6.25 The formation of a white layer by the 'abusive' cutting action of a dull drill in steel (After J. B. Kohls et al.)

The white layer is a result of a phase transformation that produces untempered matensite. It considerably increases the local hardness over that of the bulk substrate material. Fig. 6.25 shows an example of a set of microhardness readings taken through the white layer; the hardness within the white layer itself is typically of the order of 63HRC. In this example, the white layer penetrates to approximately 0.1mm. Immediately beneath the area of untempered martensite is a heat-affected zone with hardness of 44HRC; this is less than that of the bulk material, which is about 48HRC. The combination of the very hard white layer, the softer heat-affected zone and the marginally-harder substrate produces an unstable metallurgical condition.

Normally, the drill's geometry produces the well-known 'saw-tooth' effect on the machined surface as is passes through the component. Such a 'saw-tooth' profile on a surface with a white layer coupled with a softer heat-affected zone will give rise to weak areas susceptible to fatique or crack initiation and propagation when the component is subjected to high stresses in service. In some cases, the formation of a white layer at the hole's surface can be minimised by reaming or boring the component after twist drilling.

It is worth mentioning that formation of a white layer is not confined solely to drilling operations: it can occur under *any* abusive cutting conditions on steels, but is simply more difficult to detect inside a hole!

Drilling a powder-metallurgy component

In recent years, the powder- metallurgy process has been increasingly applied, particularly in the automative field. Although the process is chosen because components can be pressed to the required net shape, there are times when some secondary machining is required, for example when features such as holes, re-entrant angles or threads are required at 90° to the pressing direction. Therefore drilling, tapping, turning, milling and so on are frequently necessary. The most common secondary machining operation is drilling. Products produced by the powder-metallurgy route cause a number of problems for a drill. Some of these will now be considered, after a brief review of the range of component conditions that can arise.

The properties of a typical powder-metallurgy component can vary considerably: its structure can vary from an open aspect (porous) through to reasonably dense. The density and chemical composition can also vary markedly across a component. Therefore, when drilling a powder-metallugry component, any of these condition may be encountered, in varying degrees.

Whenever a hole is drilled in a component – regardless of whether it is a wrought or powder-metallurgy product – there is a tendency for the hole to deviate from the axial centre-line. This deviation is well-known in wrought products. It is usually caused by errors in the tool point's geometry, resulting in a bent hole. In the case of powder-metallurgy components, the problem is compounded by the lack of a preliminary centre-drilling operation or a modification to the point of the drill to increase its self-centring effect. If, as is normal, an unmodified drill has been used, the hole will not only be 'bell-mouthed' at its top and bottom, but is also likely to have a 'helical wandering' of the centre-line along its length. The reason for the 'bell-mouthed' effect at the top of the hole when using a conventional drill point as purchased from a drill manufacturer (i.e. with a standard unmodified chisel point and an 118° point angle) is that several effects occur as the drill enters the top surface of the component. The drill point will describe a circle as the flat chisel edge contacts the workpiece and 'walks' across the surface. The 'walking' tendency will decrease as the drill moves further into the component's surface, producing the 'bell-mouth' at the top of the hole. The 'wandering' along the length of the hole is caused by density variations in the workpiece materials and geometry errors on the drill, and the 'bell-mouthing' at the bottom of the hole is normally attributed to the unsupported drill being fed past the bottom face, causing a trepanning-like effect.

As well as these geometric errors in the hole, its surface topography is also modified by such factors as feedrate, smearing, tearing and rubbing, as shown in Fig. 6.26.

Owing to the open nature (porosity) of a powder-metallurgy product, a drill can be destabilised because the frequent changes from pore to particle set up an excitation vibration effect and lead to a rapid wear-rate. Depending on the component's future application, in certain instances no coolant will be used, and this further exacerbates the problem.

There is also a tendency for some surface hardening to be produced by the drill's passage through the component. It tends to affect a greater depth in powder-metallurgy components than in wrought products and to be more uniform throughout the hole's length, reinforcing the argument that abusive machining conditions are present.

Some of these problems of drilling powder-metallurgy components can be minimised by using stub drills, which are more rigid and therefore decrease the vibration and wandering. If the point angle is increased and the geometry is modified by web-thinning, and if a multi-coated drill is used, wear-rates will decrease considerably.

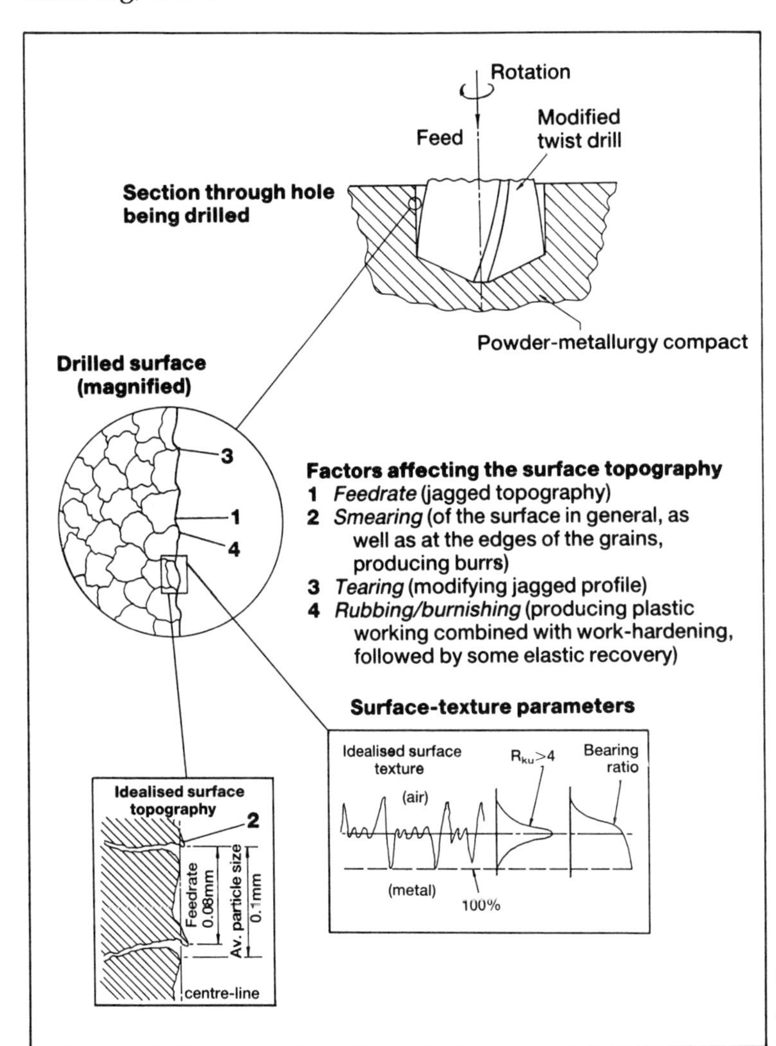

Fig. 6.26 The surface topography produced by drilling a ferrous powder-metallurgy component

6.9 Surface-texture parameters – a means of defining a machined surface

Any surface – regardless of whether or not it has been machined – can be accurately defined using surface-texture parameters. A typical machined surface has a combination of roughness, waviness and lay (Fig. 6.27a). The roughness is the primary texture, which is produced by the combination of the tool's form and the feedrate; it is a high-frequency component. The waviness results from such factors as deflections of the machine or work, vibrations, chatter, heat treatment and warping strains; it is the medium-frequency component upon which the surface roughness is superimposed. Lay, on the other hand, is a result of the production method used; it determines the direction of the predominant surface pattern. The surface texture is normally assessed by taking a sampling length at right angles to the lay.

The marked improvements in component reliability in the post-war years have been largely due to the developments in the techniques and understanding of surface-texture assessment. It is no use a designer specifying close dimensional tolerances for a specific process when the inherent surface roughness that will be produced in the process lies

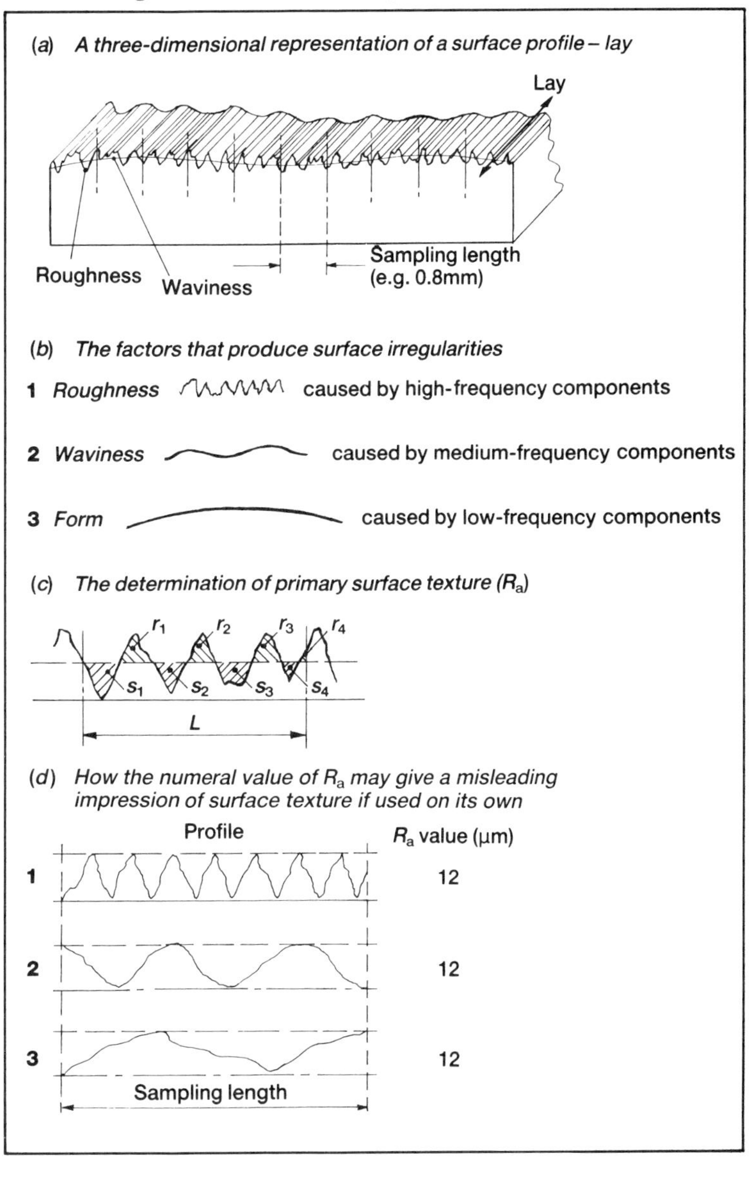

Fig. 6.27 The main parameters that define surface texture

outside this range. In fact, the 'education' of designers in terms of what is practicable in particular production processes still has a long way to go in many companies.

The service life of a component, which is affected by the wear-rate of mating surfaces, is dependent upon the specified level of surface finish. Rough surfaces, with large peaks and valleys, will of course have smaller areas of contact and, as a result, will wear more quickly than smooth mating surfaces.

The measure of the primary surface texture known as arithmetic roughness, is defined by the following 'general-case' equation.

$$R_a = \frac{\sum \text{areas } r + \sum \text{ areas } s}{L} \times \frac{100}{V}$$

where L is the sampling length, V is the vertical length, the areas r and s are as shown in Fig. 6.27b, and R_a is measured in μm. Reasonably inexpensive instruments (Figs. 6.31c and 6.32) are now available that will measure R_a automatically.

When exceptional peaks and valleys are present within the sampling length (normally 0.8mm), however, then another measure known as R_z is preferred. Methods used in determining surface parameters such as R_z are metrologically important, and are well-documented elsewhere.

The parameter R_a is given as a numerical value: the higher the R_a value, the rougher the surface texture. In practice, designers will often quote an R_a value in isolation on a drawing for a component, not realising that this information is not only misleading, but that it fails completely to describe the surface they want. The examples shown in Fig. 6.27d show how misguided such a use of this surface-texture parameter is: it shows three completely different surface textures that all have identical R_a values. Obviously, machined surface 1 would hardly be appropriate for a bearing surface, yet it has an identical R_a value to surface 3, which is much closer to what was required. This illustrates the sort of 'trap' that awaits any engineer foolish enough to use this value in such an over-simplistic and misguided way. Yet many companies still specify only this value and doggedly hang on to its use, in spite of much evidence corroborated by research and published articles to persuade them otherwise! With this being the case, what methods may be used for more fully describing the machined surface?

Whenever a surface-texture assessment is necessary, it is important to consider what sampling length should be used for the trace of the machined surface, as this choice will in part determine the magnitude of the assessed parameters. It is also crucial that the instrument is calibrated before any assessment in order to ensure repeatabilty of results. In most electronic surface-texture instruments used nowadays, a standard 'meter cut-off' or sampling length of 0.8mm is used, which is applicable to most conditions. This sampling length allows the results to be compared with other surface-texture parameters produced by different methods with some reliability. It must be long enough to include a reliable amount of roughness, yet short enough to exclude waviness from the measurement. Judicious use of the large array of surface descriptors available will result in an accurate and realistic evaluation of a machined surface. The machined surface's total geometry has components of roughness, waviness and form, and parameters for describing it fall into the following groups:

- Amplitude – these are measures of the vertical characteristics of the surface deviations.

- Spacing – these are measures of the horizontal characteristics of the surface deviations.
- Hybrid – these are a combination of both.

Depending upon the application for which the component is designed, different surface-texture parameters will be chosen. Typically they might include skewness R_{sk} and kurtosis R_{ku}, which are both amplitude parameters, the high-spot count HSC, which is a spacing parameter, the bearing ratio *tp* (or Abbott-Firestone curve, as it is sometimes known), which is a hybrid parameter, and – of course – R_a.

If one considers the typical surface-texture traces from machined surfaces shown in Fig. 6.28, it can be seen that a variety of surface topographies can be generated. The amplitude distribution curves shown on the right of Fig. 6.28 are produced from the traces of the surface and indicate the air-to-metal ratios on two axes, i.e. the skewness and kurtosis. They show clearly how the values of skewness and kurtosis change considerably, depending upon the machine surface generated. The shape of this curve gives a visual confirmation of the trace's form, with a negative skewness when the metal ratio is high, through to a positive skewness when it is low. The kurtosis value, on the other hand, indicates the sharpness of the trace in terms of the distribution of the air-to-metal ratio over the peak-to-valley height. If the curve is spiky, it means that the metal is predominantly distributed in a particular region, whereas a flattened profile will mean that the metal is more evenly distributed over the whole peak-to-valley height.

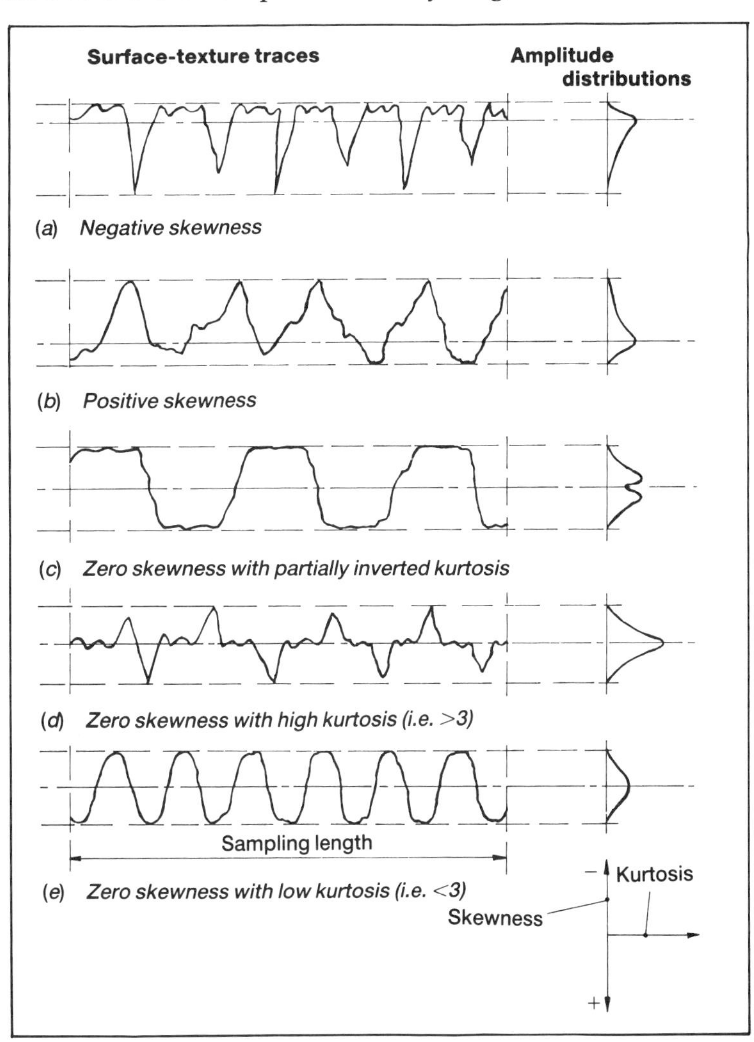

Fig. 6.28 The value of using the parameters skewness and kurtosis in defining the general shape of the surface-texture trace

The high-spot count is a count of the number of complete peaks that project above the mean line, or an arbitrary line selected at a particular depth, within the sampling length. This value allows the user to consider the peaks' plastic deformation, and can be used to specify a wear limit for a particular application, for example a transition fit between a shaft and a bush, or the wear on a bearing surface after a period of time.

Another useful method for estimating a machined surfaces's topography is the bearing ratio. This bearing ratio, or Abbott–Firestone curve, is the bearing surface's length expressed as a percentage of the sampling length at a particular depth below the highest peak. This value may be used in a similar manner to the high-spot count to determine the bearing condition at a preselected depth. An example of an Abbott – Firestone curve is shown in Fig. 6.26.

In all stylus-based instruments, the finite radius at the tip of the stylus means that a true trace of the surface texture is not produced: the stylus cannot follow the deepest valleys of the profile, and there is a certain amount of truncation of any narrow deep valleys. However, this is not too serious a problem in most machined-surface assessments. Some companies are even examining the use of light scattered off the surface to determine the true topography.

Using the surface descriptors skewness and kurtosis, it is possible to construct graphs that illustrate different machined surfaces. The envelopes of the skewness and kurtosis produced by different manufacturing process are shown on such a graph in Fig. 6.29a, which shows how diverse their topographies can become. In fact, they can be seen to fall into two main groups, a 'locking' group and a 'bearing' group. Processes falling in the locking envelope can be used when the two surfaces are required to lock together – occasionally, as in a disc-brake situation, or permanently, when adhesive bonding conditions are required. Processes falling in the bearing envelope can, however, be used satisfactorily for the bearing applications in sliding or rotating assemblies. So, when a designer requires a surface for a specific application, either bearing or locking, such manufacturing-process envelopes can be used to determine the correct machining condition, and the surface descriptors which are attained in these conditions can be defined and assessed. This is another argument for not relying on the R_a value alone!

6.10 The surface condition after machining

As can be seen from the photomicrographs in Fig. 6.30, the cutting speed plays an important role in determining the surface condition generated during machining, through its influence on the temperatures of the tool and workpiece. Furthermore, the properties of materials are dependent on the strain rate, with the type of tool wear changing according to the cutting speeds: if low cutting speeds are used, the wear on the tool is characterised by attrition (mechanical removal of surface layers), whereas at the high cutting speeds, attrition gives way to diffusion wear and Fick's laws dominate the cutting regime. These tool wear mechanisms will have some influence on the type of surface produced on the workpiece, as will the chip formation and strain behaviour.

When a carbide insert is used on an alloy steel, it has been shown that the feedrate and depths of cut have only marginal effects on the sub-surface damage, and the cutting speed is the most influential parameter in these cases. The surface shown in Fig. 6.30a was produced by a very low cutting speed, 2.6m/min. The chip formation is discontinuous and the surface shows an alternating effect of chip formation and fracture; there is some evidence of deposited residual built-up edge. This surface topography is due to a

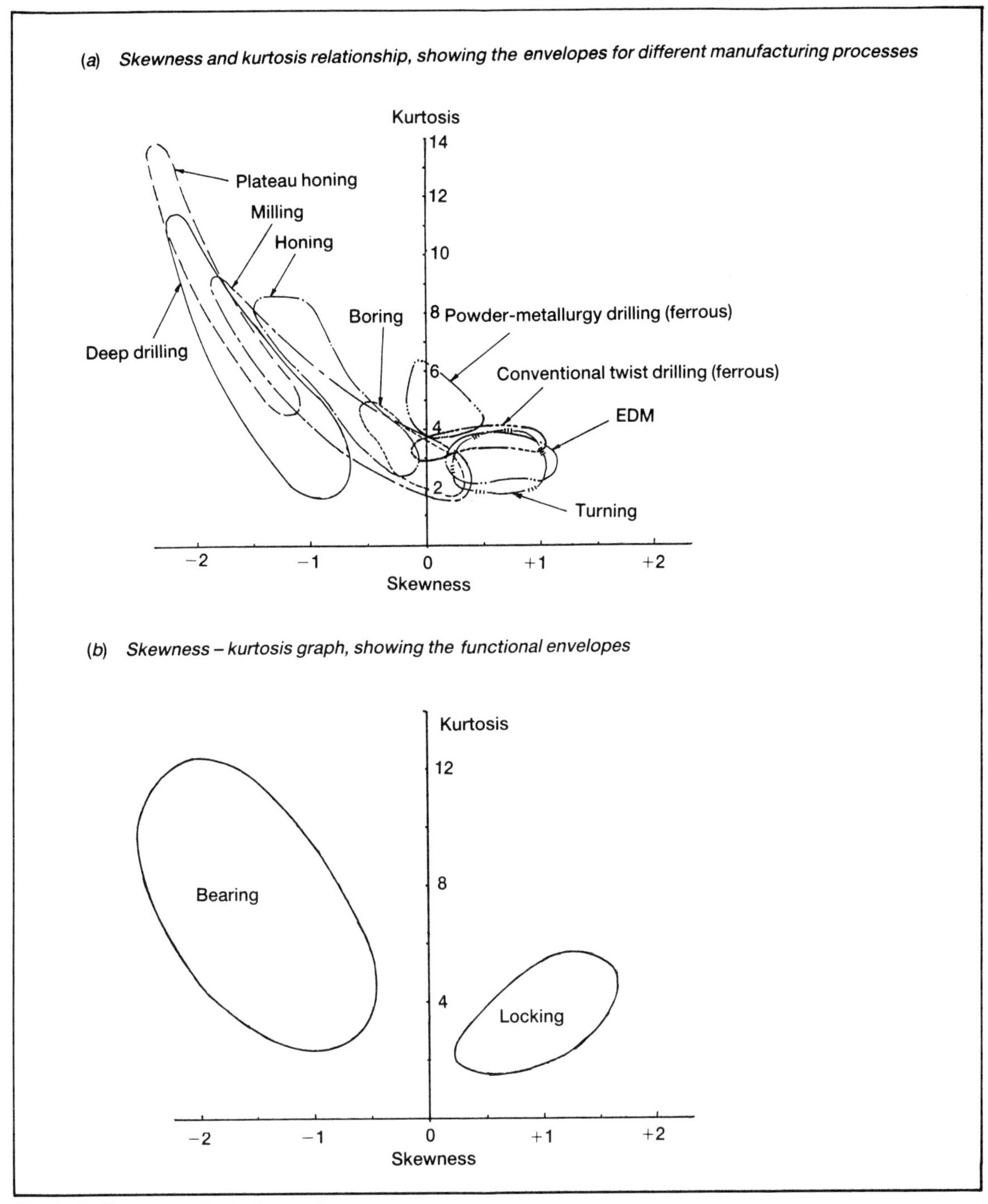

Fig. 6.29 Skewness–kurtosis envelopes

complex interaction of a number of effects, such as changes in shear angles in the contact area between the chip and tool, and straining which causes increases in the thickness of the chip. These effects produce a variety of conditions, from strain to cracking, and cause the irregular alternating topography of the surface.

Cutting speeds in the range from 11m/min to 59m/min generate a continuous chip formation. It is evident from the photomicrographs in Fig. 6.30b, c and d that the surface finish improves over this range, although there is some indication of debris from the built-up edge remaining in Fig. 6.30d. Once the 'approved' cutting speed has been reached (112m/min for this carbide insert grade), the surface finish appears to be good (Fig. 6.30e). Some isolated areas of the surface exhibit side flow (shown in Fig. 6.24). If the

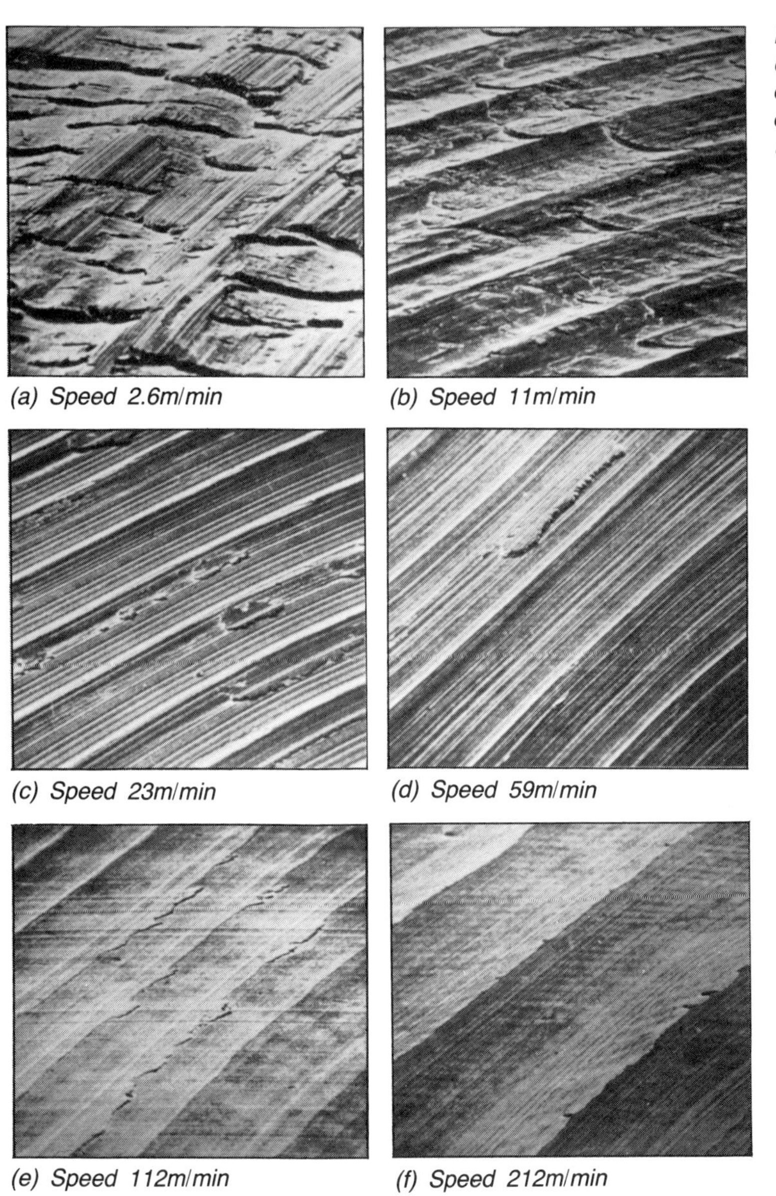
(a) Speed 2.6m/min
(b) Speed 11m/min
(c) Speed 23m/min
(d) Speed 59m/min
(e) Speed 112m/min
(f) Speed 212m/min

Fig. 6.30 Electron micrographs of component surfaces machined at different cutting speeds (Courtesy of Edgar Vaughan/D. C. Watson and M. C. Murphy)

cutting speed is raised still higher, to 212m/min, then a greater tool wear-rate occurs. This is not due to appreciable carbide breakdown, however, and an excellent surface finish is still produced (Fig. 6.30f).

The machined surfaces produced at the lower cutting speeds in Fig. 6.30 a to d show where scales of built-up edge have broken away during some revolutions of the work-piece and been deposited over several feed grooves. To obtain a clearer understanding of the surface and sub-surface effects in the extreme conditions of very low and very high cutting speeds, in Fig. 6.30 a and f respectively, the following comments can be made. If longitudinal sections are taken through the surface, they reveal its damaged nature. In the case of Fig. 6.30a, built-up edge is present on the surface, there is a cutting/fracture sequence with evidence of work-hardening, and the structure is of layered scales with cracks and crevices beneath them. In the component machined at the high cutting speed (Fig. 6.30f), a 'white layer' has formed, which is quite complex. In fact, the good surface finish disguises the fact that a white-layer underlay is present, with a hardness of 860HV. By way of comparison, if this component received conventional heat treatment at 1200°C, the hardness would only be 700HV.

It is evident from these discussions that an optimum surface finish is obtained when the cutting speed selected is in line with the manufacturer's recommendations; in this case, the speed would be 112m/min, with the surface topography similar to Fig. 6.30e. Obviously, if cutting fluids of the correct grade are used, 'abusive' cutting conditions are minimised and become closer to those in the 'gentle' machining regimes.

6.11 Metrological equipment used for assessing machined surfaces

An exhaustive account of all of the metrological equipment that is available for the assessment of machined surfaces is beyond the scope of this book. However, some examples of the inexpensive and the highly sophisticated equipment and methods will be briefly reviewed below.

6.11.1 Inexpensive and relatively unsophisticated methods

When companies cannot financially justify a large capital outlay on sophisticated equipment, or when they do not have an inspection department that would be able to use it regularly, some form of *limited* surface-texture assessment is still possible using less sophisticated equipment for very little cost.

The easiest to use but the most limited in scope is the comparator gauge method shown in Fig. 6.31a and b. These comparator gauges allow the surface finish to be determined by means of visual and tactile comparison with a number of 'standard' surfaces representative of each machining process. Gauges are available for turning, milling, shaping, etc., and contain blocks in a range of ascending surface finishes, identified by their R_a values.

Designers can use the gauges in conjunction with a comparison chart, as shown in Fig. 6.31a, to obtain a basic interpretation and visual and tactile impression of the machined surface produced by a particular method. Alternatively, the machine operator can compare the surface just produced against those on the comparator gauge. This gives a basis for at least some elementary surface-finish assessment on the machine, without removing the workpiece from its set-up. This system suffers from the problem that the interpretation or estimation of the surface condition is subjective: two people could vary considerably in their estimates of the R_a value obtained visually or by tactile assessment.

A range of easy-to-use hand-held instruments that overcome this problem are available, of which the one shown in Fig. 6.31c is typical. This instrument gives reasonably repeatable values of R_a for assessment of the machined surface. Its usefulness, of course, is subject to the proviso that use of the R_a value alone can be misleading.

The instrument shown in Fig. 6.32 provides even greater accuracy and just as much flexibilty of use. It can be used in any position, horizontal or vertical, and is small enough to be carried in the operator's pocket until needed for assessment. Its carrying case contains a reference standard on which the instrument can be calibrated before use, enabling consistent, calibrated and accurate assessments of the surface to be done in seconds. Once again, this instrument is only capable of an assessment of the R_a value, so its use must be tempered with caution. Its advantages are its portability, cheapness, ease of use and adaptability.

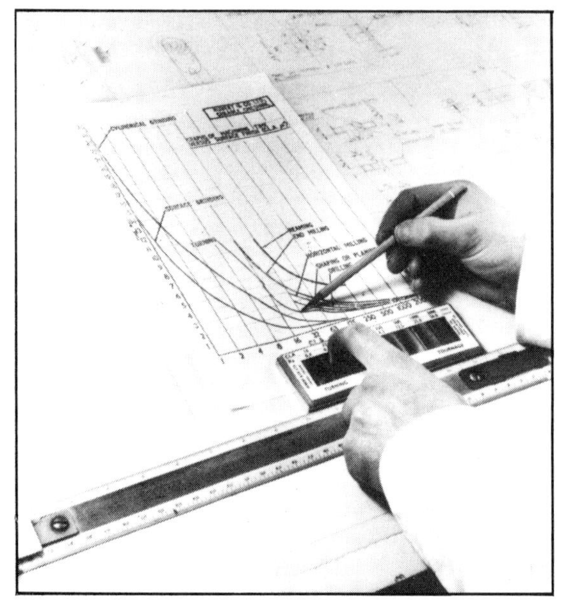

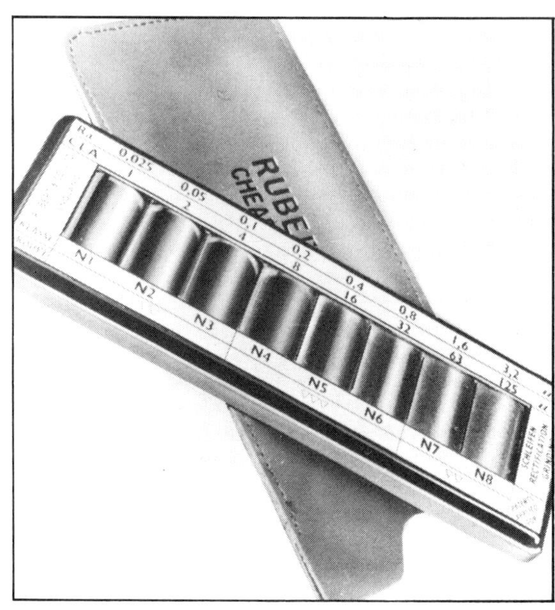

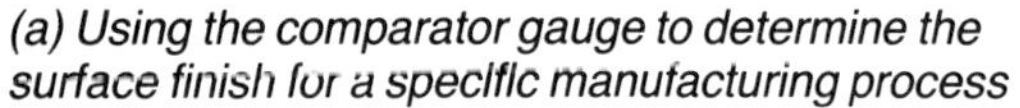

(a) Using the comparator gauge to determine the surface finish for a specific manufacturing process

(b) A typical comparator gauge

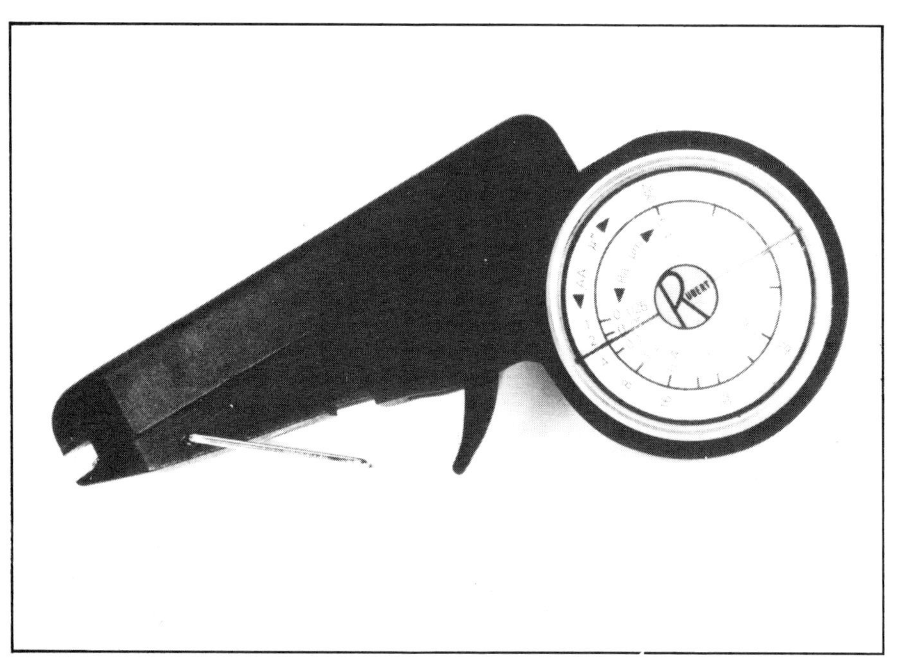

(c) A hand-held surface-texture instrument.

Fig. 6.31 Inexpensive methods for surface-finish determination (Courtesy of Rubert & Co.)

6.11.2 Highly sophisticated instruments

When a more complete interpretation of the roundness and surface texture is required, together with the ability to program a large range of parameters and store the results or programs for future use, then instruments of the kind shown in Fig. 6.33 are necessary. The range of applications of such systems is vast, and just some of their capabilities are mentioned here.

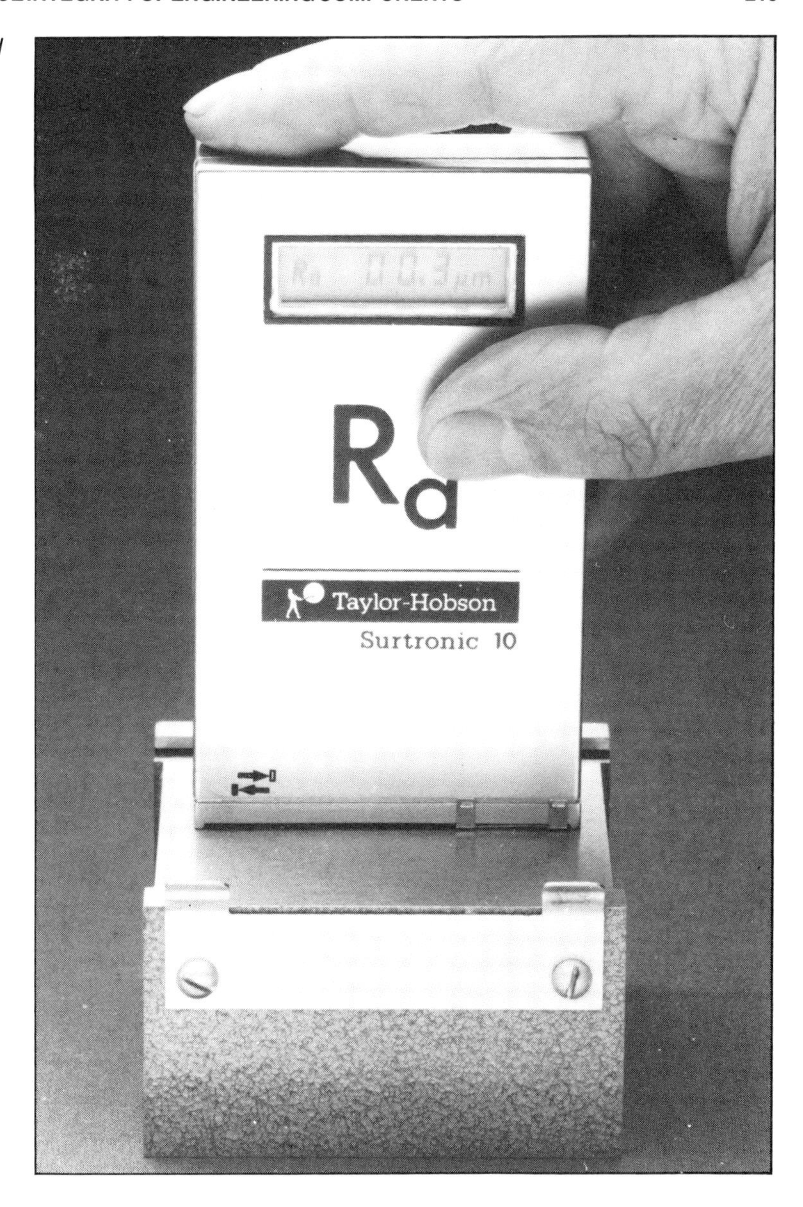

Fig. 6.32 An electronic hand-held instrument for the determination of surface texture (Courtesy of Rank Taylor-Hobson)

Assessing roundness parameters

Roundness is a geometric characteristic of cylindrical or spherical workpieces. If there are errors in the workpiece – which is almost always the case – they have been caused by the machine tool that produced the component. Obviously any deviations from the perfect circle can be considered as large-scale, rather than small-scale errors. The reason why these deviations are usually considered alongside surface-texture results – which relate to small-scale errors – is that the main factors which affect the efficiency of rotary bearings are the roundness of the mating parts and their surface texture.

Machines for assessing roundness should generally include the following features:

- An accurate measurement basis, in the form of an independent axis of rotation produced using a precision bearing.
- A method by which the axes of the work and the bearings may be aligned.
- An instrument capable of measuring and amplifying roundness errors.
- A recording device capable of transferring data to and from the computer's memory for further processing – either to a screen or to a printer to obtain a 'hard copy'.
- A method by which quantitative assessment of the errors can be made.

Fig. 6.33 Instruments for determining some aspects of surface integrity (Courtesy of Rank Taylor-Hobson)

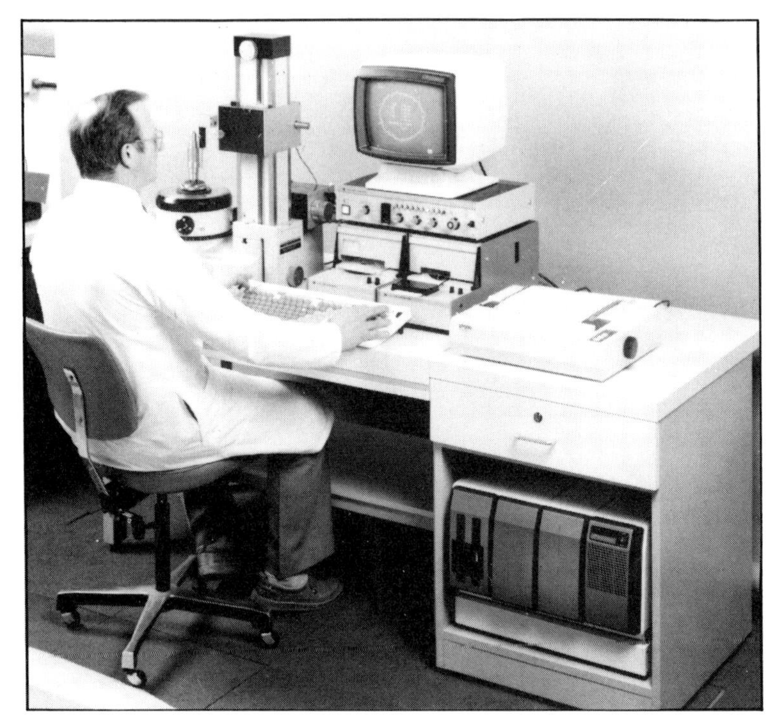

(*a*) *The Talyrond 200, for the assessment of roundness parameters*

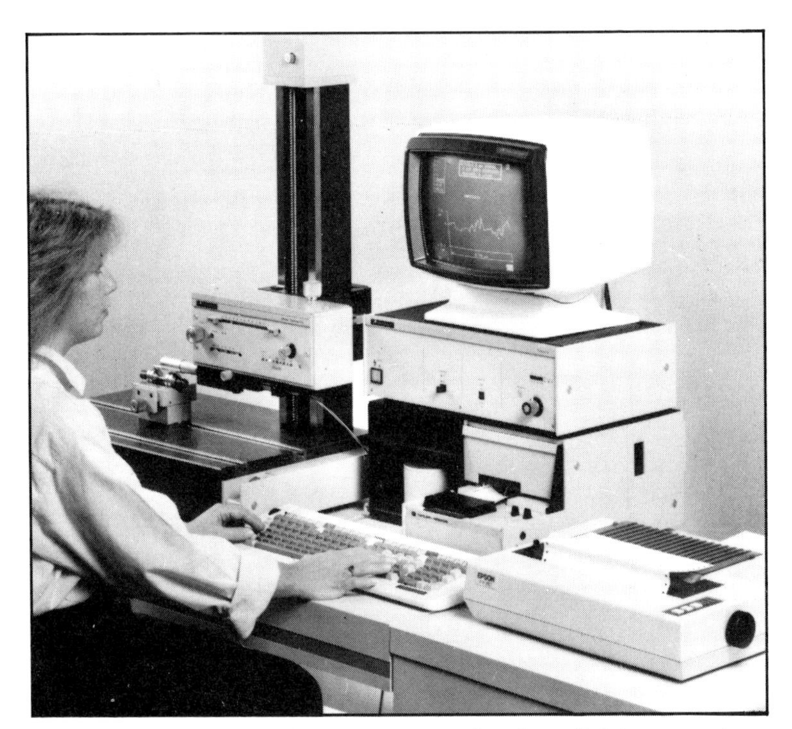

(*b*) *The Talysurf 6, for the assessment of surface-finish parameters*

Instruments with this level of adaptability and electronic sophistication, such as the one in Fig. 6.33a, can be used in a vast number of roundness applications, including the assessment of such parameters as roundness, cylindricity, co-axiality, eccentricity and straightness. The inspector can choose which parameters need to be assessed on the machined part, as well as the levels of magnification and filtering, and what method of measurement is required.

Several methods are used in the determination of roundness, four of which are discussed in the following sections. Of these, the first two are non-preferred methods, based on trial and error.

Plug-gauge centre method This method is based on determining the largest circle which may be just contained inside the trace, which touches at least two points on the trace. The deviation from roundness is the difference in radii between this circle and an outer one, which has the same centre and touches at least one point on the trace, divided by the magnification.

Ring-gauge centre method This method is based on determining the smallest circle which will just contain the trace. An inner circle touching at least one point on the trace, is drawn from the same centre, and the deviation from roundness is quoted as the difference between their radii, divided by the magnification.

Minimum-zone centre method In this case, a pair of concentric circles are found that just contain the trace in the annular zone between them. The roundness error is the width of the narrowest such zone divided by its magnification.

Least-squares centre method This is by far the most popular method of assessing the roundness of a machined part. It is based on a circle chosen so that the sum of the squares of the radial distances of all the trace points from this circle is a minimum. Having found this circle, two circles are drawn from its centre which contain the inner and outer extreme points on the trace. The radial difference between these circles, divided by the magnification, represents the roundness error. This method is analogous to finding the 'best line' on a graph by the least-squares method.

The highly sophisticated roundness machines can simply and speedily superimpose the least-squares circle on any trace and determine the roundness error. The machine in Fig. 6.33a is also capable of changing the filtering frequency automatically, and of producing magnifications of up to 20 000 times. If the straightness column is fitted (as shown in Fig. 6.33a) traces of a variety of diameters can be superimposed without moving the part, so that the degree of concentricity between the diameters can be assessed.

Assessing surface-finish parameters

A wide range of highly sophisticated electronic and computer-operated equipment is available, allowing just about every conceivable parameter of a machined surface to be assessed. Therefore, only some of the main features will be mentioned in this section. Both machines shown in Fig. 6.33 are prompted by menu-driven software. An exciting feature of the surface-texture machine (Fig. 6.33b) is its ability to produce a contour map of the machined surface. This mapping facility depends on a traversing unit, which can perform up to 64 traverses, with a minimum distance of 0.5mm between successive traces. A large area of a machined surface can be assessed by displaying all the individual traces isometrically; there is also the ability to scale the display. Other features of the display are the ability to invert the image and view the valleys as peaks, and to alter the viewing position.

By using other attachments, the surface finish of radii and curvatures can be assessed, as well as all of the normal amplitude, spacing and hybrid parameters.

Such machines are modular in construction, so once a basic machine is purchased other levels of sophistication can be added later, allowing for just about every variation of surface-texture assessment through the purchase of optional extras.

6.11.3 The latest technology – almost a complete measuring system

Such is the pace of technological advance that it is now possible to acquire machines with vast computing power and capability for automatic measurement; an example is the machine shown in Fig. 6.34. With this machine, it is possible to use computer programs to measure surface texture, roundness, radial change, straightness and many other parameters.

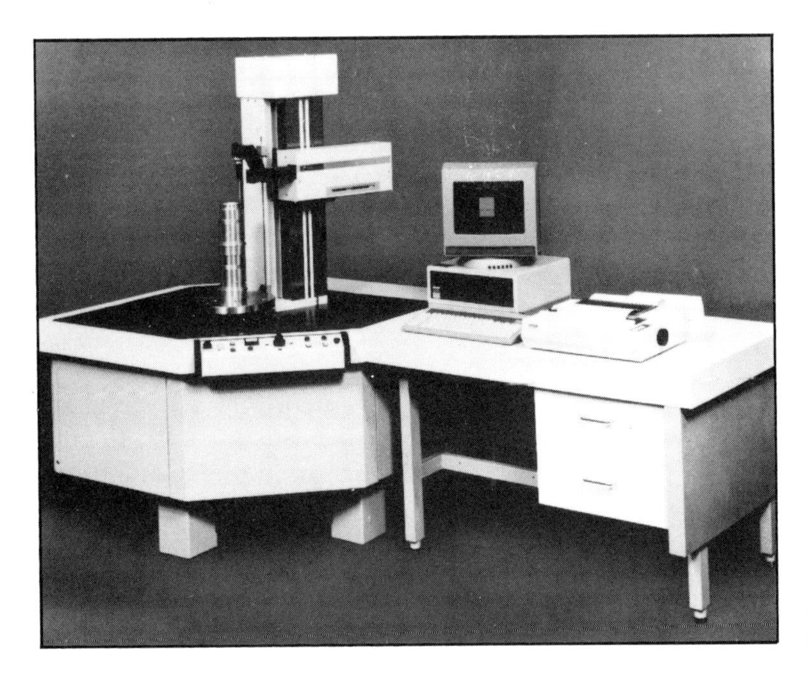

Fig. 6.34 The Talyrond 300 – a highly sophisticated instrument for determining roundness, radial change, concentricity, straightness, surface finish, etc. (Courtesy of Rank Taylor-Hobson)

Not only can the gauging head be programmed for different orientations (allowing the component to be addressed from a number of angles), but the workpiece may be positioned onto the table by either a 'pick-and-place' unit or a robot. Obviously, the positioning ability of such handling devices is not very accurate compared with the high accuracies of the instrument itself. However, the instrument will accommodated a certain amount of misalignment: so long as the centre of gravity of the workpiece is within the triangular engraved lines on the table (shown in the lower photograph in Fig. 6.34), the computer will modify its program and carry on measuring the part.

Once the part has been fully assessed, the data may be transferred to peripheral devices to control automatic updating as necessary. Machines of this type will revolutionise the way in which companies perceive the assessment of machined surfaces. However, they must be used under full metrological environmental conditions of controlled temperature, humidity and cleanliness, if their full benefits are to be obtained.

This discussion of the variety of equipment available for the determination of surface conditions after machining operations was not intended to be an exhaustive account. In the last two sections of this chapter, two extreme material types – treated steel and metal-matrix composites – are discussed. These are now becoming serious competitors of the present materials. They show considerable variation in their machining behaviours, with the steel being a material of 'ideal' machinability and the metal-matrix composites anything but easy to machine.

6.12 A treated steel with 'ideal' machining characteristics

First of all, it is worth mentioning that there are two kinds of steels with improved machinability characteristics. The first type is the so-called free-cutting steel, in which the enhanced machinability is achieved through additions of alloying elements such as sulphur, bismuth and lead (although the lead varieties are now losing favour because of the effects on the body of too much lead in the atmosphere). Elements such as these form a large number of inclusions in the steel; these aid machinability but also have the effect of impairing the mechanical properties. The second type of high-machinability steel is basically a standard-grade steel with additions of calcium. These cause the inclusions to be modified into a form more beneficial for machinability with little effect on the mechanical properties.

Fig. 6.35a illustrates the marked effect of non-metallic inclusions in steels. In general terms, inclusions such as the manganese sulphides can be considered as beneficial. Conversely, the aluminium oxides, the silicates and the complex oxides are thought of as detrimental to machinability. In calcium-treated steel, however, the oxides are modified into a more beneficial form, calcium aluminate. This is much less abrasive than the aluminium oxide and is associated with sulphides which, in turn, also improve machinability. A further advantage is that the sulphur is present as an envelope around the calcium aluminate, decreasing the abrasive effect on the tool still more.

The calcium treatment affects the steel in two ways:

- It reduces the abrasion of the tool caused by the hard oxides, especially the aluminium oxides.
- When high cutting speeds are used (when using carbide tooling, for example), a protective film forms on the tool. This arises from the inherent sulphur in the steel, as shown in Fig. 6.35b.

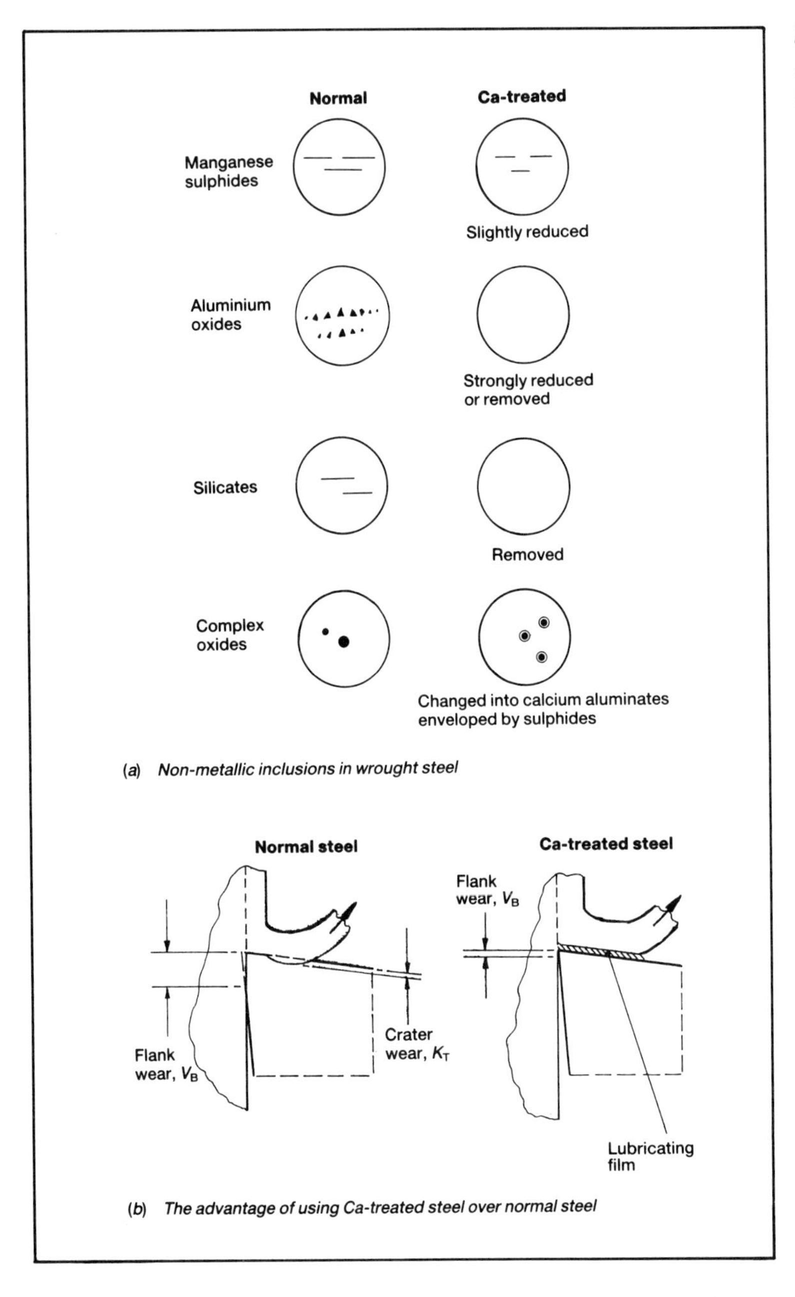

Fig. 6.35 Ca-treated steel provides excellent machinability (Courtesy of Ovako Steel)

Most modern steels are aluminium treated for grain refinement, but this inevitably leads to inclusions of alumina, Al_2O_3 . These inclusions are extremely abrasive and are a prime cause of poor machinability and high tool wear. Aluminium oxide has a higher melting point than steel, so the inclusions in the steel matrix have an angular shape, because they are already present in the steel when it solidifies.

In the calcium-treatment process, a lance is inserted into the molten steel and an inert gas is flushed through to equalise the temperature and to act as a carrier for the reactive elements. It is essential to remove any dissolved oxygen in the steel before the calcium treatment takes place, because otherwise it would react immediately with the calcium. Powdered calcium-treatment elements are injected into the steel through the gas stream. This results in all the silicates being completely removed and the calcium reacting with the alumina to form calcium aluminate. This has a lower melting point than the steel, so the inclusions become spheroidal and softer than the previous alumina ones. This treatment gives an obvious improvement during machining, in the form of less tool wear owing to modification of the abrasive inclusions. As a result, the tool life is increased.

The sulphides in the steel are modified by the process and tend to associate with the calcium aluminate, forming an envelope around the inclusions. This affinity of the calcium-treated inclusions has the added advantage of plating a sulphide layer onto the tool's surface (Fig. 6.35b) and in so doing improving the machinability characteristics.

The advantage of using calcium-treated steels is that they can be used at considerably higher cutting speeds, with longer tool life and less power consumed by the machine. This can result in productivity savings 'right across the board', including 150% increases in tool life, 30% increases in throughput and a 75% reduction in the scrap rate (as more consistent tolerances can be held). It has also been found that power savings of 20% occur, and that the carbide tools can be used without coolant. Another advantage of these steels is that their cost is similar to those of the equivalent grades processed by other routes.

The companies machining with these steels have all shown a marked improvement in their profitability since using them; it remains to be seen how long other companies will take before they start machining with them too (assuming that the engineering parameters allow).

If these currently-available commercial steels are, in a sense, a machinist's 'dream', the new metal-matrix composites are the cutting-tool manufacturers', machine-tool builders' and machinists' challenge for the future!

6.13 Metal-matrix composites – the challenge for the future

Large numbers of new materials are being developed for the 1990's. A particularly active research program at the moment is into the area of metal-matrix composites. These materials offer the designer vastly superior structural performance, with excellent strength-to-weight ratios, but can present major difficulties in machining. Some of the early researchers in the field have found that conventional tooling can be totally ineffectual in cutting some of these materials. The difficulties of machining these materials must be overcome, though, to ensure that the benefits of this research and development are transferred to the end-user. The materials have a high potential usage in replacing conventional products, particularly in the automotive and aerospace industries.

The photomicrographs in Fig 6.36 give some idea of the degree of difficulty likely to be encountered by the production engineer whose brief it is to machine them to produce components such as the ones in Fig. 6.37. Considering the first example of a metal-matrix composite, in Fig. 6.36a, it is obvious to any engineer that a mixture of tough stainless-steel wire and ductile aluminium will present considerable machining problems, even if the surface-integrity effects resulting from the machining are ignored! Similarly, the composite in Fig. 6.36b, will present another magnitude of difficulty, as in this case, the ductile aluminium has silicon carbide (grinding dust!) particles present throughout the structure. Just what this does to the tool is anybody's guess, quite apart from the problem of this dust abrading the machine tool's slideways!

All of these metallic and non-metallic additions to the metal-matrix composites produce severe tests for the engineer, but they are problems that must be overcome as the application potential of these materials is vast. They offer near-net shapes with minimal secondary machining; the components shown in Fig. 6.37, for example can be produced with no draft angles present. This means that products of the future will not bear the cost of the large increase in value added, which is known as machining.

Fig. 6.36 Metal-matrix composites (Courtesy of Hi-Tec Metals (R&D) Ltd.)

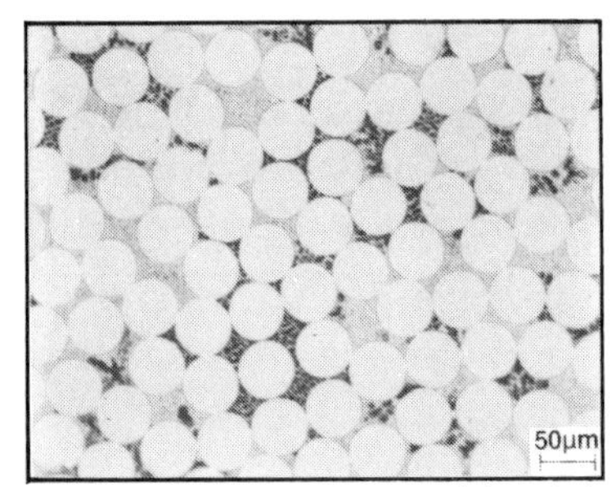

(a) Aluminium with 50μm diameter stainless-steel wire

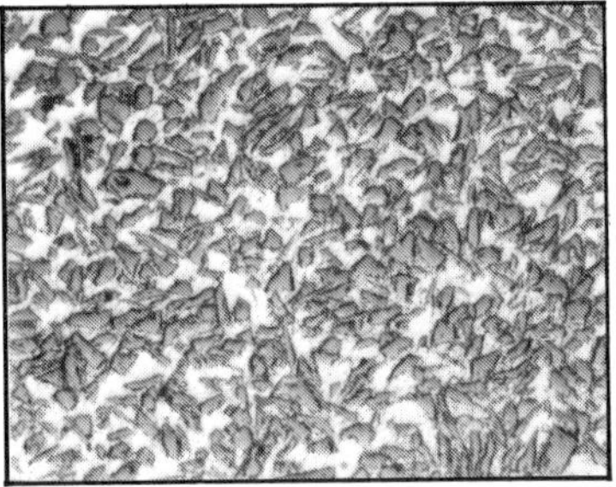

(b) Aluminium with 5μm particles of silicon carbide

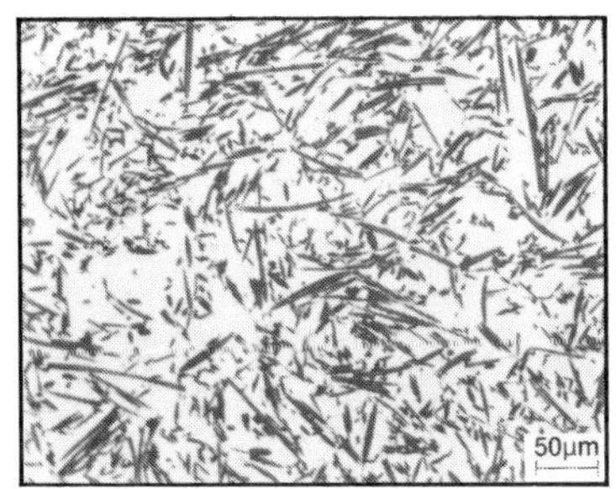

(c) Magnesium with an infiltration of Saffil particles

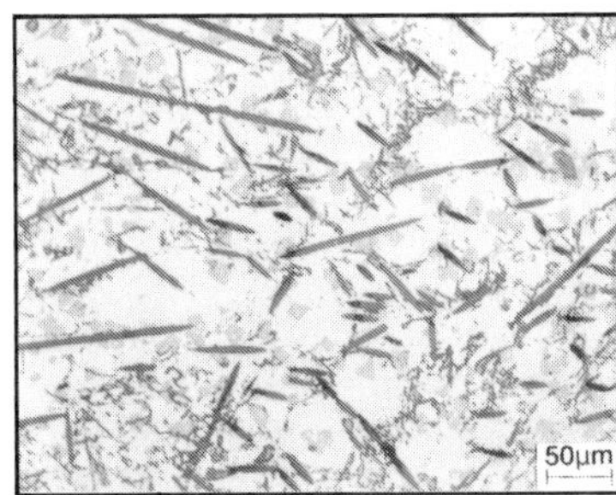

(d) Magnesium with a mixture of glass and graphite particles

(e) Detail of particulate reinforcement: Saffil (Al_2O_3)

Fig. 6.37 Components manufactured by the metal-matrix composite route (Courtesy of Hi-Tec Metals (R&D) Ltd.)

Appendix A
Turning troubleshooting

Turning operations with carbide tooling

Balanced wear

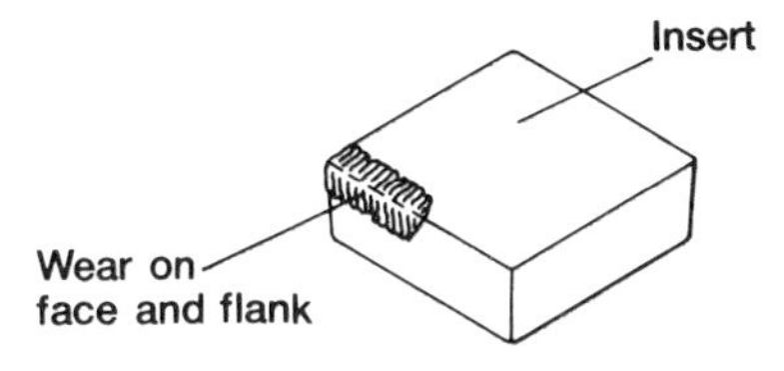

In theory, wear life is reached when the component loses 'size,' or when the component loses its surface finish.

Flank wear

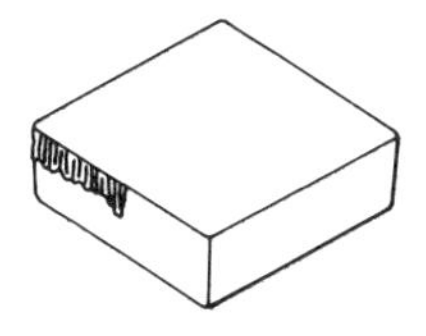

Cause	Remedy
Too high a cutting speed.	Reduce cutting speed.
Carbide grade with insufficient resistance to flank wear.	Choose wear-resistant carbide grade.
Deformation of the cutting edge, resulting in heavy flank wear.	
Insert above centre-height.	Check centre-height.
Tool-holder tilted out of position.	Check tool-holder position.

Crater wear

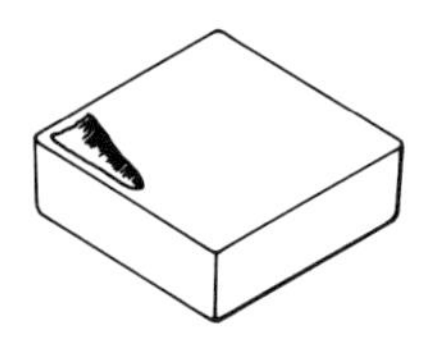

Cause	Remedy
Too high a cutting speed.	Reduce cutting speed.
Too low a feed.	Increase feedrate.
Low resistance to crater wear.	Choose a carbide grade more resistant to crater wear.

Mechancial failure

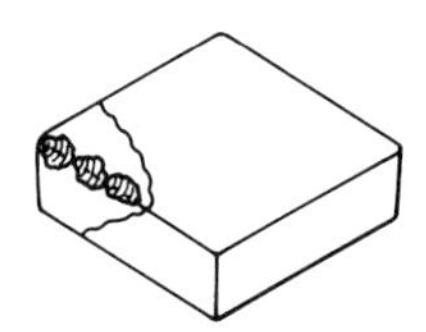

Cause	Remedy
Incorrect chip-breaker	Increase chip-breaker width.
Cutting edge too weak for the job.	Increase edge strength by honing or by a negative edge-reinforcing chamfer.
Vibration.	Improve stability.
Insufficient microtoughness of the cemented carbide.	
Intermittent supply of the coolant.	Choose coolant, or cut 'dry.'
Incorrect insert seating.	Check condition of tool-holder.

Built-up edge

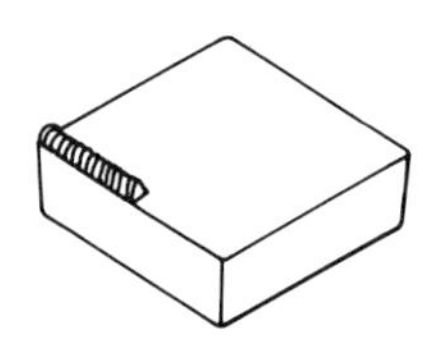

Cause	Remedy
Too low a cutting speed or feedrate or too small a cutting rake angle.	Increase cutting-edge temperature by increasing cutting speed, reducing coolant or, preferably, cutting 'dry'.

Comb cracks

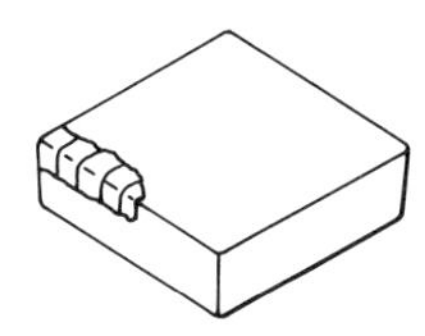

Cause	Remedy
Thermal shocks, for instance in milling and intermittent turning operations.	Choose a tougher grade of carbide.
	Cut 'dry'.

Deformation

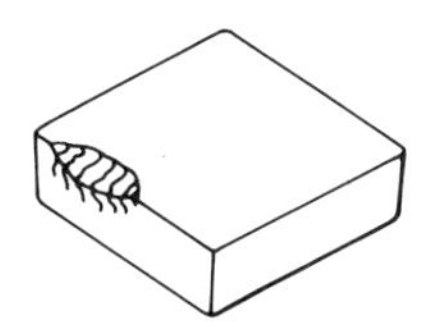

Cause	Remedy
Too high a load on the tool face.	Reduce feed.
Too high a cutting temperature.	Reduce cutting speed.
Too soft a carbide grade	Choose a harder and more wear resistant carbide grade.

(Courtesy of Seco)

Threading operations using carbide tooling

Rapid flank wear

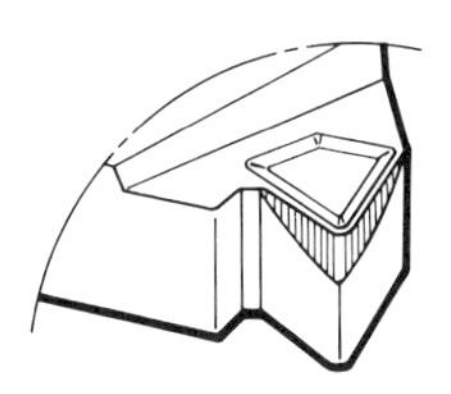

Cause	Remedy
Cutting speed too high.	Reduce the cutting speed.
Lack of coolant.	Increase coolant supply.
Infeed per pass too small – too many passes.	Increase the depth of infeed for the smallest depths – reduce the number of passes.
Incorrect grade.	Select a more wear-resistant grade.

Uneven flank wear

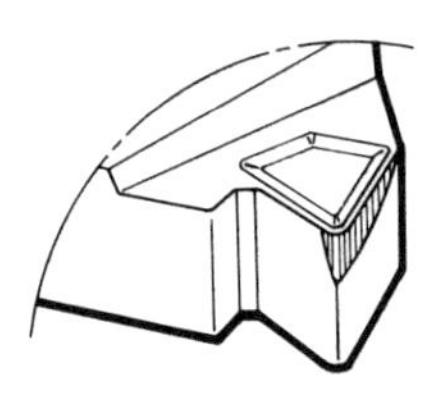

Cause	Remedy
Incorrect method of infeed.	At flank infeed, decrease infeed angle by 3°–5°.
Incorrect angle of inclination.	Correct angle of inclination.

Excessive plastic deformation

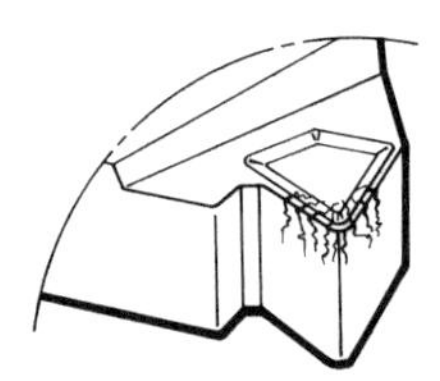

Cause	Remedy
Infeed per pass too big – too few passes.	Decrease the depth of infeed for the biggest depths – increase the number of passes.
Lack of coolant.	Increase coolant supply.
Cutting speed too high.	Reduce cutting speed.
Incorrect grade.	Select a harder grade.
Excessive stock removal from crest.	Check material stock above the crest.

Edge frittering

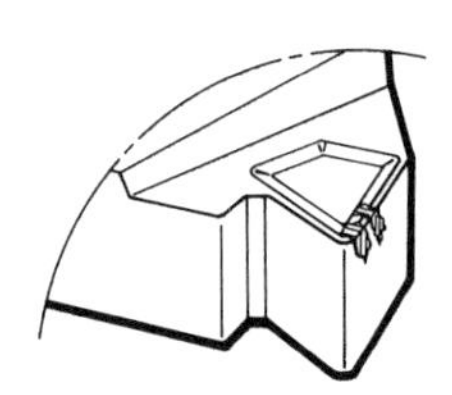

Cause	Remedy
Instability of work-holding and/or tool set-up.	Check ridigity of operation. Select a tougher grade.

Edge spalling

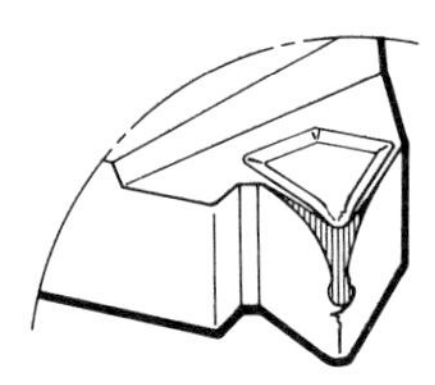

Cause

Intermittent coolant supply.

Remedy

Direct coolant flow and/or increase coolant supply.

Insert breakage

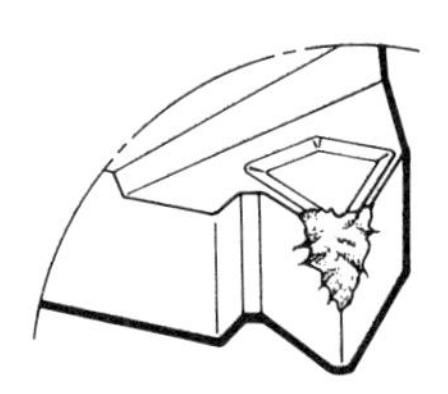

Cause	Remedy
Instability.	Check rigidity of operation.
Lack of chip control.	Select a tougher grade.
Excessive plastic deformation.	Machine with same infeed per pass.
Intermittent or inadequate coolant supply.	Direct coolant flow and/or increase coolant supply.
Incorrect preparatory operation.	Check dimension of blank.

Shallow thread profile

Cause	Remedy
Wrong centre-height.	Adjust cutting-edge height.
Insert not cresting.	Check dimension of blank.
Excessive tool wear.	Change insert earlier.

Incorrect thread profile

Cause	Remedy
Incorrect tool setting.	Correct tool setting.

Lack of chip control

Cause	Remedy
Incorrect depth of infeed per pass.	Machine with same infeed per pass. Use U-lock threading insert.

Bad surface finish

Cause	Remedy
Cutting speed too low.	Increase the cutting speed.
Incorrect angle of inclination.	Correct the angle of inclination.

(Courtesy of Sandvik (UK) Ltd.)

Parting-off operations

Insert chipping

Cause	Remedy
Built-up edge.	Try a positive geometry and/or increase the cutting speed.
Tool and/or workpiece deflection.	Try a positive geometry. Reduce feed and overhang.
Feed movement accelerates at breakthrough (machine play).	Adjust the feed mechanism (pneumatic or hydraulic feed is not desirable in parting operations). If this fails, fit a shock absorber on the slide.
Parted-off pieces rebound against the cutting edge.	Reduce machine speed. Try a negative insert and/or a tougher grade. Fit a component-catcher unit.
Residual pipe or ring damages the cutting edge at breakthrough.	Reduce feed just before breakthrough and stop the feed movement as quickly as possible.

Vibrations or poor surface finish

Cause	Remedy
Tool incorrectly positioned or mounted.	Check correct positioning or mounting.
Excessive overhang.	Reduce overhang.
Insufficient feed.	Increase feed.
Excessively high cutting speed.	Reduce cutting speed.
Excessive cutting forces.	Try a positive geometry, if possible with reduced part-off insert width.
Built-up edge.	Try a positive geometry and/or increase cutting speed.
Machine play.	Adjust.

Convex or concave machined surface

Cause	Remedy
Right or lefthand insert.	Try a positive geometry, if possible in a neutral insert.
Built-up edge.	Try a positive geometry and/or increase the cutting speed.
Excessive feed.	Try a positive geometry or reduce feed.
Worn or chipped insert.	Change the insert.
Insufficient lateral rigidity.	Try an increased insert width if possible and/or reduce overhang.
Chip jamming.	Use a positive geometry. Adjust feed to achieve an acceptable chip form. Increase the cutting fluid flow.

Short tool life

Cause	Remedy
Chipping or breakage of insert.	Select tougher grade.
Excessive flank wear and/or crater wear.	Select a more wear-resistant grade and/or reduce the cutting speed.
Cutting-tip height not according to recommendations.	Adjust cutting-tip height.

(Courtesy of Sandvik (UK) Ltd.)

Vibrations produced in turning operations

Vibrations or chatter marks caused by the tooling or the tool mounting (typical of internal machining with boring bars)

Cause	Remedy
High radial cutting forces, due to:	
Unsuitable entering angle.	Select as large an entering angle as possible (K=90°).
Nose radius too large.	Select a smaller nose radius.
Unsuitable edge roundness, or negative chamfer.	Select an uncoated grade with a sharp cutting edge.
Dull edge through excessive flank wear.	Select a more wear-resistant grade or reduce speed.
High tangential forces, due to:	
Insert geometry creating high cutting forces.	Select a positive insert geometry.
Chip breaking is too hard, giving high cutting forces and inducing vibration.	Reduce the feed or select a geometry for higher feeds.
Varying or too low cutting forces due to small depth of cut and insert rubbing.	Increase the depth of cut slightly to make the insert cut, not rub.
Tool incorrectly positioned.	Check the centre-height.
Instability in the tool due to long overhang.	Reduce the overhang if possible. For boring- bars use the largest-diameter bar possible. If the overhang is over four times the diameter, use an antivibration bar.
Unstable clamping gives insufficient rigidity.	Improve the clamping of the boring-bar. The clamping length should be at least three times the diameter. Rebush mounting. Use a rigid clamping method.

Vibration or chatter due to poor stability in the machine or vibration-sensitive parts like long slender shafts or thin-walled workpieces

Cause	Remedy
Machine problems.	Check the machine for play in the bearings, transmission and slides, and adjust or repair if needed.
Vibration-sensitive workpieces.	Check the set-up and make it as rigid as possible. Try to dampen or support the workpiece with a steady, or by using dampening materials such as rubber or plastics.

(Courtesy of Sandvik (UK) Ltd.)

Surface-finish problems in turning operations

The surface looks dull and grey and does not meet surface-finish requirements

Cause	Remedy
Cutting speed too low, causing built-up edge.	Increase cutting speed to reduce the risk of built-up edge.
Dull cutting edges due to excessive flank wear.	Change the cutting edge. Select a more wear-resistant grade. Reduce the cutting speed.
Large edge roundness of a coated grade or a negative cutting-edge geometry.	Select a uncoated grade with high wear resistance. Select a positive cutting geometry.
Depth of cut too small, making the insert rub against the workpiece, not cut.	Increase the depth of cut slightly.
Too small a clearance angle between the trailing edge of the insert and workpiece (especially for incopying operations).	Reduce the incopy angle to increase the clearance. Select a positive tool system.

The surface looks and feels 'hairy' and does not meet the tolerance requirements

Cause	Remedy
The chips are breaking against the turned shoulder, marking the finished surface.	Select an insert geometry with a wavy cutting edge to guide the chips away. Change the entering angle. Reduce the depth of cut. Select a positive tool system with a neutral angle of inclination.
Hairy surface caused by excessive notch wear on the trailing edge.	Select a grade with better resistance to oxidation wear (Al_2O_3-coated or a TiC-Ni-based grade). Reduce the cutting speed.

The surface finish is too rough after turning

Cause	Remedy
Too high a feed, in combination with too small a nose radius, generates a rough surface.	Select a larger nose radius. Reduce the feed.

(Courtesy of Sandvik (UK) Ltd.)

Appendix B
Turning data determination

The machine power requirement

Turning data determination

The power required for a metal-cutting operation is usually of most concern when roughing, when it is essential to ensure that the machine will produce sufficient power for the operation. The main parameters to be considered in the power calculation are the tangential cutting force F_t and the cutting speed V. Also of great importance in the overall calculation is the efficiency factor, normally denoted by the Greek symbol η. This factor will depend upon the type of transmission system fitted to the machine, and its overall condition. In practice, this means that the calculated cutting data is often irrelevant and cannot be used because of losses in machine transmission from the motor to spindle. The machine's efficiency, η, is normally between 0.6 and 0.9, depending upon the machine's condition. Usually a value of $\eta = 0.7$ may be used in the power calculation to give a reasonable approximation.

The equation is

$$\text{Power, } P = \frac{P_{net}}{\eta}$$

where P = available motor effect (in kW), and
P_{net} = net effect for metal cutting (in kW).

$$P_{net} = \frac{V \times F_t}{60 \times 1000}$$

$$P = \frac{V \times F_t}{\eta \times 60 \times 1000}$$

The tangential cutting force can be determined using the equation

$$F_t = k_s \times a \times s$$

where k_s is the specific cutting force (obtained from the manufacturer's data sheets for specific materials, insert geometry, plan approach angle and feedrate correction factors), a is the depth of cut and s the feed.

Therefore, the complete formula for the power is:

$$P = \frac{V \times a \times s \times k_s}{\eta \times 60 \times 1000}$$

where V = cutting speed (in m/mm),
a = depth of cut (in mm),
s = feed (in mm/rev), and
k_s = specific cutting force (in N/mm^2).

Example

V = 130m/min
a = 8mm
s = 0.8mm/rev
k_s = $1430N/mm^2$ } or F_t = 9150N
η = 0.7

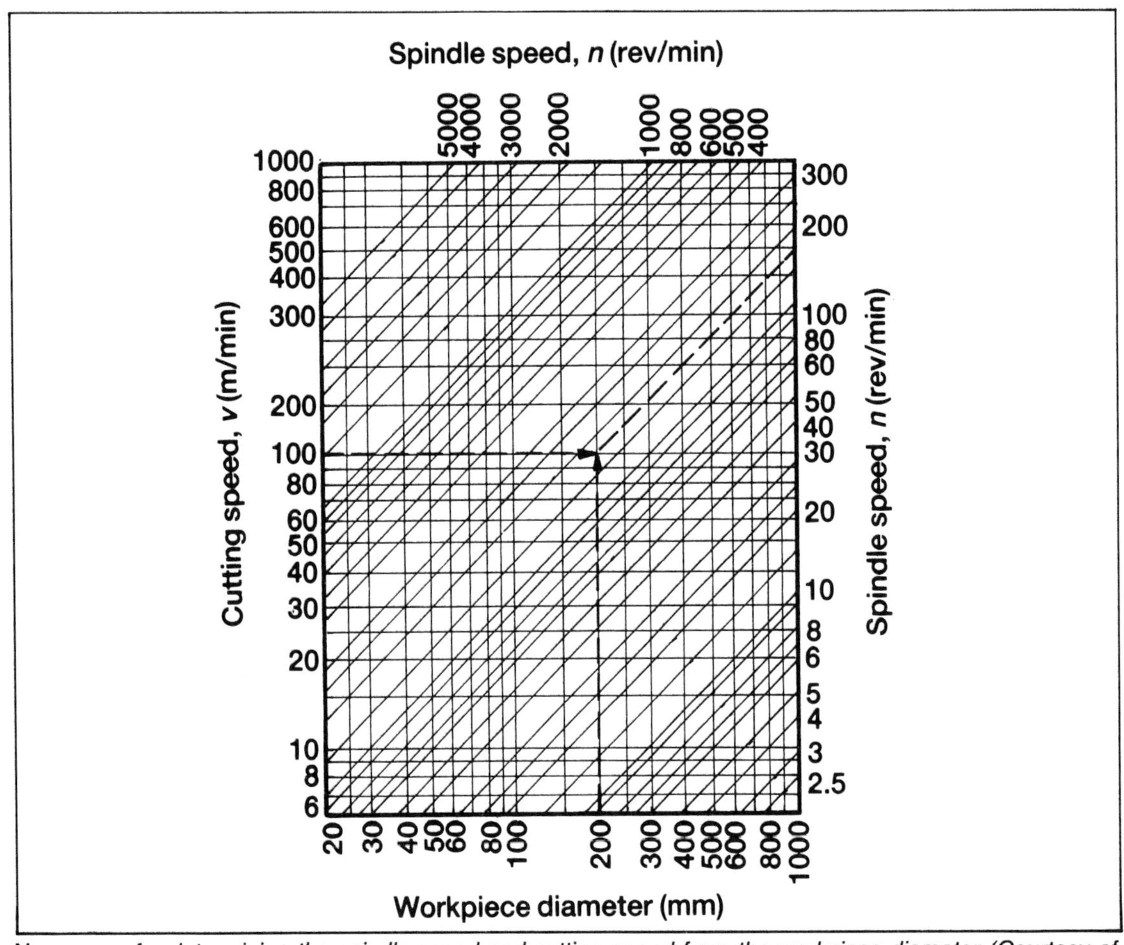

Nomogram for determining the spindle speed and cutting speed from the workpiece diameter (Courtesy of Sandvik (UK) Ltd.)

The power required for the roughing-out operation is therefore

$$P = \frac{130 \times 8 \times 0.8 \times 1430}{0.7 \times 60 \times 1000} \quad \text{or} \quad \frac{130 \times 9150}{0.7 \times 60 \times 1000}$$

$$\approx 28\text{kW}$$

Suppose that the calculated power (28kW in this case) is just below the maximum power available on the lathe. Then it is possible to increase the effective cutting performance of the machine by increasing the depth of cut (if possible), the feedrate or the cutting speed. If, however, the calculated power came out to be above the maximum power of the machine, it would be necessary to decrease the values.

The graph below provides speedy and reasonably accurate approximations of the machine power required for a specific cutting operation (roughing). The same data has been used for the example given on the graph as in the 'long-hand' method above.

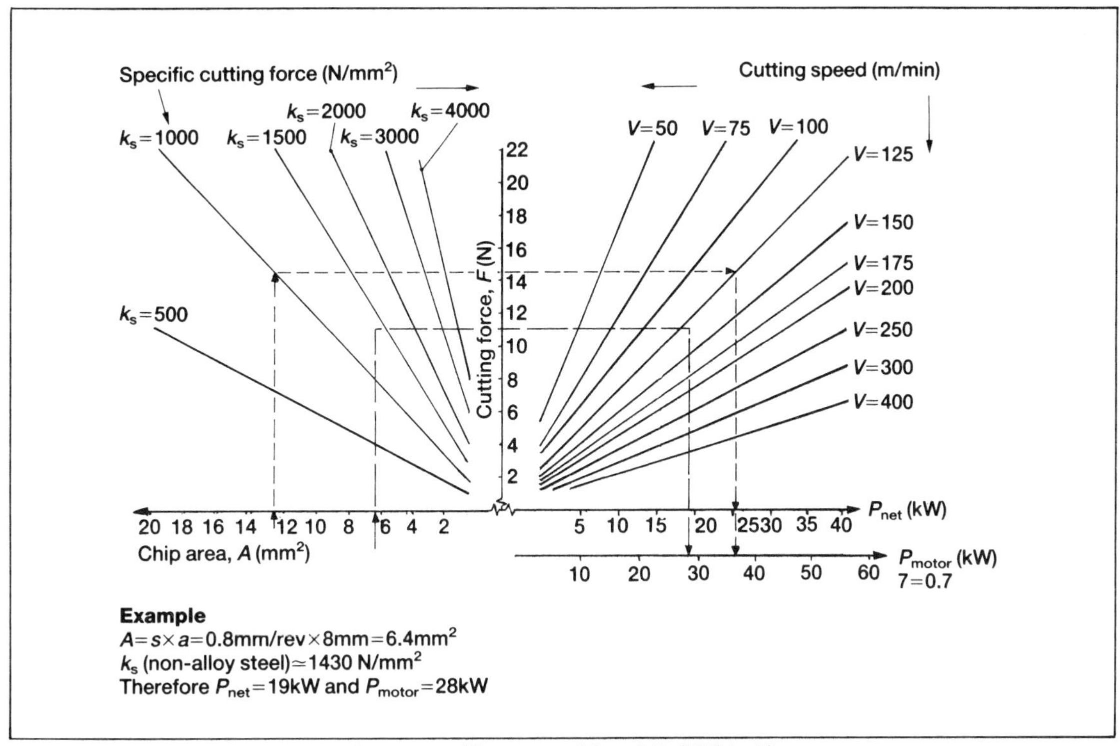

Diagram for determining the machine power (Courtesy of Sandvik (UK) Ltd.)

Appendix C
Milling troubleshooting

Milling operations with carbide tooling

Flank wear

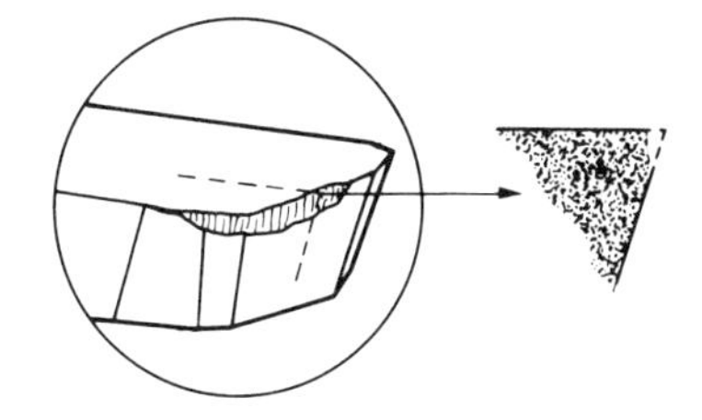

A uniform flank wear is the optimum type of wear. It is often combined with some chipping of the cutting edge towards the end of the tool life. If the flank wear increases too quickly, the cutting speed should be lowered without changing the table feed, or a more wear-resistant grade should be used.

Crater wear

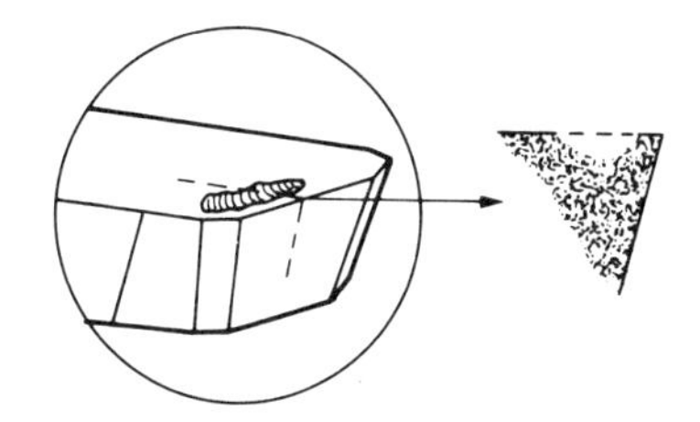

Crater wear can appear in steel milling, but as long as there is a balance between flank wear and crater wear, nothing should be changed. If the crater wear becomes too great there is a risk that the cutting edge will break.

- Reduce the cutting speed.
- Select a more wear-resistant grade.

Chipping

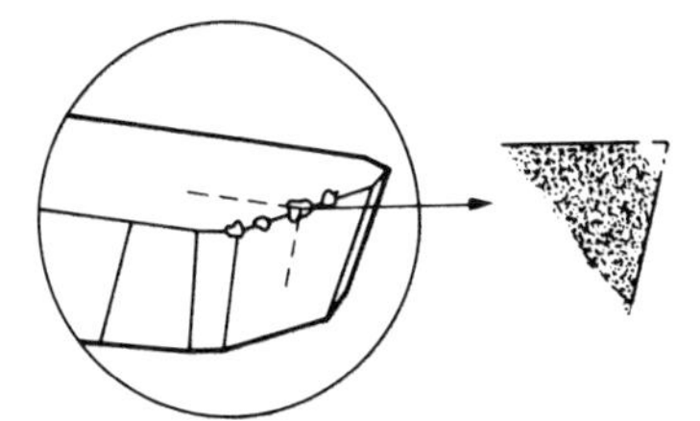

Some chipping of the cutting edge (small pieces of carbide knocked out) can be accepted as long as the tool life is unaffected.

If chipping is a problem:

- Increase the cutting speed.
- Reduce the feed per tooth.
- Select a tougher carbide grade or an insert with reinforced edges.
- Improve the stability of the machining conditions.

Thermal cracks

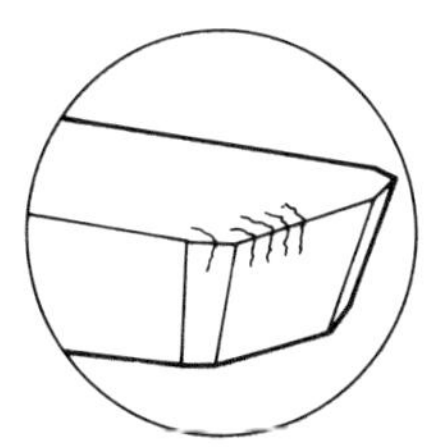

Cracks perpendicular to the cutting edge are caused by the temperature variations involved in milling. The tool life can be affected if the cracks develop early: pieces of carbide will be knocked out between the cracks.

When thermal cracking becomes a problem:

- Change to a smaller-diameter cutter.
- Reduce the cutting-edge temperature by reducing the cutting speed and, if necessary, the feed per tooth.
- Choose a tougher grade.

Note: Do *not* use coolant.

Mechanical cracks

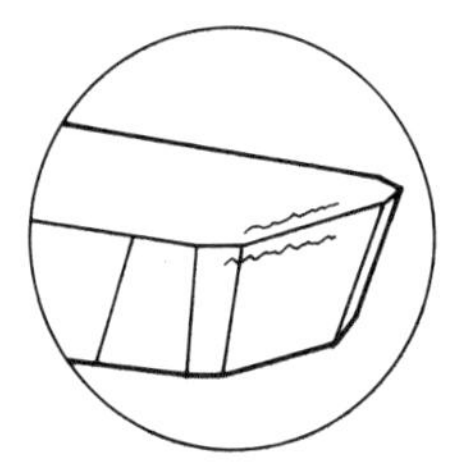

Cutting forces vary during a cut, and milling inserts especially are subjected to varying mechanical stresses which, in some cases, result in cracks parallel to the cutting edges being produced. Edge fracture is often the result of mechanical cracking, and a thick exit chip especially critical.

Deformation

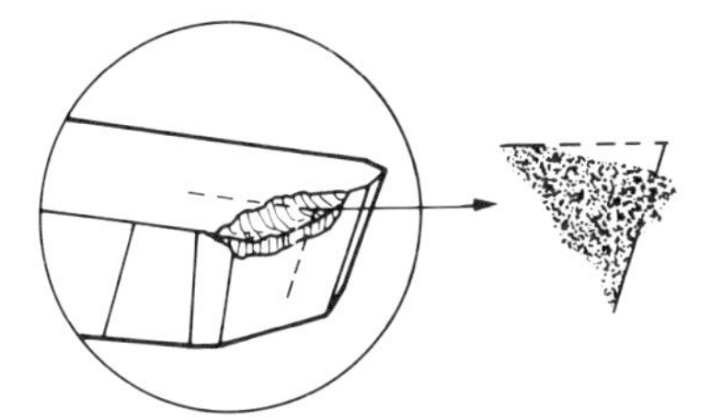

The combination of high edge temperature and high cutting forces can result in plastic deformation of the cutting edge. To reduce this, lower cutting data has to be applied, or a more wear-resistant (harder) carbide grade should be used.

Notch wear

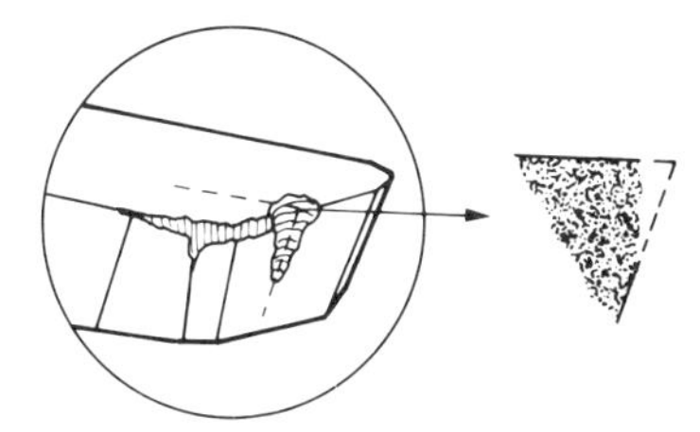

An abrasive workpiece skin may cause excessive wear where the skin rubs on the cutting edge.

A more wear-resistant grade improves the situation, but if the notch wear is caused by sand inclusions, a coated insert should be chosen.

Edge build-up

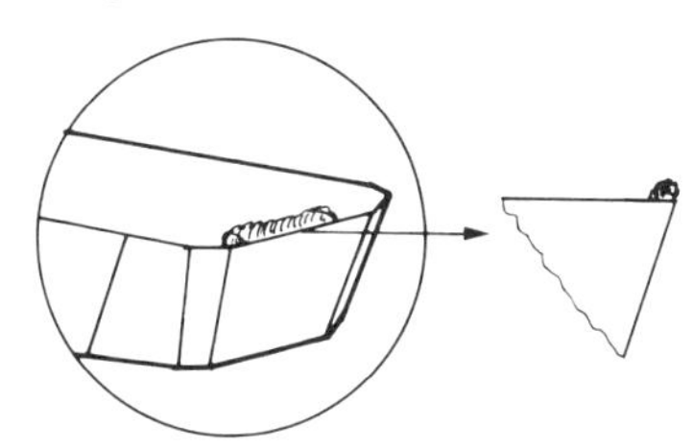

Milling of 'sticky' materials, e.g. some stainless steels, can cause edge build-up. Workpiece material is 'welded' onto the edge, and when the build-up breaks off, small pieces of carbide may follow. This condition is fatal to the tool life.

Edge build-up is due to low cutting speeds; it normally occurs at cutting speeds between 80m/min and 160m/min. To avoid edge build-up, increase the cutting speed. For materials giving too short a tool life at the high cutting speed, apply a very low speed.

Note: Do *not* use coolant on stainless or low-carbon steels.

In aluminium milling the edge build-up will only affect the surface finish, not the carbide. Edge build-up can occur even at cutting speeds up to 800m/min. In the case of aluminium milling a spray-mist or flood coolant may improve the surface finish.

(Courtesy of Sandvik (UK) Ltd.)

Summary

Cause								
Rapid flank wear	Rapid crater wear	Insert chipping	Thermal cracks	Mechanical cracks	Plastic deformation	Notch wear	Edge build-up	*Remedy*
X	X				X	X		Select a more wear-resistant carbide grade
		X	X	X				Select a tougher carbide grade
		X					X	Increase the cutting speed
X	X		X		X			Reduce the cutting speed
X							X	Increase the feed per tooth
		X	X	X	X			Reduce the feed per tooth
				X				Change the cutter position
					X		X	Change the cutter geometry
			X					Use a smaller cutter diameter
		X						Use a fly wheel to absorb peak power consumption
		X						Improve the stability

Vibrations produced in milling operations

Cause	Remedy
Weak fixturing.	Improve the fixture. Put fixed supports in the direction of the cutting forces. Change cutter geometry. Reduce the feed per tooth.
Axially-weak component. 	Use a square-shoulder face mill.
Radially-weak spindle, e.g. long spindle overhang. 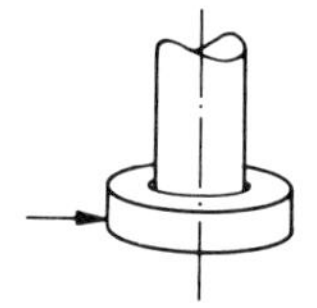	Use a differential pitch cutter. Increase the feed per tooth, e.g. by reducing the cutting speed. Use a 45° entering angle.
Square-shoulder face milling – radially-weak spindle.	Use smallest-possible diameter cutter. Use a differential pitch cutter (for diameters greater than 125mm). Change to up-milling.

Erratic table feed.	Adjust the back-lash eliminator (when fitted, on a conventional mill). Reduce the feed per tooth, e.g. by increasing the cutting speed. Change to up-milling.
Back-lash in the spindle (heavy banging when a tooth enters into cut).	Repair the spindle and the transmission. Use a fly wheel (when horizontal milling on a conventional machine). Reduce the feed per tooth, e.g. by increasing the speed (this tends to give a fly-wheel effect to the cutter). Change to up-milling.
Vibrations due to other causes.	Check cutter mounting. Change cutter geometry. Change rpm to break the harmonics.

Note: Do not reduce the feed per tooth as soon as vibrations occur. More problems are solved by *increasing* the feed.

(Courtesy of Sandvik (UK) Ltd.)

Surface-finish problems in milling operations

Cause	Remedy
Axial run-out too large.	Set or repair the cutter. Use inserts with parallel lands. Check the cutter mounting. Check the spindle run-out.
Feed per rev too high.	Increase the cutting speed. Use a wiper insert. Set the cutter accurately.
Vibrations.	Check cutter mounting. Reduce the depth of cut. Change geometry. Do not use wiper. Take out every second insert.
Edge build-up.	Increase cutting speed. Change to positive geometry.
Back-cutting.	Reduce the depth of cut. Check the spindle. Use smaller-diameter cutters.
Workpiece frittering.	Reduce the feed per tooth. Use a close-pitch cutter. Reposition the cutter to achieve a thin exit chip. Change cutter geometry.

(Courtesy of Sandvik (UK) Ltd.)

Appendix D
Milling data determination

The machine power requirement

A number of factors affect the power consumed when milling; these include the work-piece material, the material-removal rate, the tool geometry and the chip thickness. When milling, a thick chip is more power-efficient than a thin chip.

For example, when the metal-removal rate is kept constant and the cutting speed is decreased from $V = 180$m/min to $V = 120$m/min, the result will be to increase the feed per tooth by a third. This effective increase in the feed per tooth from the original value $s_z = 0.1$mm, say to $s_z = 0.15$mm will mean that there will be a 12% drop in the power consumed. This implies that it is important to avoid too small a feed per tooth, if the machine's available power is limited.

In fact, however, it is the average chip thickness h_m which has the greatest influence on the power consumed when milling. The table below lists the power correction factors that need to be applied in the calculation of milling power for different values of the average chip thickness.

Average chip thickness, h_m (mm)	0.05	0.10	0.15	0.20	0.25	0.30	0.35	0.40
Power correction factor	1.50	1.23	1.10	1.00	0.94	0.89	0.85	0.81

If a face-milling operation is carried out, it is usually accurate enough to assume that the chip thickness is the same as the feed per tooth. However, exceptions to this occur when smaller approach angles are used, as this reduces the chip thickness. To illustrate the extent of the reduction effect caused by the smaller approach angle, three popular approach angles are listed below, together with their respective correction factors.

Approach angle, k_r	90°	75°	45°
Correction factor	1.00	0.97	0.71

The expression for the power consumed during milling is

$$P_{net} = \frac{a_a \times a_r \times s'}{1000 \times V_p}$$

where-

P_{net} = net power (in kW),
V_p = metal-removal (in cm^3/min) per kW,
a_a = axial depth of cut (in mm), and
a_r = radial depth of cut (in mm), and
s' = table feed (in mm/min).

Note that P_{net} is the power *of the cutter.* In order to determine the machine power required it is necessary to *divide* this value by the machine's efficiency η, which might typically be around 0.8.

In order to minimise insert wear when milling, the power utilised should not be more than 70% of the available machine power.

The net power calculation above is valid when the insert used in the milling operation has a 0° top rake. For every degree that the rake is greater or smaller than 0°, the power consumption will change by 1.3%. If a cutter has an insert geometry with a positive top rake, the power consumed will be decreased, whereas a negative top rake will increase the power consumption. For example if the cutter geometry has a positive 15° top rake, it will require approximately 20% less power than a cutter with a 0° top rake.

The table below lists some typical data for determining the net power, which are valid for a 0° top rake geometry and a chip thickness of 0.2mm.

Material		*Hardness Brinell HB*	V_p *(cm³/min kW)*	*Specific cutting force, k_s (kp/mm²)*
Non-ally carbon steel	C< 0.25%	125	25	245
	C< 0.25%	150	23	270
	C< 0.25%	250	21	295
Low alloy steel	Annealed	125-200	21	285
	Hardened	200-450	17	350
High alloy steel	Annealed	150-250	19	315
	Hardened	250-500	17	365
Stainless steel	Ferritic/martensitic	175-225	19	325
	Austenitic	150-200	17	350
Steel castings	Non-alloy	225	27	230
	Low alloy	150-250	24	250
	High alloy	150-300	21	285
Hardened steel		> 50 HRC	10	600
Malleable cast iron	Short-chipping	110-145	31	195
	Long-chipping	200-250	34	180
Grey cast iron	Low-tensile	150-225	49	125
	High-tensile	200-300	38	160
Nodular cast iron	Ferritic	125-200	45	135
	Pearlitic	200-300	31	200
Chilled cast iron		40-60 HRC	14	425
Aluminium		100	82	75

The cutting forces in milling

The determination and calculation of the cutting forces in milling is outside the scope of this book: it is enough to be able to give rough estimates of the forces involved, for example, to find out if the fixturing is strong enough.

F_R F_A F_T

F_T = Tangential cutting force
F_A = Axial cutting force
F_R = Radial cutting force

The total tangential cutting force, for all the inserts in cut, is related to the power consumption by the expression

$$\sum F_t = \frac{P_{net}\ (\text{kW}) \times 60 \times 1000}{V\ (\text{m/min})}$$

where V is the cutting speed (in m/min). In the situation illustrated in the diagram, $\sum F_t = F_{t1} + F_{t2} + F_{t3}$. However, the tangential forces are acting in different directions, so the load on the fixture is less than $\sum F_t$.

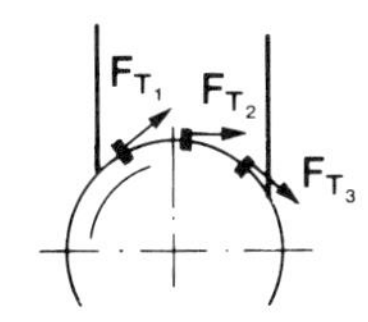

The radial cutting force is related to the tangential force.* The expression

$$\sum F_r = 0.6 \times \sum F_t$$

is valid for new inserts, but the radial force increases dramatically when the inserts are worn: increases of 100% or more are normal.

The axial cutting force depends on both the tangential force and the entering angle, as shown below.

Entering angle, K_r	90°	75°	45°
$\sum F_a$ (approximate)	0	$0.2 \times \sum F_t$	$0.5 \times \sum F_t$

The feed force F_m is a combination of the tangential and radial cutting forces, and varies as shown in the diagram below.

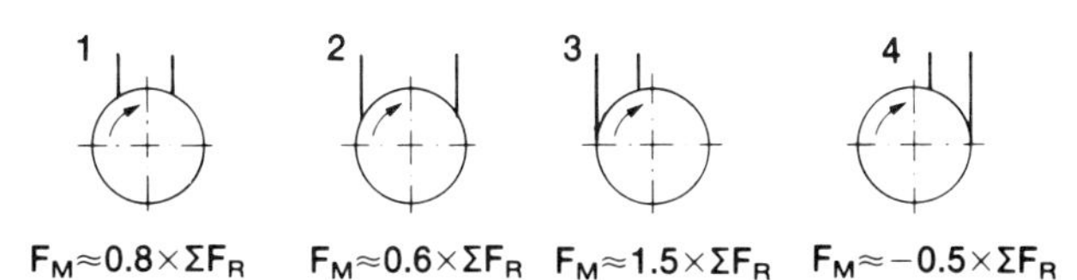

(Courtesy of Sandvik (UK) Ltd.)

---

*Note that relationships between the cutting forces can differ by as much as 100% from those quoted in this section, depending on the number of teeth in cut, insert edge treatment, etc.

Speed conversion chart

Cutting speed v (m/min)	Cutter diameter (mm) 10	12	16	20	25	32	40	50	63	80	100	125	160	200	250	315
	Spindle speed (rev/min)															
10	318	265	199	159	127	99	80	64	51	40	32	25	20	16	13	10
20	636	531	398	318	255	199	159	127	101	80	64	51	40	32	25	20
30	954	796	597	477	382	298	239	191	152	119	95	76	60	48	38	30
40	1 270	1 060	796	637	509	398	318	255	202	159	127	102	80	64	51	40
50	1 590	1 330	995	796	637	497	398	318	253	199	159	127	99	80	64	51
60	1 910	1 590	1 190	955	764	597	477	382	303	239	191	153	119	95	76	61
70	2 230	1 860	1 390	1 110	891	696	557	446	354	279	223	178	139	111	89	71
80	2 550	2 120	1 590	1 270	1 020	796	637	509	404	318	255	204	159	127	102	81
90	2 870	2 390	1 790	1 430	1 150	895	716	573	455	358	286	229	179	143	115	91
100	3 180	2 650	1 990	1 590	1 270	995	796	637	505	398	318	255	199	159	127	101
110	3 500	2 920	2 190	1 750	1 400	1 090	875	700	556	438	350	280	219	175	140	111
120	3 820	3 180	2 390	1 910	1 530	1 190	955	764	606	477	382	306	239	191	153	121
130	4 140	3 450	2 590	2 070	1 660	1 290	1 040	828	657	517	414	331	259	207	166	131
140	4 460	3 710	2 790	2 230	1 780	1 390	1 110	891	707	557	446	357	279	223	178	141
150	4 780	3 980	2 960	2 390	1 910	1 490	1 190	955	758	597	477	382	298	239	191	152
160	5 090	4 240	3 180	2 550	2 040	1 590	1 270	1 020	808	637	509	407	318	255	204	162
180	5 730	4 780	3 580	2 870	2 290	1 790	1 430	1 150	909	716	573	458	358	286	229	182
200	6 370	5 310	3 980	3 180	2 550	1 990	1 590	1 270	1 010	796	637	509	398	318	255	202
220	7 000	5 840	4 380	3 500	2 800	2 190	1 750	1 400	1 110	875	700	560	438	350	280	222
250	7 960	6 630	4 970	3 980	3 180	2 490	1 990	1 590	1 260	995	796	637	497	398	318	253
300	9 550	7 960	5 970	4 780	3 820	2 980	2 390	1 910	1 520	1 190	955	764	596	477	382	303
350	11 100	9 280	6 960	5 570	4 460	3 480	2 790	2 230	1 770	1 390	1 110	891	696	557	446	354
400	12 700	10 600	7 960	6 370	5 090	3 980	3 180	2 550	2 020	1 590	1 270	1 020	796	637	509	404
450	14 300	11 900	8 950	7 170	5 730	4 480	3 580	2 870	2 270	1 790	1 430	1 150	895	716	573	455
500	15 920	13 300	9 950	7 960	6 370	4 970	3 980	3 180	2 530	1 990	1 590	1 270	995	796	637	505
550	17 510	14 600	10 900	8 750	7 000	5 470	4 380	3 500	2 780	2 190	1 750	1 400	1 090	875	700	556
600	19 100	15 900	11 900	9 550	7 640	5 970	4 780	3 820	3 030	2 390	1 910	1 530	1 190	955	764	606
700	22 300	18 600	13 900	11 100	8 910	6 960	5 570	4 460	3 540	2 790	2 230	1 780	1 393	1 110	891	707
800	25 500	21 200	15 900	12 700	10 200	7 960	6 370	5 090	4 040	3 180	2 550	2 040	1 590	1 270	1 020	808
1000	31 800	26 500	19 900	15 900	12 700	9 950	7 960	6 370	5 050	3 980	3 180	2 550	1 990	1 590	1 270	1 010

(Courtesy of Sandvik (UK) Ltd.)

Appendix E
Drilling troubleshooting

Delta short-hole drilling

Over-size hole diameter

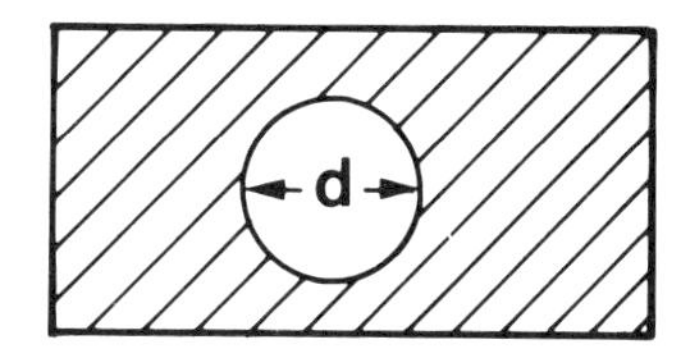

Cause	Remedy
Drill not central.	Check the alignment.
Machine spindle not central.	Change or repair spindle.
Machine spindle dirty.	Clean spindle (drill location).
Too high feed per rev.	Lower the feed per rev.
Insufficient coolant flow (chip blockage).	Increase coolant flow (check pump pressure). Clean filter.
Lack of rigidity of machine, tool or component.	Clamp tool and/or component better. Repair or change machine spindle.

Machine stops

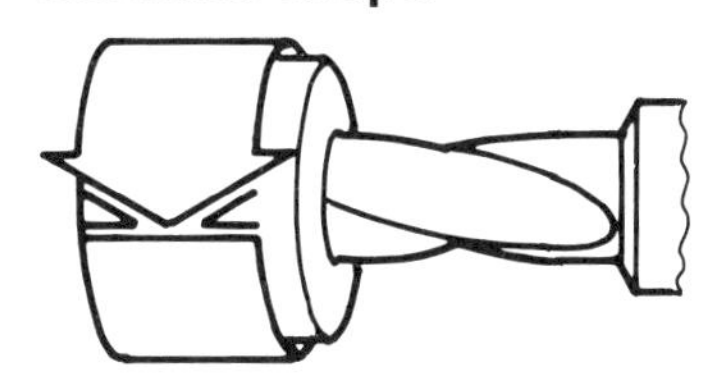

Cause	Remedy
Drill not central.	Check the alignment.
Machine spindle not central.	Change or repair spindle.
Machine spindle dirty.	Clean spindle (drill location).
Wrong cutting data.	Change cutting data.
Wrong coolant mix.	Check solution, add more oil, EP additives (15% – 25%) or water.
Lack of rigidity of machine, tool or component.	Clamp tool and/or component better. Repair or change machine spindle.

Hole not symmetrical

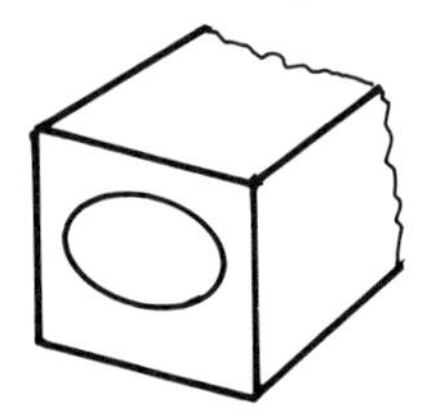

Cause	Remedy
Lack of rigidity of machine, tool or component.	Clamp tool and/or component better. Repair or change machine spindle.
Wrong cutting data for the material being drilled.	Lower the feed per rev. If uncertain, check material cross-reference list and cutting data against the material group.

Poor tool life

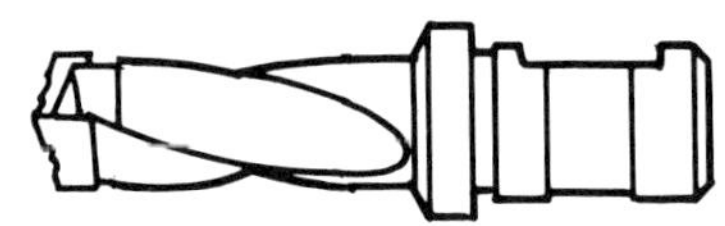

Cause	Remedy
Wrong cutting data.	Check material cross-reference list and cutting data against material group.
Insufficient coolant flow.	Increase coolant flow (check pump pressure). Clean filter. Add more oil, EP additives or water.
Lack of rigidity of machine, tool or component.	Clamp tool and/or component better. Repair or change machine spindle.

(Courtesy of Sandvik (UK) Ltd.)

'U' short-hole drilling

Front face of drill broken and fused in the workpiece

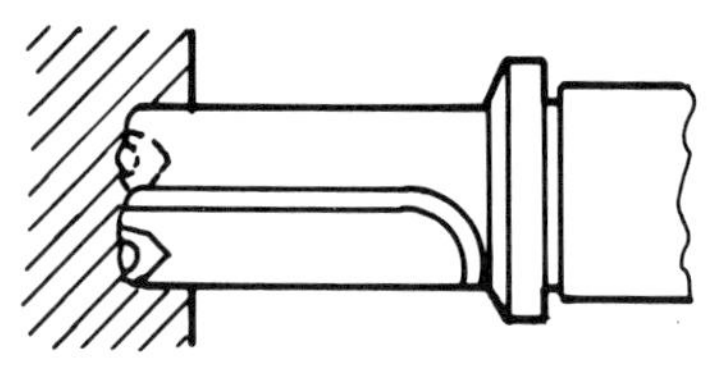

Cause	Remedy
Drill too much off-centre.	Realign drill.
Insufficient coolant volume and pressure.	Increase coolant flow, clean filter and coolant holes in drill.
Wrong cutting data, leading to jamming of both coolant holes and chip channels with chips.	Check data for material being drilled, to change chip size.
Wrong insert grade.	Check and change insert grade.

Wear on outside diameter of drill

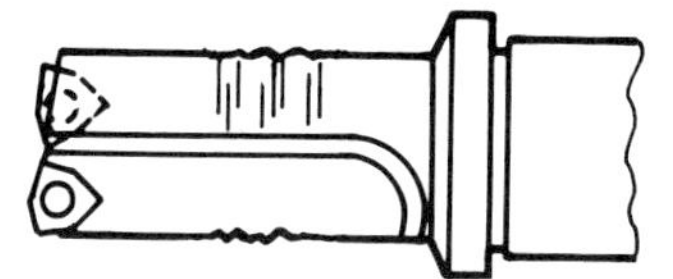

Cause	Remedy
Drill too much off-centre.	Realign drill.
Wrong grade for peripheral insert, leading to spalling and failure on the insert radius.	Change peripheral insert to tougher grade.

Over-size or under-size hole

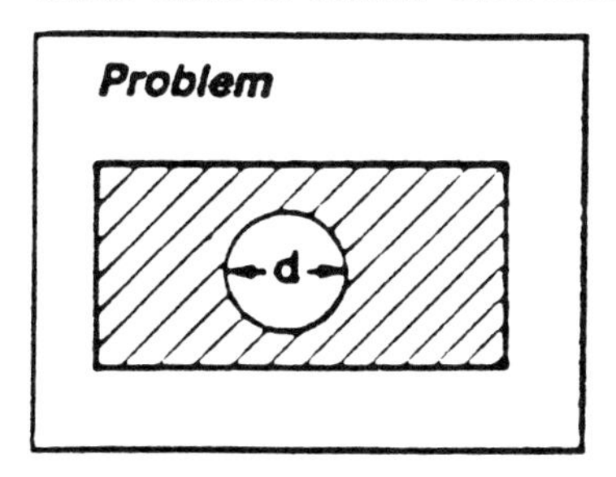

Cause	Remedy
Drill off-centre.	Realign drill.
Feed per rev too high.	Reduce feed.

Chip jamming in the drill flutes

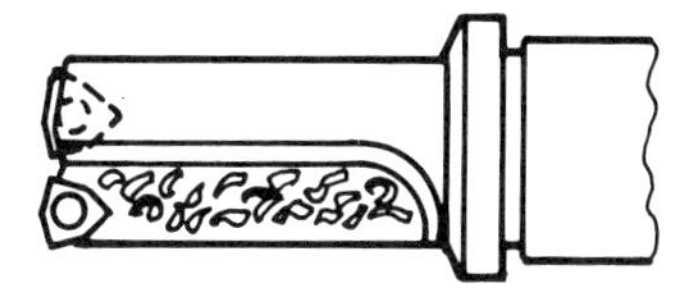

Cause	Remedy
Insufficient coolant volume and pressure.	Increase coolant pressure, clear coolant holes in drill.
'Sticky' material.	Change insert geometry, increase feed and lower cutting speed.
Machine not sufficiently rigid.	Check rigidity, change to more rigid machine, check data.

Vibration

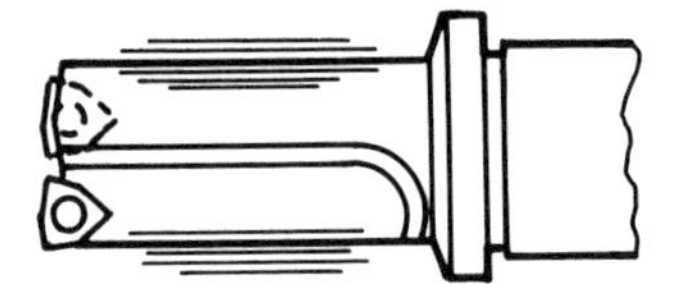

Cause	Remedy
Too big an overhang.	Refixture component and shorten drill overhang.
Starting face in material not flat – more than 2° angle.	Reduce feed.
Lack of rigidity of machine or component.	Check fixture, change to more rigid machine, check data.

Machine stops

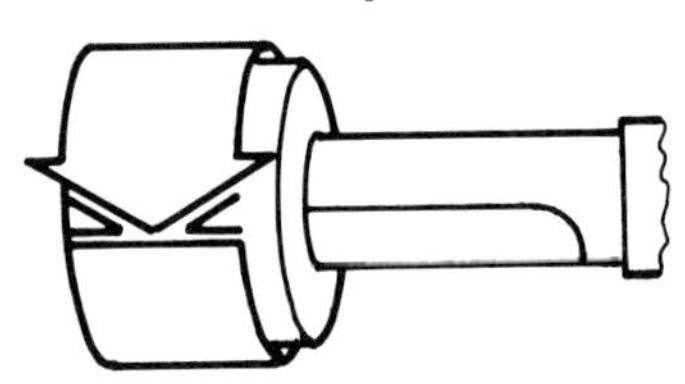

Cause	Remedy
Lack of power.	Check the available machine power.
Too high cutting data.	Adjust data to material and power available.

Chipped inserts

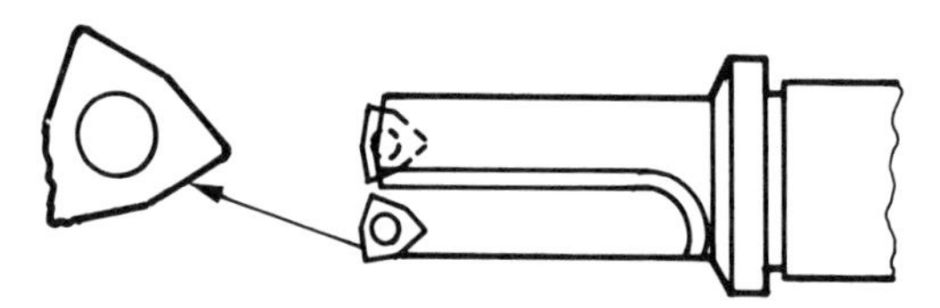

Cause	Remedy
Wrong grade.	Check carbide grade and geometry is suitable for workpiece material and cutting data.
Drill off-centre.	Realign drill.

Insert incorrectly located in seat

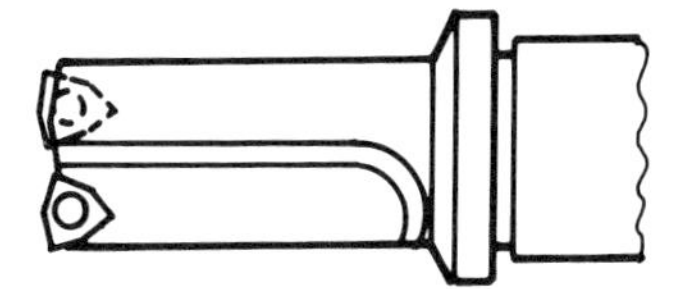

Cause	Remedy
Dirt in insert seat.	Clean insert seat.
Maltreatment of drill has damaged insert seats.	Check and replace drill if necessary.

(Courtesy of Sandvik (UK) Ltd.)

Trepanning operations

Chattering

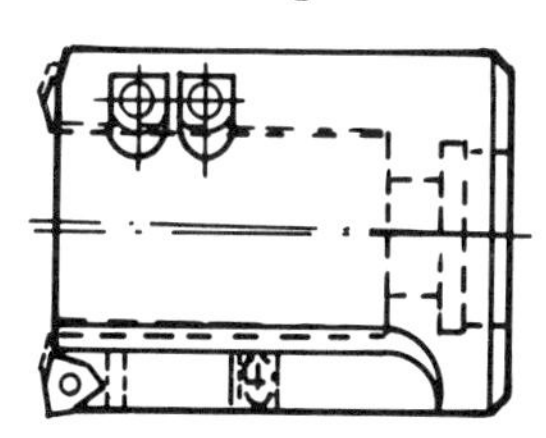

Cause	Remedy
Overhang of trepanning head too great.	Reduce overhang.
Peripheral cartridge not positioned correctly, axially.	Reposition cartridge.
Starting face in material not flat – more than 2° angle.	Reduce feed.
Unstable machine.	Check machine and component fixture.

Machine stops

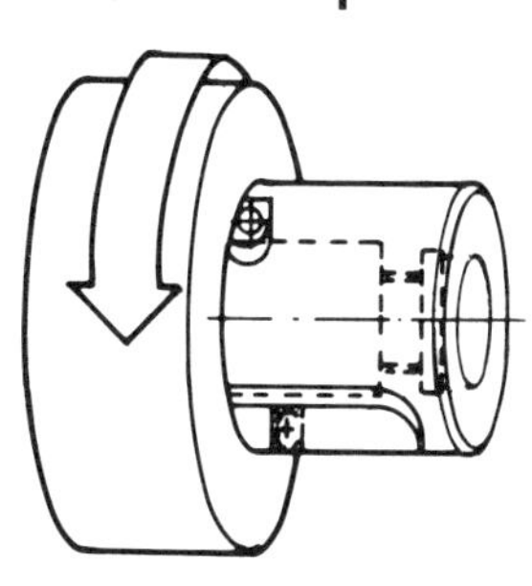

Cause	Remedy
Lack of power.	Check available machine power.
Incorrect data.	Reduce cutting speed. Check data with material and power available.

Insert chipped

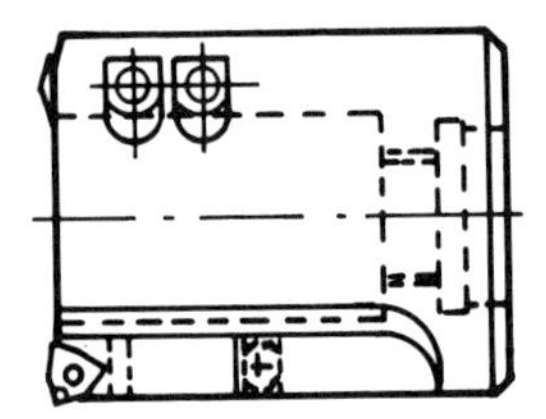

Cause	Remedy
Wrong carbide grade.	Use tougher grade.
Incorrect cutting data.	Check data.
Trepanning head off-centre.	Realign trepanning head.
Lack of rigidity.	Check machine and component fixture.

Insert failure

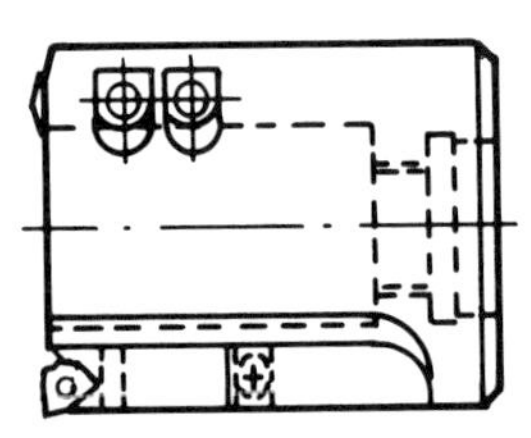

Cause	Remedy
Core falls down and breaks the insert.	Check the position of the trepanning head. On the stationary drill, the cartridge should be at 12 o'clock and 6 o'clock, with the peripheral insert at 6 o'clock.

Insert not seated properly

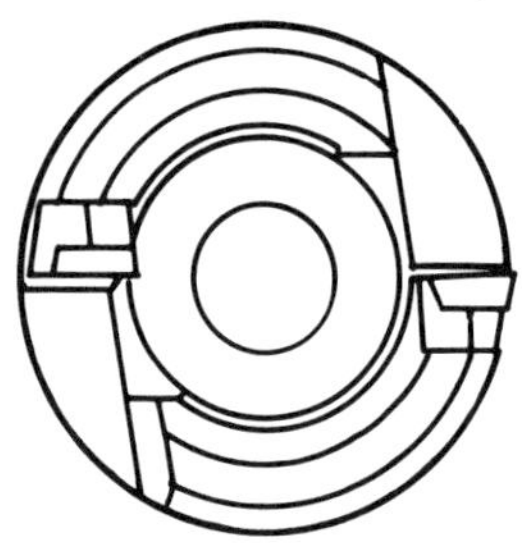

Cause	Remedy
Dirt or swarf in seat.	Clean insert seat.
Insert locking screw damaged.	Replace screw.
Seat damaged.	Replace cartridge.

Trepanning head fails or becomes 'welded' to workpiece

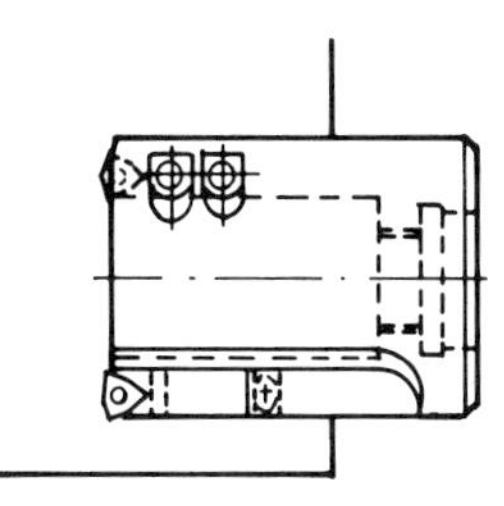

Cause	Remedy
Trepanning head not on centre.	Realign trepanning head.
Poor coolant pressure, and insufficient quantity.	Increase coolant pressure, clean filter and coolant supply holes in trepanning head.
Wrong cutting data, causing chips to clog chip ducts.	Check data to alter chip size.

Wear of outside diameter of trepanning head body

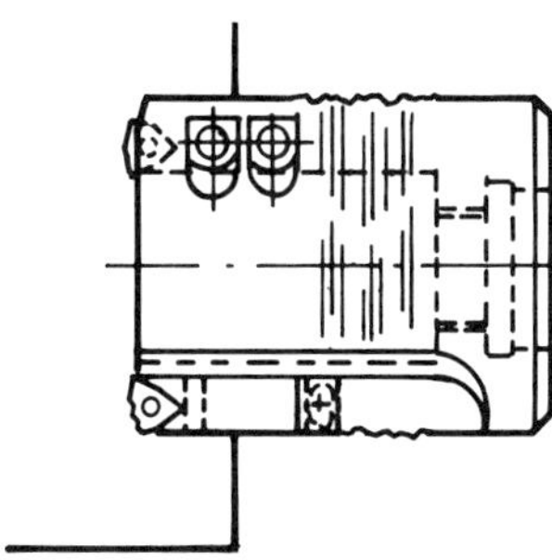

Cause	Remedy
Trepanning head not on centre.	Realign trepanning head.
Incorrect chip breaking.	Check data to alter chip size.
Frittering of peripheral insert.	Change insert grade to tougher carbide.
Diameter of trepanning head incorrectly set.	Reset trepanning head.

Over-size or under-size hole

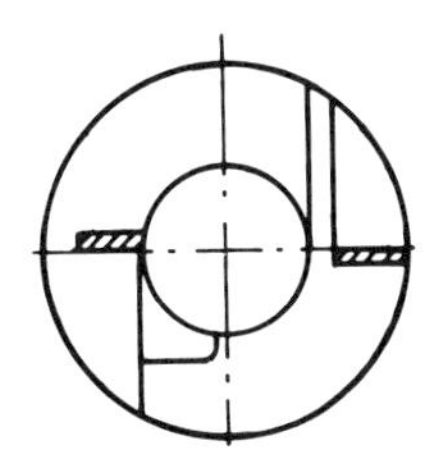

Cause	Remedy
Trepanning head not on centre.	Realign trepanning head.
Excessive feed pressure.	Reduce feed.
Trepanning head not mounted to drill holder correctly, causing deflection.	Check trepanning head mounting.

Chips jamming

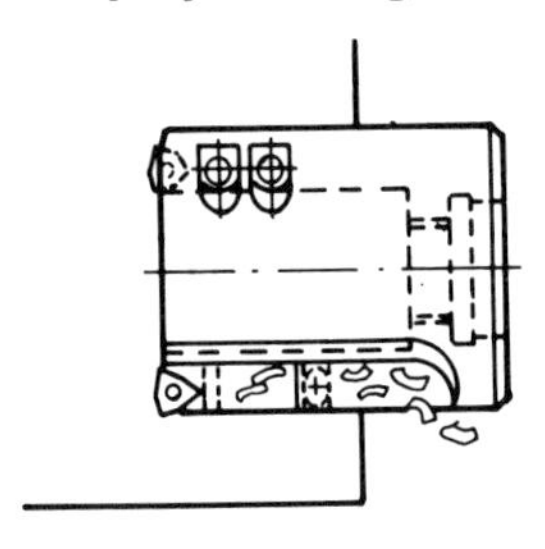

Cause	Remedy
Insufficient coolant pressure and volume.	Increase coolant pressure, clean filter and coolant supply holes.
Wrong cutting data.	Check data for material being trepanned to alter chip size.

(Courtesy of Sandvik (UK) Ltd.)

Appendix F
Cutting speeds using carbide inserts – a basic guide

Material	*Rake*	*Use*	*Grade (ISO)*	*SFPM*	*Speed (m/min)*
Armour-plate steel	Double-negative or positive with negative land	Long continuous cuts	P15	125–175	38–53
		Shock applications	P25, P30	80–125	24–38
Resulphurised steel and leaded steel	Positive	General purpose	P15, P20	500–800	152–244
		Milling	P20	400–600	122–183
		High-speed finishing on rigid parts	P01, P05	1000–1600	304–488
		High-speed shock applications	P10, P15	800–1200	244–366
		Severely interrupted cuts	P25, P30	250–500	76–152
		Broad range and higher speeds	P10, P30	500–1200	152–366
Low-carbon steel	Positive	General purpose	P15, P20	300–500	91–152
		Milling	P20	400–600	122–183
		High-speed finishing on rigid parts	P01, P05	600–1200	183–366
		High-speed shock applications	P10, P15	400–800	122–244
		Severely interrupted cuts	P25, P30	200–400	61–122
		Broad range and higher speeds	P10, P30	300–900	91–274
Medium-carbon steel	Negative for roughing, positive for finishing	General purpose	P15, P20	200–400	61–122
		Milling	P20	300–500	91–152
		High-speed finishing on rigid parts	P01, P05	400–1000	122–304
		High-speed shock applications	P10, P15	300–600	91–183
		Severely interrupted cuts	P25, P30	200–400	61–122
		Broad range and higher speeds	P10, P30	300–800	91–244
High-carbon steel	Negative	General purpose	P15, P20	150–350	46–107
		Milling	P20	200–400	61–122
		High-speed finishing on rigid parts	P01, P05	200–600	122–183
		Severely interrupted cuts	P25, P30	150–350	46–107
		Broad range and higher speeds	P10, P30	300–500	91–152

Material	*Rake*	*Use*	*Grade (ISO)*	*SFPM*	*Speed (m/min)*
Cast steel	Positive	General purpose	P15, P20	200–300	61–91
		Milling	P20	200–450	61–137
		High-speed finishing on rigid parts	P01, P05	300–700	91–213
		Severely interrupted cuts	P25, P30	180–250	55–76
		Broad range and higher speeds	P10, P30	200–500	61–152
High manganese steel	Negative	General purpose	P15, P20	50–100	15–31
		Milling	P20, M10–20	50–125	15–38
		Severely interrupted cuts	P25, P30	35–80	11–24
High-speed steel	Double-negative	General purpose	P15	120–220	37–67
		High-speed finishing on rigid parts	P01, P05	300–400	91–122
Hot-work die steel	Negative	General purpose	P15	120–220	37–67
		Milling	P20, M10–20	250–350	76–107
Tool steel	Positive	General purpose	P15, P20	200–275	61–84
		Milling	P20, M10–20	200–300	61–91
		High-speed finishing on rigid parts	P01, P05	250–400	76–122
		Severely interrupted cuts	P25, P30	150–230	46–70
Ferritic stainless steel	Positive	General purpose	P15, P20	175–300	53–91
		Milling	P20, M10–20	400–500	122–152
		Broad range and higher speeds	P10, P30	200–700	61–213
Martensitic stainless steel	Negative	General purpose	P15, P20	175–300	53–91
		Milling	P20, M10–20	400–500	122–152
		Prevent build-up	P01, P05	300–800	91–244
		Heat-treated material	K10	200–400	61–122
		Broad range and higher speeds	P10, P30	200–700	61–213
Austenitic stainless steel	Positive	General purpose	K10	200–400	61–122
		Broad range and higher speeds	K10–K20 M10–M20	400–500	122–152
Precipitation-hardening stainless steel	Negative or positive with negative land	General purpose	K10	200	61
		Heavy cuts	P15	175–300	53–91
		Very light finishing cuts	P01, P05	300–800	91–244
		Broad range and higher speeds	P10, P30	200–700	61–213
Wrought aluminium	12°–15° positive with 20° positive clearance	General purpose	K10	750–2400	228–732
		Interrupted cuts	K20	750–1600	228–488
		High-speed finishing	P01, P05	1800–3000	549–914
Cast aluminium	12°–15° positive with 20° positive clearance	General purpose	K10	750–1800	228–549
		Interrupted cuts	K20	750–1200	228–366
		High-speed finishing	P01, P05	1400–2400	427–732
Beryllium*	12°–15° positive with 20° positive clearance	Continuous cuts	K10	125–250	38–76
		Interrupted cuts	K30	125–150	38–46
Grey iron	Positive	General purpose	K10	350–500	107–152
		Interrupted cuts	K20	225–400	69–122
		Broad range and higher speeds	K10–K20	500–850	152–259

**Warning: Machining exotic materials*

Beryllium: beryllium metal and its compounds are toxic

Material	*Rake*	*Use*	*Grade (ISO)*	*SFPM*	*Speed (m/min)*
Nodular or malleable cast iron low alloy	Positive	General purpose	K10	350–500	107–152
		Interrupted cuts or milling	K20	225–400	69–122
		Broad range and higher speeds	K10–K20 M10–M20	500–850	152–259
High alloy	Positive	General purpose and milling	P20, M10–20	225-350	69–107
		Heavy cuts	P15	225-300	69–91
		High-speed finishing	P01, P05	600–1000	183–305
		Broad range and higher speeds	P10, P30	300–700	91–213
Chilled cast iron	Negative or positive with negative land	General purpose	K10	350–500	107–152
		Heavy cuts	P15	250–400	76–122
		Long rigid cuts	P10, M10	200–400	61–122
		Broad range and higher speeds	P10, P30	400–800	122–244
Cobalt-based alloys	Positive	General purpose	K10	50–130	15–40
		Interrupted cuts or milling	K20	50–100	15–30
Copper, zinc and leaded copper	Positive	General purpose	K10	400–1600	122–488
		Interrupted cuts or milling	K20	300–600	91–183
High-strength bronze	Negative	General purpose	K10	200–600	61–183
		Interrupted cuts or milling	K20	200–400	61–122
		High-speed finishing cuts			
Cast copper alloys	12°–15° positive with 20° positive clearance	General purpose	K10	600–900	183–274
		Interrupted cuts or milling	K20	600–800	183–244
		High-speed finishing cuts	P01, P05	700–1300	213–396
Fibreglass*	Positive	General purpose	K10	180–250	55–76
Graphite	Positive	General purpose	K10	200–250	61–76
Lava	Positive	General purpose	K10	500–750	152–229
Magnesium*	12°–15° positive with 20° positive clearance	General purpose	K10	500–2400	152–732
		Interrupted cuts or milling	K20	500–1000	152–305
Magnetic alloys	Positive	General purpose	K10	75–120	23–46
Manganese	Positive	General purpose	K20	50–65	15–20
Molybdenum	12°–15° positive with 20° positive clearance	General purpose	K20	300–350	91–107
Nickel-based alloys	Positive	General purpose	K10	40–350	12–107
		Heavy cuts or conventional milling	K20	40–100	12–30
		Heavy cuts and climb milling	P20, M10–20	40–100	12–30
		Interrupted cuts	P15	40–100	12-30
Plastic	Positive	General purpose	K10	250–1500	76–457
Tantalum	12°–15° positive with 20° positive clearance	General purpose	K10	200–800	61–244
		Milling	K20	200–400	61–122
Tantalum–tungsten	12°–15° positive with 20° positive clearance	General purpose	K01, K05	180–250	55–76

**Warning: Machining exotic materials*

Fibreglass: dust control is of the utmost importance

Material	*Rake*	*Use*	*Grade (ISO)*	*SFPM*	*Speed (m/min)*
Thallium*	Positive	General purpose	K01, K05	150–200	46–61
		Interrupted cuts	K05	150–200	46–61
Thorium*	25°–30° positive, special bronzed tool	General purpose	K01, K05	500–600	152–183
		Interrupted cuts	K05	500–600	152–183
Tin–bismuth alloys	12°–15° positive with 20° positive clearance	General purpose	K01, K05	200–500	61–152
		Interrupted cuts	K20	200–300	61–91
Titanium	Positive	General purpose	K20	225–350	69–107
		Finishing cuts	K20	300–400	91–122
Titanium alloys	Positive	General purpose	K20	75–150	23–46
		Finishing cuts	K20	125–200	38–61
Tungsten	12°–15° positive with 20° positive clearance	General purpose	K20	60–70	18–21
		Finishing cuts	K20	60–70	18–21
Tungsten–rhenium alloys	Positive	General purpose	K20	80–100	24–30
		Milling	K20	80–100	24–30
Zirconium*	12°–15° positive	General purpose	K20	225–250	69–76
		Finishing cuts	K01, K05	225–250	69–76

**Warning: Machining exotic materials*

Thallium: thallium oxide (produced in machining) is poisonous
Thorium: thorium is pyrophoric, use water as a coolant. (This also applies when machining magnesium and zirconium)

(Courtesy of Seco)

Appendix G
Hardness conversion table

Applies for plain carbon and low-alloy steels and cast steel, and to a limited extent for high-alloy and/or work-hardened steel.

DIN 50 150 contains hardness value and conversion details.

Tensile strength R_m (N/mm²)	*Brinell hardness BHN*	*Vickers hardness HV*	*Rockwell hardness HRB*	*Rockwell hardness HRC*	*Tensile strength R_m (N/mm²)*	*Brinell hardness BHN*	*Vickers hardness HV*	*Rockwell hardness HRC*
285	86	90			1190	352	370	37.7
320	95	100	56.2		1220	361	380	38.8
350	105	110	62.3		1255	371	390	39.8
385	114	120	66.7		1290	380	400	40.8
415	124	130	71.2		1320	390	410	41.8
450	133	140	75.0		1350	399	420	42.7
480	143	150	78.7		1385	409	430	43.6
510	152	160	81.7		1420	418	440	44.5
545	162	170	85.0		1455	428	450	45.3
575	171	180	87.1		1485	437	460	46.1
610	181	190	89.5		1520	447	470	46.9
640	190	200	91.5		1555	456	480	47.7
675	199	210	93.5		1595	466	490	48.4
705	209	220	95.0		1630	475	500	49.1
740	219	230	96.7		1665	485	510	49.8
770	228	240	98.1		1700	494	520	50.5
800	238	250	99.5		1740	504	530	51.1
820	242	255		23.1	1775	513	540	51.7
850	252	265		24.8	1810	523	550	52.3
880	261	275		26.4	1845	532	560	53.0
900	266	280		27.1	1880	542	570	53.6
930	276	290		28.5	1920	551	580	54.1
950	280	295		29.2	1955	561	590	54.7
995	295	310		31.0	1995	570	600	55.2
1030	304	320		32.2	2030	580	610	55.7
1060	314	330		33.3	2070	589	620	56.3
1095	323	340		34.4	2105	599	630	56.8
1125	333	350		35.5	2145	608	640	57.3
1155	342	360		36.6	2180	618	650	57.8

Brinell hardness: Degree of loading $0.102 \times F/D^2 = 30 N/mm^2$, where F=testing force in N, D=diameter of ball in mm
Vickers hardness: Diamond pyramid 136°, testing force $\geqslant$ 98N
Rockwell hardness B: Diameter of ball 1/16 inch, total testing force 980 ± 6.5N
Rockwell hardness C: Diamond cone 120°, total testing force 1471 ± 9N

(Courtesy of Krupp Widia)

Appendix H
The characterisation of chip forms

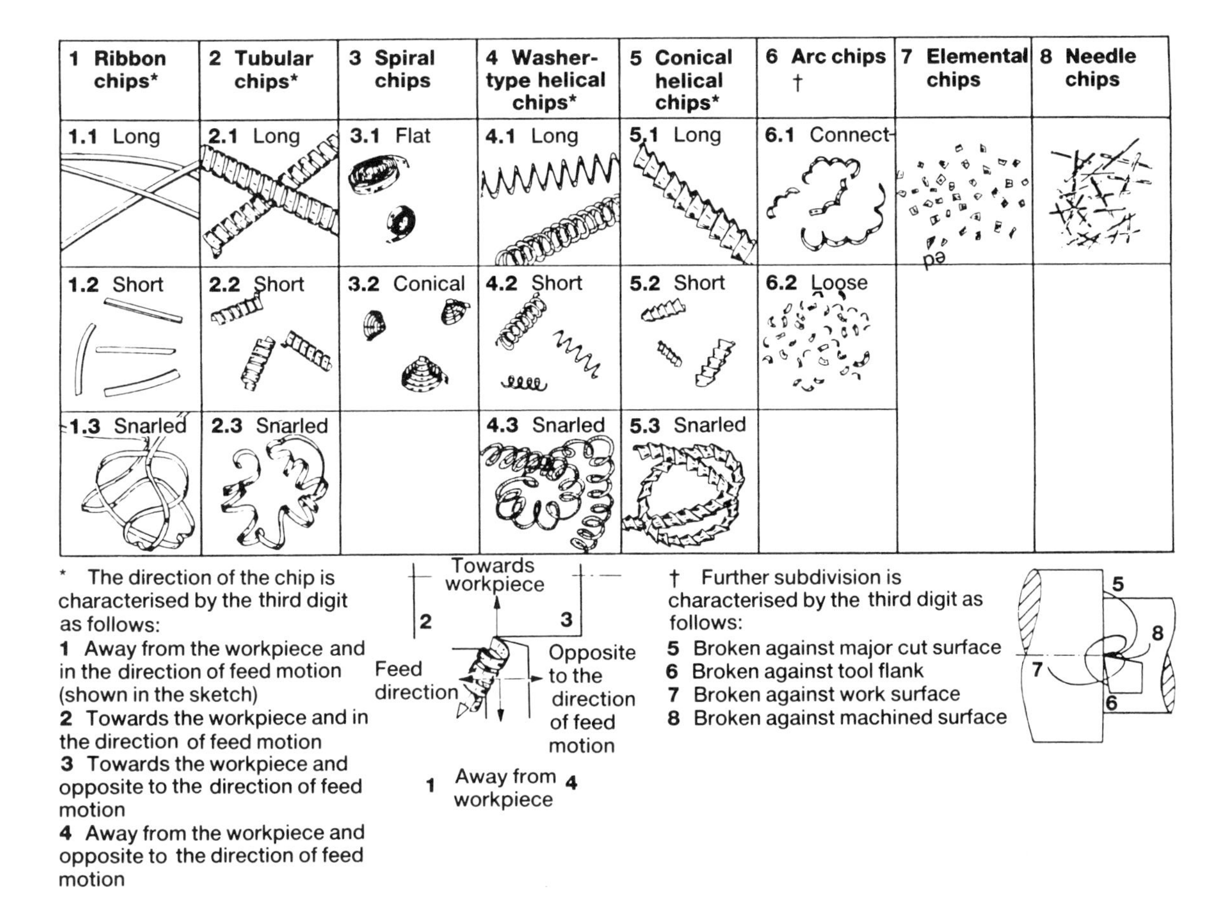

(Courtesy of ISO: extract from ISO 3685 : 1977 (E))

Appendix I
Standards relating to tooling for CNC machining and turning centres

British Standard	*ISO Standard*	*Subject*
—	243 : 1975	External turning tools using carbide inserts
—	504 : 1975	Carbide inserts for turning tools, their designation and marking
—	513 : 1975	Carbide groupings with their applications for machining
—	514 : 1975	Internal turning tools using carbide inserts
122 : 1980 Part 3	523 : 1974	Outside diameters recommended for the specification of milling cutters
4193 : 1980 Part 2	883 : 1976	Carbide inserts – indexable, without a fixing hole
4193 : 1980 Part 1	1832 : 1977	Inserts – indexable (throwaway), designation for cutting tools
4193 : 1980 Part 3	3364 : 1977	Carbide inserts – indexable, with a cylindrical fixing hole
4193 : 1980/2 Parts 4 and 5	3365:1977/80 Parts 1 and 2	Carbide inserts – indexable, for milling cutters: square (Part 1) and triangular (Part 2)
5623 : 1979	3685 : 1977	Tool-life testing with single-point turning tools
4193 : 1982 Part 6	5608 : 1980	Tool-holders for turning/copying and cartridges for inserts – indexable
4193 : 1982 Part 7	5610 : 1981	Tool-holders for single-point turning/copying for inserts – indexable
4193 : 1982 Part 8	5611 : 1981	Dimensions of cartridges using inserts – indexable
4193 : 1984 Part 14	6261 : 1983	Boring bars for use with inserts – indexable
4193 : 1982 Parts 9 and 10	6262 : 1982 Parts 1 and 2	End mills with inserts – indexable, using flatted parallel shank (Part 1) and morse-taper shank (Part 2)
4193 : 1983 Part 11	6462 : 1983	Face-milling cutters using inserts – indexable
4193 : 1984 Part 12	6986 : 1983	Side and face-milling cutters using inserts – indexable
4193 : 1984 Part 13	6987 : 1983 Part 1	Carbide inserts – indexable, with a fixing hole partly cylindrical

Bibliography

Recent books

Gorczyca, Fryderyk E. (1987) *Application of Metal Cutting Theory.* Industrial Press, New York.

Gough, P. J. C. (1970) *Swarf and Machine Tools – a Guide to the Methods used in the Handling and Treatment of Swarf and Cutting Fluids.* Hutchinson, London.

Juneja, B. L. and Sekhon, G. S. (1987) *Fundamentals of Metal Cutting and Machine Tools.* Wiley, New York.

Metal Cutting Tool Institute (1987) *Metal Cutting Tool Handbook.* Industrial Press, New York.

Mills, B. and Redford, A. H. (1983) *Machinability of Engineering Materials.* Applied Science, London.

Shaw, Milton C. (1984) *Metal Cutting Principles.* Clarendon Press, Oxford.

Trent, E. M. (1984) *Metal Cutting.* Butterworth, London.

Venkatesh, V. C. and Chandrasekaran, H. (1987) *Experimental Techniques in Metal Cutting.* Prentice-Hall, India.

Conference papers

Cutting Tools International Colloquium: Short Papers, Saint Etienne, France, November 1981. Société Française de Metallurgie.

Cutting Tool Materials: International Conference Papers and Discussions, Fort Mitchell, Ky., USA, September 1980. American Society for Metals, Working and Forming Division.

Influence of Metallurgy on Hole Making Operations: Proceedings from an International Symposium, Metals Park, Ohio, USA, 1978. American Society for Metals, Mechanical Working and Forming Division.

Influence of Metallurgy on Machinability: Proceedings from an International Symposium, Ohio, USA, October 1975. American Society for Metals, Mechanical Working and Forming Division.

Influence of Metallurgy on the Machinability of Engineering Materials: Papers from an International Conference, Rosemont, Ill., USA, September 1982. Institution of Production Engineering Research.

Machinability Testing and Utilization of Machining Data: International Conference Papers, Oak Brook, Ill., USA, September 1978. Society of Manufacturing Engineers, Materials Removal Council.

Materials for Metal Cutting: Conference Proceedings, Scarborough, UK, April 1979. British Iron and Steel Research Association.

New Tool Materials, Cutting Techniques and Metal Forming: Fifth International Conference Papers, Monte Carlo, Monaco, March 1983. Engineer's Digest.

New Tool Materials, Metal Cutting and Forming: International Conference, London, March 1981. Automation (Leatherhead, UK).

Periodicals

American Machinist and Automated Manufacturing
Monthly (USA)
Industrial Engineering
Monthly (USA)
The International Journal of Advanced Manufacturing Technology
Quarterly (UK)
International Journal of Production Research
Quarterly (UK)
Machinery and Production Engineering
Fortnightly (UK)
Metalwork Production
Monthly (UK)
Precision Toolmaker
Bi-monthly (UK)
Production Engineer
Monthly (UK)

Note: Cutting-tool companies also produce literature, and in some cases magazines, in this field of technology.

Company addresses

UK Head Office

Bristol Erickson Ltd.
Tower Road North
Warmley
Bristol
BS15 2XF
Tel: 0272 677571

Cincinnati Milacron Co.
PO Box 505
Kingsbury Road
Birmingham
B24 0QU
Tel: 021-351 3821

Devlieg Microbore Tooling Co.
Leicester Road
Lutterworth
LE17 4HE
Tel: 04555 3030

Edgar Vaughan
Legge Street
Birmingham
B4 7EU
Tel: 021-359 6100

International Head Office

American Machinist
Penton Publishing Inc.
1100 Superior Avenue
Cleveland
Ohio 44114
USA
Tel: 216-867-9191

Kennametal Inc.
International Group
PO Box 231
Latrobe
PA 15650
USA
Tel: 412-539-4700

Cincinnati Milacron Co.
4701 Marburg Avenue
Cincinnati
Ohio 45209
USA
Tel: 513-841-8294

UK Head Office	International Head Office
Gildemeister (UK) Ltd. Unitool House Camford Way Sundon Park Luton LU3 3AN Tel: 0582 570661	**Gildemeister Automation GmbH** Max-Müller-Strasse 24 3000 Hanover 1 West Germany Tel: 0511-6707-1
Hi-Tec Metals (R&D) Ltd. Unit B4 Millbrook Close Chandlers Ford Southampton SO5 3BZ Tel: 0703 255510	
Karl Hertel Ltd. Corporation Street Nuneaton Warks CV11 5AJ Tel: 0203 386279	**Karl Hertel GmbH** Wehlauer Strasse 73 Postfach 1751 D-8510 Fürth/Bay West Germany Tel: 0911-7370-0
Krupp Widia UK Ltd. 5 The Valley Centre Gordon Road High Wycombe Bucks HP13 6EQ Tel: 0494 451845	**Krupp Widia GmbH** Münchener Strasse 90 D-4300 Essen 1 West Germany Tel: 0201-725-0
Kennametal Erickson UK Ltd. PO Box 29 Kingswinford West Midlands DY6 7NP Tel: 0384 401000	**Kennametal Inc. International** PO Box 231 Latrobe PA 15650 USA Tel: 412-539 4700
LK Tool Company Ltd. East Midlands Airport Castle Donington Derby DE7 2SA Tel: 0332 811349	**Cincinnati Milacron Co.** 4701 Marburg Avenue Cincinnati Ohio 45209 USA Tel: 513-841 8294
Morris Tooling Ltd. 304 Bedworth Road Longford Coventry CV6 6LA Tel: 0203 367548	

UK Head Office	International Head Office
NCMT Ferry Works Thames Ditton Surrey KT7 0QQ Tel: 01-398 3402	
Ovako Steel Ltd. Neachells Lane Wolverhampton West Midlands WV11 3QF Tel: 0902 307437	**Ovako Steel Imatra** 55100 Imatra Finland Tel: 954 6021
Polstore (Materials Handling) Ltd. Brooklands Works Wintersells Road Byfleet Weybridge Surrey KT14 7LQ Tel: 09323 40666	
Rank Taylor-Hobson Ltd. PO Box 36 New Star Road Thurmaston Lane Leicester LE4 7JQ Tel: 0533 763771	
Renishaw Metrology Ltd. New Mills Wotton-under-Edge Gloucestershire GL12 8JR Tel: 0453 844211	
Rocol Ltd. Rocol House Swillington Leeds LS26 8BS Tel: 0532 866511	
Rubert & Co Ltd. Acru Works Demmings Road Cheadle Cheshire SK8 2PG Tel: 061-428 6058	

UK Head Office

Sandvik Coromant UK
Manor Way
Halesowen
West Midlands
B62 8QZ
Tel: 021-550 4700

International Head Office

Sandvik AB
81181 Sandviken
Sweden
Tel: 026-260000

Scanatron
Poole House
Castle Street
Nether Stowey
Somerset
TA5 1LN
Tel: 0278 732491

Seco Tools (UK) Ltd.
Alcester
Warwickshire
B49 6EL
Tel: 0789 764341

Seco Tools AB
773 01 Fabest
Sweden
Tel: 022-340000

Swedish Machine Group Ltd.
4 The Argent Centre
Silverdale Road
Hayes
Middlesex
UB3 3BL
Tel: 01-573 3337

SMG Machine Company AB
Box 800
721 22 Västerås
Sweden
Tel: 021-120340

Stellram Ltd.
Avro Way
Bowerhill Industrial Estate
Melksham
Wiltshire
SN12 6TP
Tel: 0225 706882

Stellram SA
1260 Nyon
Switzerland
Tel: 022-613101

'Toolware'
Isis Informatics Ltd.
8 Croft Road
Godalming
Surrey
GU7 1BY
Tel: 04868 24340

Index

Page numbers in *Italics* refer to figures.